Free Radicals in Food

ACS SYMPOSIUM SERIES **807**

Free Radicals in Food

Chemistry, Nutrition, and Health Effects

Michael J. Morello, Editor
Quaker Oats Company

Fereidoon Shahidi, Editor
Memorial University of Newfoundland

Chi-Tang Ho, Editor
Rutgers University

American Chemical Society, Washington, DC

Library of Congress Cataloging-in-Publication Data

Free radicals in food : chemistry, nutrition, and health effects / Michael J. Morello, Fereidoon Shahidi, Chi-Tang Ho, editors.

p. cm.—(ACS symposium series ; 807)

Includes bibliographical references and index.

ISBN 0–8412–3741–7

1. Free radicals (Chemistry)—Pathophysiology—Congresses. 2. Free radicals (Chemistry)— Congresses. 3. Food—Composition—Congresses.

I. Morello, Michael J., 1953- II. Shahidi, Fereidoon, 1951- III. Ho, Chi-Tang, 1944. IV. Series.

RB170 .F685 2002
616.07—dc21 2001053394

The paper used in this publication meets the minimum requirements of American National Standard for Information Sciences—Permanence of Paper for Printed Library Materials, ANSI Z39.48–1984.

Distributed by Oxford University Press

PRINTED IN THE UNITED STATES OF AMERICA

Foreword

The ACS Symposium Series was first published in 1974 to provide a mechanism for publishing symposia quickly in book form. The purpose of the series is to publish timely, comprehensive books developed from ACS sponsored symposia based on current scientific research. Occasion-ally, books are developed from symposia sponsored by other organiza-tions when the topic is of keen interest to the chemistry audience.

Before agreeing to publish a book, the proposed table of contents is reviewed for appropriate and comprehensive coverage and for interest to the audience. Some papers may be excluded to better focus the book; others may be added to provide comprehensiveness. When appropriate, overview or introductory chapters are added. Drafts of chapters are peer-reviewed prior to final acceptance or rejection, and manuscripts are prepared in camera-ready format.

As a rule, only original research papers and original review papers are included in the volumes. Verbatim reproductions of previously published papers are not accepted.

ACS Books Department

Contents

Free Radicals and Food Chemistry

Natural Antioxidants

Nutritional Biochemistry and Health

Preface

Epidemiological studies point to the link between dietary intake, health, and disease. These studies are driving the demand for foods that are consistent with long-term health, have the potential to prevent or ameliorate disease, and at the same time deliver desirable hedonic characteristics. This in turn is driving efforts to identify foods that provide these hypernutritional benefits, specific substances within the food that impart the efficacy, and the biochemical basis for their efficacy. As this body of information expands, it is increasingly evident that foods affecting free radical reactions play a pivotal role in providing these benefits. This book presents state-of-the-art contributions by researchers from around the world in an attempt to unravel factors involved in the chemistry, nutrition, and health effects of free radicals in foods. The intent of the book is to serve as a focal point for dissemination of recent developments in this rapidly expanding field.

The book starts with an introductory overview, which is intended to refresh the reader's memory on key aspects of free radicals and reactions thereof. Subsequent sections cover free radicals and food chemistry, natural antioxidants, and nutritional biochemistry and health. In the food chemistry section, topics range from analysis of free radicals within food matrices to Maillard reactions, emulsions, dairy, and meat products. In the antioxidant section, results are presented on the efficacy of antioxidants from tea, seeds, and selected naturally occurring compounds. Finally, in the nutritional biochemistry and health section, free radical inhibition is discussed in relationship to biochemical paths and cancerous cell cultures.

Finally, we thank all the authors for their contributions and efforts in the preparation of this book. May their insight and research lead us all to longer and healthier lives!

Michael J. Morello
The Quaker Oaks Company
617 West Main Street
Barrington, IL 60010

Fereidoon Shahidi
Department of Biochemistry
Memorial University of Newfoundland
St. John's NF, A1B 3X9
Canada

Chi-Tang Ho
Department of Food Science
Rutgers, The State of New Jersey
New Brunswick, NJ 08903

Chapter 1

Free Radicals in Foods: Chemistry, Nutrition, and Health Effects

Michael J. Morello[1], Fereidoon Shahidi[2], and Chi-Tang Ho[3]

**[1]The Quaker Oaks Company,
617 West Main Street Barrington IL 60010-4199
[2]Department of Biochemistry, Memorial University of Newfoundland,
St. John's, Newfoundland A1B 3X9, Canada
[3]Department of Food Science, Rutgers University,
65 Dudley Road, New Brunswick, NJ 08901-8520**

An overview of the free radicals and reactions thereof is presented. Free radicals are atoms or groups having an unpaired electron and hence are paramagnetic. Electron paramagnetic resonance spectroscopy (EPR) and trapping methods are used to analyze radicals. In lipids, radical reactions lead to autoxidation and hence flavor reversion. Reactive oxygen species are key components involved in such reactions. Finally, descriptions for phenolic, sequesterant and enzymatic antioxidants and their mode of action are provided.

Free radical reactions are ubiquitous in food and biological systems. They play major roles in biochemical pathways and food degradation. In addition, there is escalating evidence for the fundamental role free radicals play in disease. The pervasive interest in free radical chemistry is documented by over three hundred reviews in the last eighteen months. Consequently, a comprehensive review is beyond the present scope. The intent of this chapter is to provide an overview of the fundamental aspects of free radical chemistry and to set the stage for research reported in subsequent chapters.

Free radicals are defined as atoms or compounds that have at least one unpaired electron. Radicals follow the Pauli exclusion principle: no two electrons may have the same set of four quantum numbers, and Hund's rule of maximum spin multiplicity: electrons tend to avoid being in the same orbital and have paired spin when possible. Consequently, radicals are typically paramagnetic. Examples of free radicals include ground state O_2 ($^3\Sigma_g^- O_2$), superoxide ($O_2^{\bullet -}$), singlet O_2 ($^1\Sigma_g^+ O_2$: electrons with antiparallel spin in orthogonal π*2p orbitals), hydroxyl radical ($OH^\bullet$), peroxy and alkoxy radicals ($RO_2^\bullet$ and $RO^\bullet$, respectively), thiyl and perthiyl radicals ($RS^\bullet$ and $RS_2^\bullet$, respectively), halogen atoms ($X^\bullet$), nitric oxide and nitrogen dioxide ($NO^\bullet$ and $NO_2^\bullet$), and numerous carbon-centered radicals (*1*). Transition metals can also be considered radicals. They are extremely important in many reactions as they have the ability to accept/donate single electrons. It is also important to note that peroxide ion (O_2^{2-}), hydrogen peroxide (H_2O_2), and singlet oxygen ($^1\Delta_g$ O_2: paired electrons in one of the π*2p orbitals) do not have unpaired electrons and are therefore not radicals.

Analysis of Radicals

There are three techniques commonly used to detect free radicals: electron paramagnetic resonance (EPR) spectroscopy – this is also referred to as electron spin resonance (ESR) spectroscopy, spin trapping, and reaction fingerprinting (*2*). EPR detects radicals directly but is limited to relatively stable radicals. EPR detects the rate of absorbance between the two spin energy levels associated with an unpaired electron; this absorbance is induced when the substance with the unpaired electron is placed in a magnetic field. Additionally, the electron spin can couple to nuclei, leading to splitting patterns analogous to those observed for NMR. It also should be noted that non-radicals, that is substances with paired electrons, are not detected by EPR, because the magnetic effects on the electron spin cancel each other out. Spin trapping is a technique in which reactive radicals react with non-radicals thereby forming more stable radicals that can be detected by EPR. The key features of spin traps are that they react rapidly and form stable – less reactive – radicals that give strong EPR response. Spin traps discussed later in this volume include 5,5-dimethyl-1-pyrroline-*N*-oxide (DMPO), *N*-*t*-butyl-α-phenylnitrone (PBN) and, 2,2,6,6-tetramethyl-piperidine-*N*-oxide iodoacetamide (TEMPO-IA). Reaction fingerprinting is specific for radicals and substrates. Halliwell and Gutteridge (*2*) have summarized fingerprints for many radical reactions. Fingerprints for lipid peroxidation illustrates the range and specificity of this technique.

- Loss of substrates: GC or HPLC analysis of fatty acids, oxygen loss measured by an oxygen specific electrode.
- Peroxide assays – peroxide measurement: Iodine formation – titration, ferrous oxidation xylenol orange (FOX) in vitro assay for LDL peroxidation, glutathione peroxidase (GPX) specific for fatty acid peroxides, Cyclooxygenase (COX) used to measure trace peroxides in biological fluids
- Separation of reaction products: HPLC or GC analysis of aldehydes, lipid hydroperoxides, cholesterol esters, and phospholipids.
- Miscellaneous: conjugated dienes, 2-thiobarbituric acid reactive substances (TBARS), and aldehydes, among others.

Radical Reactions

Formation and reaction of radicals are closely linked. The general scheme of lipid autoxidation – initiation, propagation, and termination – provides an outline of typical radical reactions.

- Initiation: formation of the primary source of radicals is usually brought about through homolytic fission, photo-excitation, and transition metal ion assisted redox reactions.
- Propagation: radical – molecule reactions generate the characterizing reaction products and include abstraction, substitution, addition, and fragmentation.
- Termination: radical – radical reactions remove radicals from the overall scheme and include combination and disproportionation.

Initiation

Initial formation of radicals can result from homolytic fission and electron transfer reactions (*3*). Homolysis can be induced both thermally and photochemically. Peroxy and azo compounds are particularly susceptible to homolysis. Examples of homolysis reactions are given below.

$$CH_3CH_2OOCH_2CH_3 \rightarrow 2\ CH_3CH_2O^{\bullet} \quad \text{ca. } 80\ ^{\circ}C$$

$$H_2O_2 \rightarrow 2\ HO^{\bullet} \quad h\nu \text{ (sunlight or 254 nm)}$$

$X_2 \rightarrow 2\,X^{\bullet}$ $h\nu$ (wavelength is halogen specific)

Transition metal ions are commonly involved in electron transfer reactions. Iron, copper, zinc and manganese are important in these reactions. Redox potentials are important when evaluating the viability of these reactions. Examples of electron transfer reactions discussed in this volume include: Fenton reaction and transformation of oxymyoglobin to metmyoglobin with the release of superoxide.

$$Fe^{2+} + H_2O_2 \rightarrow Fe^{3+} + HO^{\bullet} + HO^{-}$$

$$MbFe^{2+}O_2 \rightarrow O_2^{\bullet -} + MbFe^{3+}$$

Propagation

Characteristic products of free radical reactions are generated through abstraction, substitution, addition, and fragmentation (*3*). A common feature of these reactions is that radical reactants lead to radical products. This feature is key to the autocatalytic nature of radical reactions.

Hydrogen abstraction reactions are very important in lipid autoxidation and the efficacy of phenolic antioxidants. Ease of hydrogen abstraction from fatty acids is related to the degree of allylic stabilization. That is, the bond dissociation energy of hydrogens adjacent to two double bonds is less than that for hydrogens adjacent to one double bond, which is less than that for saturated fatty acids. The ability of phenolic antioxidants to donate hydrogen atoms (that is, the ease that hydrogen atoms are abstracted from phenols) is related to the degree of resonance stabilization provided to the oxygen-centered radical by the aromatic ring and its substituent groups.

Substitution reactions may serve an antioxidant function. Halogenation and nitration of isoflavones is discussed later in this volume. These reactions imply that the proinflamitory oxidants (hypobromous acid, hypochlorous acid, and peroxynitrite) might be reduced or moderated by reaction with isoflavones in vivo.

The most notable addition reaction is that of ground state oxygen with carbon centered radicals. This leads to peroxy radicals that can abstract hydrogens from other molecules regenerating carbon-centered radicals. Additionally, the peroxide so formed can then undergo homolysis to yield alkoxy and hydroxy radicals. Consequently, this homolytic initiation leads to additional radicals that propagate and accentuate the autocatalytic nature of the reaction.

$$R\bullet + O_2 \rightarrow RO_2\bullet$$

$$RO_2\bullet + RH \rightarrow ROOH + R\bullet$$

$$ROOH \rightarrow RO\bullet + HO\bullet$$ (homolytic initiation, leads to propagation)

At this point, it should be noted that ground state oxygen does not undergo direct addition to olefins. Ground state oxygen has two unpaired electrons with parallel spins. Electrons in the π–bond of the olefin have paired spins. According to the Pauli principle this reaction is prohibited.

Another important addition reaction is addition of hydroxyl radicals to DNA bases. This initiates a series of redox and rearrangement reactions that lead to fragmentation and cross-linking that ultimately impair replication and transcription.

An example of a radical fragmentation reaction discussed in this volume is the metal ion catalyzed degradation of Amadori compounds. This reaction provides an alternate path to osones, which are key intermediates in the formation of nitrogen heterocycles via the Maillard reaction.

Termination

When radicals combine or disproportionate this effectively serves to terminate the overall reaction (*3*).

$$2\,R\bullet \rightarrow R\text{–}R$$

$$2\,CH_3CH_2^{\bullet} \rightarrow CH_2{=}CH_2 + CH_3\text{–}CH_3$$

Combination reactions have low activation energies and are typically very fast. However, steric effects can inhibit these reactions. Additionally, these reactions are often diffusion controlled. An example of combination reactions is the cross-linking of triacylglycerols. Disproportionation reactions are also fast. Additionally, rates of these reactions imply they do not involve normal hydrogen abstraction. Both combination and disproportionation reactions are influenced by resonance stabilization.

Reactivity and Stability

Reactivity and stability of radicals are related to structure. Delocalization of the unpaired electron can stabilize the radical, reducing reactivity. That is, delocalization reduces spin density with a concomitant reduction in reactivity (*3*). Steric effects can both inhibit reactivity and destabilize radicals. Large groups in proximity to the radical center can hinder the approach of reactant molecules. Alternately, structures that force the radical center out of the preferred planar conformation destabilize the radical (*4*).

Reactive Oxygen Species

Oxygen is extremely important when discussing free radical reactions, and it is central to substances collectively referred to as "reactive oxygen species (ROS)", which include both radical and non-radical substances. Radical ROS include hydroxyl, superoxide, peroxy, and alkoxy radicals. Non-radical ROS include hydrogen peroxide, hypochlorous acid, singlet oxygen, peroxynitrite, and ozone (*1*).

Hydroxyl radicals ($OH^{\bullet}$) react very rapidly. They extract hydrogen atoms and undergo addition and electron transfer reactions (*5*).

Superoxide ($O_2^{\bullet-}$), the one electron reduced form of ground state oxygen, is less reactive than the hydroxyl radical, and reactivity depends on pH: pKa = 4.8. Consequently, at physiological pH there is a 100 to 1000 fold excess of superoxide relative to hydroperoxide (*5*).

$$HO_2^{\bullet} \leftrightarrow H^+ + O_2^{\bullet-}$$

Superoxide rapidly disappears in aqueous solution. It is believed to dismutate through reaction with hydroperoxide.

$$HO_2^{\bullet} + O_2^{\bullet-} + H^+ \rightarrow H_2O_2 + O_2$$

Direct reaction of superoxide with DNA, lipids, amino acids, and metabolites is very slow. However, superoxide reacts readily with other radicals and iron ions. It can accelerate Fenton reactions and thereby generation of hydroxyl radicals. Reaction of superoxide with ascobate also accelerates generation of hydroxyl radicals.

Peroxy and alkoxy radicals ($RO_2^{\bullet}$ and $RO^{\bullet}$, respectively) are good oxidizing agents. Both can abstract hydrogen atoms from other molecules. However, this ability is affected by resonance effects; resonance stabilization reduces the rate of hydrogen abstraction. Peroxy radicals can also react with each other to generate singlet oxygen (*4*).

$$2\ R_2CHOO^{\bullet} \rightarrow R_2CHOH + R_2C{=}O + {}^1O_2$$

Hydrogen peroxide (H_2O_2) is not very reactive, yet it can inactivate enzymes with essential thiol groups. It can also react directly with α-keto acids (pyruvate, 2-oxoglutarate, etc.). Hydrogen peroxide is cytotoxic. Some cell damage may result from direct reaction. However, it can cross cell membranes rapidly and once inside it can react with metal ions to generate hydroxy radicals, which are more generally reactive (*5*).

Hypochlorous acid (HOCl) is highly reactive and capable of damaging biomolecules both directly and through formation of chlorine. Thiols and ascorbate can be oxidized by hypochlorous acid and it can chlorinate DNA. Reaction with superoxide results in generation of hydroxy radicals (*5*).

Two states of singlet oxygen exist: the free radical, electrons with antiparallel spin in orthogonal π^*2p molecular orbitals, and the non-radical, paired electrons in one of the π^*2p molecular orbitals. The free radical rapidly relaxes to the non-radical, and for practical purposes, it is only necessary to consider the non-radical. It should be noted the spin restriction for reaction of ground state oxygen with non-radicals does not apply to either state of singlet oxygen. Singlet oxygen can form endoperoxides through reaction with conjugated dienes and hydroperoxides via ene-reactions with allylic compounds (*4*). These reactions are important for olefinic compounds such as carotenoids and lipids.

Antioxidants

The primary role of antioxidants is to prevent or inhibit degradation induced by free radical reactions. Antioxidants function via the reactions discussed earlier. The two reactions that are most prevalent in antioxidant function are hydrogen abstraction and metal ion assisted electron transfer.

For hydrogen abstraction, the antioxidant is the source of the hydrogen being abstracted; that is, the antioxidant donates the hydrogen atom to the free radical, removing a reactive radical and forming a more stable, less reactive radical. This essentially inhibits the propagation of the autocatalytic chain reaction. Compounds that fall into this group of antioxidants are the phenols and polyphenols, thiols, uric acid, ascorbic acid, and carotenoids, among others.

Phenols and polyphenols warrant extra attention as there is considerable ongoing effort to isolate and characterize these compounds from different sources. Included in this group are the tocopherols and tocotrienols, phenolic acids and esters, flavonoids, isoflavones and chalcones. Compounds in these groups are represented by a wide range of positional isomers and glycones (*6*). It would be expected that substituent effects should play a role in the ability of these compounds to donate hydrogen atoms. It is known that Hammett correlations can be applied to free radical reactions. Although ρ values are less than those for heterolytic cleavage, this does indicate the hydrogen donation and radical spin density should be influenced by substituent effects (*7*).

Metal ion assisted electron transfer reactions are important when considering the roles of sequesterants and enzymes in antioxidant protection (*8*). Compounds like ethylenediaminetetraacetic acid (EDTA), phytic acid, and polyphenols can sequester transition metal ions and thereby inhibit oxidation. Binding the metal ions essentially prevents them from regenerating hydroxy radicals through the Fenton reaction.

Superoxide dismutase (SOD), catalases, and peroxidases inhibit oxidation in vivo by inactivating ROS. There are multiple forms of SOD and all contain metal ions, which are integral to the dismutation reaction. For FeSOD, the dismutation is attributed to the following reactions.

$$Fe^{3+}\text{–enzyme} + O_2^{\bullet-} \rightarrow Fe^{2+}\text{–enzyme} + O_2$$

$$Fe^{2+}\text{–enzyme} + O_2^{\bullet-} + 2\,H^+ \rightarrow Fe^{3+}\text{–enzyme} + H_2O_2$$

$$\text{Net reaction: } O_2^{\bullet-} + O_2^{\bullet-} + 2\,H^+ \rightarrow H_2O_2 + O_2$$

Catalases directly convert hydrogen peroxide to ground state oxygen and water. Peroxidases also convert hydrogen peroxide to water, but also oxidize another substrate.

Thus, antioxidants in food and biological systems play an important role in neutralizing radicals, hence extending the shelf-life of food and preventing diseases in humans. While in many systems endogenous antioxidants can deliver a desired effect, in others, use of exogenous antioxidants is necessary.

References

1. Halliwell, B.; Gutteridge, J. M. C. *Free Radicals in Biology and Medicine, 3rd Edition*; Oxford University Press, Inc., New York, NY, 1999, pp 1-35

2. Halliwell, B.; Gutteridge, J. M. C. *Free Radicals in Biology and Medicine, 3rd Edition*; Oxford University Press, Inc., New York, NY, 1999, pp 351-425
3. Sharp, J. T. in *Comprehensive Organic Chemistry: The Synthesis and Reaction of Organic Compounds*; Barton, D. and Ollis, W. D., Eds.; Volume 1 Stereochemistry, Hydrocarbons, Halo Compounds, Oxygen Compounds; Stoddart, J. F., Ed.; Pergamon Press Ltd., Oxford, England, 1979, pp 455-467.
4. Cary, F. A.; Sundberg, R. J. *Advanced Organic Chemistry, Part A;* Plenum Press: New York, NY, 1977; pp 501-555
5. Halliwell, B.; Gutteridge, J. M. C. *Free Radicals in Biology and Medicine, 3rd Edition*; Oxford University Press, Inc., New York, NY, 1999, pp 36-104
6. Shahidi, F.; Wanasundara, P. K. J. P. D.; *Crit. Rev. Food Sci. Nutr*. **1992**, *32*, 67-103
7. Johnson, C. D. *The Hammett Equation*; Cambridge University Press: New York, NY, 1973; pp 63
8. Halliwell, B.; Gutteridge, J. M. C. *Free Radicals in Biology and Medicine, 3rd Edition*; Oxford University Press, Inc., New York, NY, 1999, pp 105-245

Free Radicals and Food Chemistry

Chapter 2

EPR Methods for Studying Free Radicals in Foods

K. M. Schaich

Department of Food Science, Rutgers University, 65 Dudley Road, New Brunswick, NJ 08901-8520

Although some roles for free radicals in foods have been proposed, relatively little research to document suspected radicals has been undertaken. EPR is a highly specific and sensitive technique for direct detection of free radicals; application of EPR to foods can reveal important information about radical reactions that may be responsible for food qualities and deterioration. To excite interest and encourage more definitive studies of free radicals in foods, this paper will introduce EPR spectroscopy, covering basic considerations necessary to obtain, interpret, and quantitate EPR spectra. Topics covered include calibration of magnetic fields and microwave frequencies for accurate determination of splitting constants and *g*-values (radical identification factors), quantification of EPR signals, methods for eliminating interference of water, instrumental and environmental conditions that must be controlled to obtain accurate and reproducible EPR spectra, and some pitfalls to avoid. Three guidelines are offered that, if followed, will ensure that data obtained reflects the radicals and reactions being studied rather than secondary or artifact radicals, that the quantitation is accurate, and that the interpretations are sound.

It is indeed encouraging that interest in free radicals in foods has finally developed to the point that this symposium has been organized to discuss various aspects of free radical applications. At least some of the current frenzy of interest in free radicals has been stimulated by a widely-held view that all free radicals are bad, that *any* free radicals in food will cause damage or initiate free radical reactions *in vivo* after consumption of the food *(1)*: "In this age...all things radical bear a bad connotation...Surely nothing can match the ill repute of free radicals." Whether this assumption is correct remains to be demonstrated. What is clear from this attitude is that there is a greatly increased impetus for measuring free radicals in foods and determining their roles and implications for food quality and safety as well as human health.

Free radicals in foods result from both physical forces during processing and chemical reactions occurring in foods, as shown below. Spin traps and spin labels, stable free radicals that are not present normally but may be added to foods for analytical purposes, should also be included in this list.

Chemical reactions	Physical forces	Added stable free radicals
lipid oxidation -- chain reactions	heat	spin labels
lipid oxidation -- co-oxidations	shear stress	spin traps
non-enzymatic browning	freezing	
enzyme reactions	freeze-drying	
ascorbic acid autoxidation	irradiation	
metal-catalyzed redox reactions		
redox reactions		
antioxidants		

Although generation of free radicals during processing and storage of foods may be reasonably anticipated, there is very little actual data (other than for oxidizing lipids) documenting the levels or types of radicals produced or the impact of the radicals on food quality or safety.

A toxic or pathological role for free radicals in foods has been proposed by extrapolation from known toxicity of free radicals in a wide range of pathological conditions in vivo *(2)*. In the absence of systematic studies on free radicals in foods, however, this contention must remain mere speculation. An opposite position deserves equal consideration, namely that free radicals generated during processing may play key functional roles in texturization, flavor formation, and other reactions contributing to food properties and characteristic qualities. Addressing this controversy, identifying and understanding reactions that are free radical-mediated in foods, and learning to control radical reactions associated with food qualities and shelf stability give us ample reason to be interested in detecting and studying radicals in foods.

The only direct and definitive method for studying free radicals is Electron Paramagnetic Resonance, EPR, also known as electron spin resonance, ESR. EPR is a technique with tremendous power but also many pitfalls. It is *not* a technique in which a sample simply can be packed into a tube and placed in a spectrometer to obtain a guaranteed spectrum. When a spectrum is fortunately detected, one cannot assume *a priori* that the radical came from the expected source. Only *stable* radicals, not all the radicals, are detected in EPR spectra, and they may or may not be from the reaction being tested. The purpose of this paper, therefore, is to provide a general introduction to what EPR can do and what experimental considerations are key and crucial for obtaining, interpreting, and quantitating EPR spectra. Comprehensive details of EPR spectroscopy and practical guidelines for methods are available in Swartz et al. *(3)*, Wertz and Bolton *(4)*, Borg *(5)*, Bersohn and Baird *(6)*, Catoire *(7)*, and Poole *(8)*.

EPR is specific for free radicals and other paramagnetic species such as transition metals while being blind to other molecules. Quite sensitive, detecting 2×10^{11} spins ($\sim 3 \times 10^{-9}$ M) at the lower limit *(3)*, under appropriate conditions EPR can reveal what kind of radicals are present, how many there are, and how fast they react. Unpaired electrons of free radicals have magnetic moment and thus interact with a magnetic field. Manipulating and detecting this interaction is the basis for EPR.

A simplified schematic diagram of an EPR system is shown in Figure 1. A typical EPR system has as its first main component a klystron, which delivers a fixed microwave frequency. EPR experiments are performed most commonly at X-band (9.5 GHz), although Q (32 GHz), L (1-2 GHz), S (2-4 GHz), K (24 GHz), and W (95 GHz) bands are also used for various special applications. The sample is placed in a quartz tube or cell that is then inserted into the EPR cavity where the microwaves are directed. The cavity is situated between two magnet poles where a magnetic field can be applied perpendicularly to the microwave frequency. The free electrons align in the magnetic field with their spin axes parallel (low energy) or perpendicular (high energy) to the field. At a field characteristic for each type of atom, low energy electrons interact with the magnetic field, absorb tiny amounts of energy and jump to their higher energy level, flipping their spins from parallel to perpendicular (Figure 2). The actual amounts of energy absorbed are quite small, so phase-sensitive amplification is applied to the signals. Hence, detectors commonly present EPR signals are first derivatives of the original energy absorptions.

Depending on the nature of the radical and its environment, EPR signals range from simple singlets to moderately complex to quite complex, as is illustrated in Figure 3. Methods for completely deciphering complex signals is beyond the scope of this paper. Several excellent references cover this in detail *(3-6)*. However, there are standard guidelines used for basic identification of radicals which should always be followed to avoid one the classic pitfalls of free radical research, namely ascribing EPR signals to whatever radical is expected rather than what is actually detected in a reaction system.

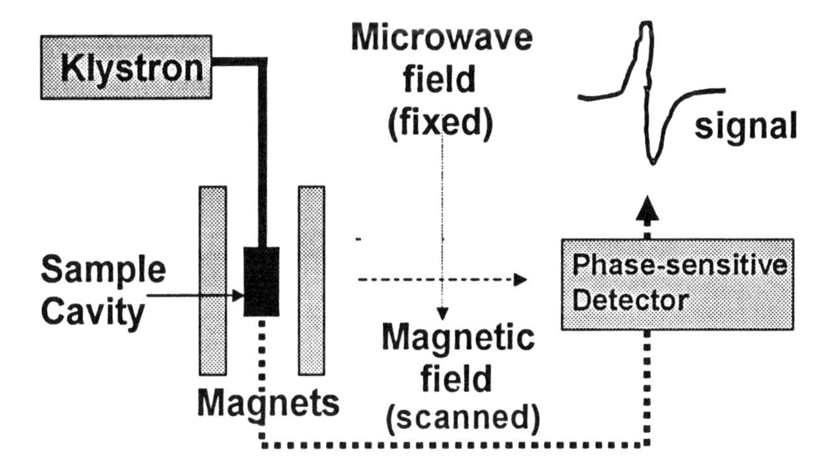

Figure 1. Schematic diagram of EPR instrumention for detection of free radicals. A sample is placed in the cavity in a fixed microwave field. A magnetic field is applied perpendicular to the microwave field and scanned. Signals produced when energy is absorbed at the resonant magnetic field are amplified and presented as first derivatives.

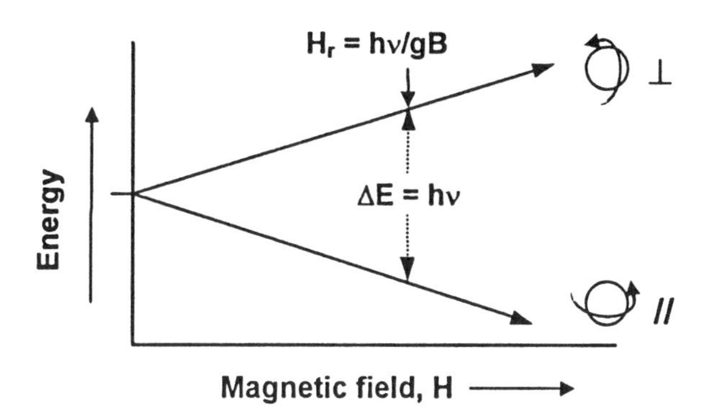

Figure 2. The origin of EPR signals, simplified. Unpaired electrons act as spinning magnets. When placed in a magnetic field, they spin in the lowest energy orientation with their axes parallel to H. As H is scanned, a field H_r is reached at which the energy [$\Delta E = h\nu = gBH_r$] equals the energy difference between low (//)and high ($\perp$) energy states. The electron absorbs the energy, flips its spin, and jumps to the high energy state. Each type of radical has a characteristic H_r. Detection of ΔE is the basis for EPR signal detection.

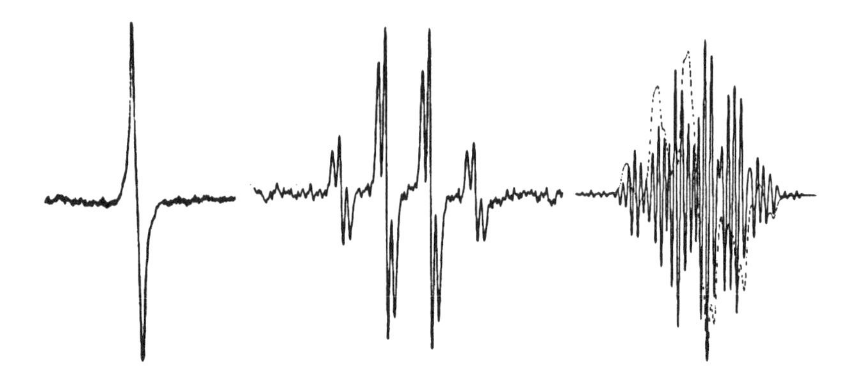

Figure 3. EPR signals of free radicals vary from simple singlets (left) to quite complex with many tens of lines (right). Adapted from (3) with permission. Copyright Wiley-Interscience, 1972.

1. Never Assume *a priori* That An Epr Signal Obtained Comes From the Source Expected Or That Lack Of a Signal Means No Radical Was Formed!!!

For detection by EPR, at least about 2 x 10^{11} spins (3 x 10^{-9} M) of a species must be present for the seconds to minutes required for a conventional EPR analysis *(3)*. Radicals produced in lower quantities or with shorter lifetimes so they do not survive the analysis period are not detected. Neither are radicals that interact with the magnetic field or the environment so that the spectral lines are tremendously broadened, spread over hundreds to thousands of gauss. However, stable radicals from secondary reactions, or even unrelated processes, may be produced in high quantities and be detected instead of the primary radicals of interest.

For example, in studying antioxidant effectiveness against Fenton-catalyzed lipid oxidation, phenoxyl radicals are very short-lived, while lipid alkoxyl and peroxyl radicals are even shorter-lived, and additionally are broadened by the oxygen. Both antioxidant and lipid radicals are produced in the reactions, but neither can be detected without special techniques. However, lipid radicals can react with any other molecules having abstractable hydrogens, such as the alcohol used as a solvent, generating stable radicals. At the same time, reactions of the iron catalyst (especially autoxidation to produce $O_2^{-\bullet}$, H_2O_2, and •OH) can initiate reactions unrelated to lipid oxidation that generate other secondary or adventitious radicals. Thus, in this system, any number of species other than lipids or antioxidants are likely to be the stable radicals detected by EPR. On the other hand, if an aprotic solvent is used and no other susceptible molecules are present, no EPR signal may be detected at all in standard measurements. Attributing EPR signals to antioxidants or lipids in the first cases would lead to derivation of incorrect reactions, mechanisms, and reaction kinetics. Equating lack of signals with lack of reaction in the latter case similarly leads to erroneous conclusions.

Hence, two general rules of thumb in EPR research are: 1) definitive assignment of EPR spectra to specific radical species requires initial identification from spectral characteristics and confirmation by generating the same radical and EPR signal via at least two different and independent reactions; and 2) multiple generation and detection modes with different specificities and lifetimes must be tested before lack of radical production or reaction can be concluded.

2. Signal Assignments Require Measurement of g-Values and Hyperfine Splittings To Interpret Spectra Accurately.

For each type of radical, there is a constant relationship for its interaction between the electromagnetic frequency and the magnetic field at which it absorbs energy. This is called the resonance condition,

$$E = h\nu = gBH_r \tag{1}$$

where E is the energy difference between the two spin states of the free electron in a radical, h is Planck's constant, ν is the microwave frequency of the spectrometer, g is the proportionality factor characteristic of each type of radical, B is Bohr's magneton, and H_r is the magnetic field at which resonance (spin-flipping) and detection occurs. Since h and B are constants, it can be seen clearly from this equation that the first requirement for identification of radicals is careful measurement and calibration of both the microwave frequency and the magnetic fields and scans. This information is then used to determine g-values (also called g-factors) and hyperfine splittings.

Microwave Frequency.

The frequency meter connected to the klystron gives an approximate reading of the resonant frequency in the cavity, but it is not sufficiently accurate for calculations of spectral parameters. Recent models of EPR spectrometers have built-in frequency counters with accuracy to 1 Hz; independent frequency counters need to connected between the klystron and cavity in older spectrometers to provide an accurate measure of frequency for each sample.

Magnetic Field.

A crude measure of magnetic field can be calculated from the field scan width scan rate, but very precise measurements require a gaussmeter connected to the field scan. Gaussmeters offer the advantage of providing absolute field values, even when the scan is non-homogeneous. For most routine analyses, however, chemical standards (compounds that have splitting constants and g-values known very precisely) are used to internally calibrate fields without using gaussmeters. The splittings provide a measure of the field sweep, while absolute or relative field positions can be calculated from the *g*-values. Some of the most commonly used field standards are listed in Table I.

Table I. Standards used for calibration of magnetic fields and determination of g-values in EPR spectroscopy

	Magnetic field standards		
	g	scan width, gauss	Characteristics
Fremy's salt[a]	2.0056	26.18	3 lines
Wurster's blue perchlorate[b]	2.0030	86.37	39 lines
Mn^{2+} in SrO powder[c]	2.0012	419.1	6 lines, 96 G apart
	***g*-value standards**		
	g at center line crossover		Characteristics
DPPH[d]	2.0036	± 0.0003	1 line, solid 5 lines soltn.[d]
Wurster's blue perchlorate[b]	2.003037	± 0.000012	op. cit.
Perylene cation (98% H_2SO_4)[e]	2.002583	± 0.000006	Many lines
Tetracene cation (98% H_2SO_4)[e]	2.002604	± 0.000007	
p-Benzosemiquinone anion[f]	2.004665	± 0.000006	5 lines, a_H=2.35 G

[a] Faber, R.J. and Fraenkel, G. K., *J. Chem. Phys.* **1967**, 2462; (3).
[b] Knolle, W. R. Ph.D thesis, Univ. Minn., 1970; *(3,4).*
[c] Rosenthal, J. and Yarmus, L., *Rev. Sci. Instr.* **1966**, 37, 381; (3).
[d] α,α'-diphenyl-β-picrylhydrazyl, stable radical; sol. bz, unstable in soltn.; Weil, J. A. and Anderson, J. K., *J. Chem. Soc.* **1965**, 5567; *(3,4)*
[e] Segal, B. G., Kaplan, M., Fraenkel, G. *J. Chem. Phys.* **1965**, 4191; (3,4).
[f] in BuOH with KOH at 23 °C; varies with temperature; Venkataraman, B., Segal, B. G., Fraenkel, G. K., *J. Chem. Phys.* **1959**, 1006; (4)

g-Values.

The constant proportionality between the frequency and magnetic field at resonance expressed in the resonance equation is designated the *g*-value. *g*-values are specifically characteristic of each type of paramagnetic species, including metals, and for free radicals are close to 2.00 (Table II). *g*-values are measured at the point where the center line of the spectrum crosses the baseline (best determined from the second derivative of the EPR signal), and can be calculated directly from the resonance equation or by comparison with standards. Some of the standards most commonly used in EPR studies are listed in Table I.

Table II. *g*-values of common free radicals and metals.

Paramagnetic species	Range of *g*	Ref.
C – centered radicals		
Hydrocarbons, straight chain	2.0022 - 2.0026	6
Hydrocarbons, aromatic	2.0025 - 2.0029	6
Hydrocarbons w/ -OH	2.0022 - 2.0033	
Porphyrin cations	2.0021 - 2.0028	3
Flavosemiquinones	2.0030 - 2.0040	3
Benzosemiquinones	2.0040 - 2.0050	3
N – centered radicals		
Protein peptide	2.0045 - 2.0060	9
Nitroxides	2.0050 - 2.0060	3
O– centered radicals		
Phenoxyl	2.01 - 2.02	3
Sulfoxyl	2.018 - 2.022	9
Cumyl peroxyl	2.0155	6
S – centered radicals		
-S• (3 anisotropic lines)	~1.99, 2.025, 2.035	10
-(S•-S)- species	2.003, 2.025, 2.032, 2.05-2.06	10,11
-S-S•	~1.99, 2.024, 2.05-2.07	10
Metal complexes		
Cu^{2+}	2.0 - 2.09	3,6
Fe^{3+} (high-spin)	2.0 - 9.7	3
Fe^{3+} (low-spin)	1.4 - 3.1	3
Mn^{2+}	2.0012	3
Mn^{5+}	2.0259	4
MgO^{-}	2.0033, 2.0386	4
Ti^{2+}	1.9280	4

When used, *g*-value standards in each application should have *g* as close as possible to the *g*-value of the experimental sample. *g*-values are calculated by comparing the field position of the test sample with that of the standard, using standard equations *(3,4)*:

$$g_x = \frac{g_s H_s}{H_x} = \frac{g_s H_s}{H_s - \Delta H} \cong g_s \left[1 + \frac{\Delta H}{H_s} \right] \tag{2}$$

where H is the resonant field, and subscripts x and s refer to the sample and standard, respectively. H_s is calculated at the cross-over point of the standard spectrum (determined from the second derivative), using the measured microwave frequency in the resonance equation.

Hyperfine Splittings

g-values reveal whether the electron sits on a carbon or a nitrogen in a radical, but this information alone is insufficient for full molecular identification of the free radical species. Exactly where the free electron resides in a molecule can be determined from hyperfine structure of a spectrum, which includes both the pattern of lines (*hfs*) and the distances (*a*) between them in a spectrum. A general introduction to hyperfine structure will be given here to demonstrate how *hfs* can help identify radicals. More detailed and rigorous development of the origins and determination of hyperfine structure can be found in References *3-5*.

Hyperfine structure arises from splitting of the energy levels shown in Figure 2 when a free electron interacts with neighboring nuclei, and electrons at each of the levels then resonate at slightly different magnetic fields. The patterns of splittings follow the same basic rules as in NMR *(4,6)*, i.e.

$$N_{hfs} = 2In + 1 \tag{3}$$

where N_{hfs} is number of lines per nuclei, I is the nuclear spin of the interacting nuclei, and *n* is the number of equivalent nuclei. Thus, for the nuclei most prevalent in foods, hydrogen (I = ½) splits lines into doublets, while nitrogen (I = 1) splits lines into triplets (Figure 4). Carbon and oxygen have no magnetic moments so do not induce hyperfine splittings, although oxygen can cause line broadening, leading to loss of resolution of neighboring narrow lines. The magnitude of the splittings is also characteristic of the type of interacting nuclei:

	lines	a (G)
C•	1	
α H	n+1	18
β H	n+1	5-10
γ H	n+1	2- 5
N •	3	15

The overall *hfs* of a radical develops as interacting nuclei split the initial electron absorbance into multiple energy levels in order of proximity to the electron, up to two groups away from the site of electron localization in a linear molecule. Each of the energy levels then contributes a line to the final spectrum. For example, a radical localized on the terminal amine of a lysine would have three initial dominant splittings from the nitrogen. Each of these would then be split into two by the amino proton, and then into three by the two protons on the β-CH_2. In some cases, three additional weak splittings from protons on the γ-CH_2 may also be resolvable.

The total number of lines contributed by an interacting nucleus or group of nuclei is $(1+2I)^n$. When splittings from multiple nuclei are non-equivalent, the lines usually do not superimpose, and consequently the full number of predicted lines are present and line intensities in the final spectrum are equal, as shown in the following stick spectra.

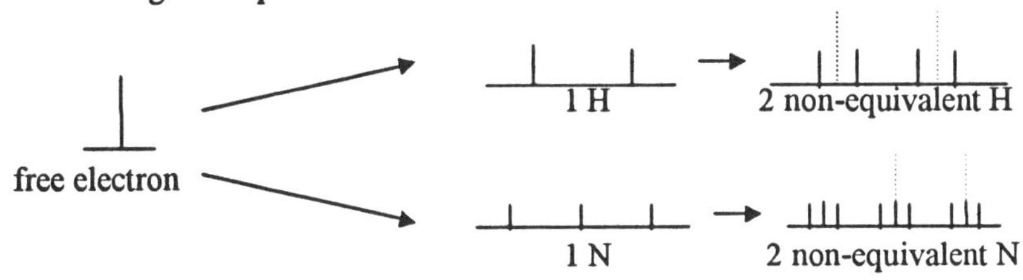

However, when multiple nuclei are equivalent (e.g. the three protons in a methyl group), some of the successive splittings from different lines fall on the same position (called degeneracy), resulting in a build up in intensity of those lines and a decrease in the total number of lines (A and B, Figure 4). The proportional intensities of the lines then follow the binomial expansion of $(1+2I)^n$ in a Pascal's triangle *(4,6)* (Figure 5). Thus, 2 equivalent protons give a spectrum with three lines with relative intensity 1:2:1 rather than the expected four lines of equal intensity. How this occurs is shown in the simple hyperfine splitting patterns below. Splittings are added in sequence, starting with the nucleus on which the free electron resides and adding additional splittings according to proximity to the electron. Dotted lines indicate positions of previous lines.

Spectra A and B in Figure 4 show the normal pattern of overlapping lines leading to the relative line intensities predicted in Pascal's triangles. C shows adventitious degeneracy when splittings from two different nuclei (e.g. nitrogen and hydrogen) are nearly identical. D shows the splitting pattern that would be seen at a terminal amine radical (not to scale). The two sets of very close lines would probably appear as single lines with twice the intensity. If resolvable, two γ-protons would split each line into a 1:2:1 pattern again.

When multiple nuclei contribute splittings to a spectrum and lines pile up on each other, it is not uncommon for several different combinations of nuclei to

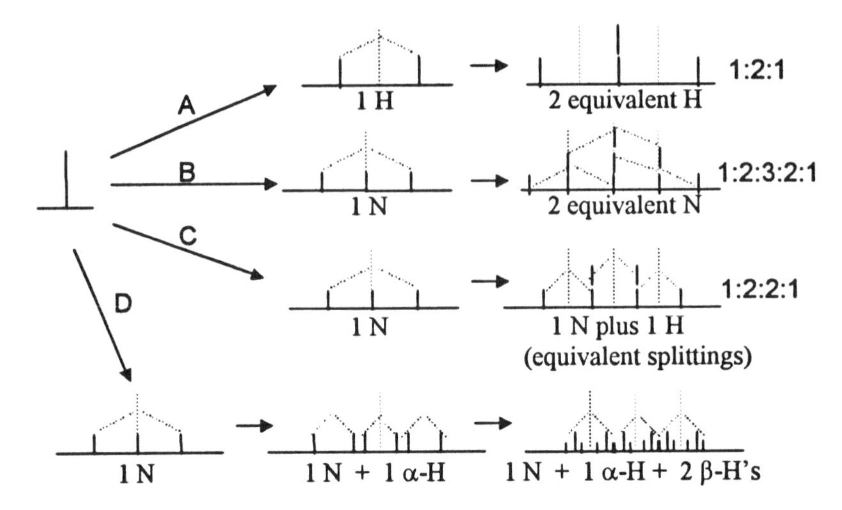

Figure 4. Hyperfine splitting patterns resulting from multiple interacting nuclei. Splittings are not drawn to scale.

# equiv. H's(*n*)		# lines, $(1+2I)^n$
1	1 1	2
2	1 2 1	4
3	1 3 3 1	8
4	1 4 6 4 1	16
5	1 5 10 10 5 1	32
6	1 6 15 20 15 6 1	64
7	1 7 21 35 35 21 7 1	128
8	1 8 28 56 70 56 28 8 1	256
1	1 1 1	3
2	1 2 3 2 1	9
3	1 3 6 7 6 3 1	27
4	1 4 10 16 19 16 10 4 1	81

Figure 5. Top. Pascal's triangle giving the relative intensities and number of lines expected for a given number of equivalent hydrogens (or other atom with spin = ± ½). Bottom. A comparable triangle starting with three lines can be constructed for nitrogen (spin = ± 1). Only the first four levels are shown.

yield very similar spectra. For example, from the Pascal triangles above it can be seen that four hydrogens on carbons or two nitrogens both give five lines with comparable proportions. Visually the spectra will appear the same, so an erroneous assignment could easily be made. Only careful measurements of line heights and *g*-values will distinguish between the two radicals. This potential conflict underlines a key point: *accurate determination of hyperfine splittings and patterns, and validation by computer simulation, combined with g-values, are critical for identification of radical species.*

Straight chain molecules usually give relatively simple spectra in which hyperfine patterns can be determined visually and then measured. However, radicals with rings or π-bonds where electrons become delocalized and, in effect, shared by multiple atoms, give somewhat more complicated hyperfine structure. Complex spectra with many lines require computer deconvolution to identify the splitting patterns, combined with spectral simulation to confirm the assignments. Excellent guidelines for deciphering and corroborating complex splitting patterns are presented by Wertz and Bolton *(4)*, Bersohn and Baird *(6)*, and Swartz et al *(3)*.

The discussion above applies primarily to solution spectra of single species where radicals can freely tumble and resulting lines are sharp. However, solid or powder materials rarely exhibit well-resolved hyperfine splittings because their signals are actually envelopes of many overlapping spectra from single radicals in multiple orientations or from multiple radicals in non-homogeneous materials. In crystals, e.g. sugar, lines are broadened so that γ- and sometimes β-splittings may be obscured, but the major splittings are still resolvable. However, in mixed solids such as flour or lyophilized materials, usually only broad singlet signals are observed. In such cases, *g*-values can be of initial help in identifying the class of radicals, and power saturation studies can reveal multiple species (see below), but specific information about electron localization and radical structures is difficult, if not impossible, to obtain. Secondary analyses must usually be combined with EPR spectra to decipher contributing radicals.

3. Control Instrumental Conditions To Avoid Signal Distortion, Maximize Detection And Resolution, And Provide a Common Basis For Quantitation.

The accuracy of measurements for *g*-values and hyperfine splittings depends on lineshape, and to a lesser extent on signal intensity, both of which are controlled by instrumental parameters. Signals need to be strong enough to give clear lines, but instrumental settings used to increase signal intensity can also broaden and distort spectra, thus altering apparent *g*-values and *a's*. Lineshapes also affect the quantification of radical concentrations. Thus, the goal of EPR measurements is to maximize signal detection and resolution while minimizing signal distortion to provide a complete spectrum that is accurate qualitatively and quantitatively. This section will discuss how instrumental parameters and some environmental factors affect EPR spectra and why it is so important that they be controlled for accurate EPR analyses.

Quantitative Analyses of EPR

The total number of measured spins (N_x) in a sample as reflected in the full integrated absorption including wings (Σ_x),

$$N_x = M \sqrt{P} \, G \, \Sigma_x \tag{4}$$

depends on modulation amplitude (M), microwave power (P), gain (G), and any other factors that affect the signal size and shape *(3)*. The latter include temperature, presence of moisture and oxygen, filter and time constants, filling factor and Q for the cavity. Although absolute numbers of spins are rarely calculated, this equation has two important implications for experimentalists: 1) accurate determination of radical concentrations requires optimization of measurement conditions; and 2) when conditions are held constant, the spin concentrations are proportional to some measure of signal intensity. Measurement of signal intensity will be considered first, and then some key factors affecting optimization will be discussed.

Signal amplitude (A) gives a first approximation of relative signal intensity, and is useful for single narrow lines. For broad signals with multiple components, a more accurate relative intensity can be determined from the product of A and the peak-to-peak linewidth squared (Figure 6):

$$\Sigma_{x\,(approx.)} = A * \Delta H^2 \tag{5}$$

This measurement can be converted to an order of magnitude estimate by comparison with a spin standard under identical conditions. However, double integration of the first-derivative EPR signal (Figure 6) is the best quantitation method for determining signal intensity, and for signals with multiple lines, complex line shapes, or wide wings this is the only accurate method. Double integration requires correction for baseline drift and careful determination of the positions at which wings end, but contemporary computers and software considerably simplify the process.

Signal intensities and spin concentrations are not identical. For most purposes, relative signal intensities are sufficient to determine whether a reaction or treatment produces more or fewer radicals. When actual concentrations of unpaired spins are needed to calculate stoichiometry or yields, signal intensities must be converted to spin concentrations. Determination of absolute spin concentrations from first principles using Equation 4, requires accurate knowledge of several additional factors, including the amplitude of H_r, sample temperature, radical *g*-value, filling factor, dielectric loss, and microwave frequency. In practice, this is exceedingly difficult, so spin concentrations are most often calculated by comparison with stable free radical standards whose spin concentrations per unit weight are known *(3,4)*:

$$N_X = \frac{M_S \sqrt{P_S}\, G_S\, \Sigma_X}{M_X \sqrt{P_X}\, G_X\, \Sigma_S} * N_S \qquad (6)$$

The most familiar standard is DPPH, but peroxylamine disulfonate (PADS), $CuSO_4{\cdot}5H_2O$ and $MnSO_4{\cdot}H_2O$ crystals, and cyclic nitroxides are also used.

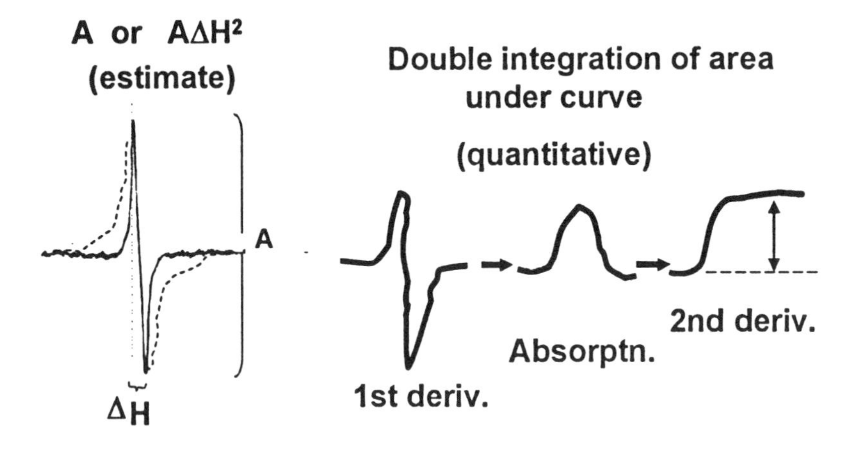

Figure 6. Methods for calculating EPR signal strength. Left: Estimates of signal intensity are provided by the signal amplitude (A) or amplitude times linewidth squared. However, these methods do not account for broad absorbance or line wings (dotted lines). Right. Double integration of EPR signals accounts for the complete area under the EPR signal.

Power Effects On Signal Amplitude And Shape

In general, EPR signal amplitude increases with the square root of the microwave power. Simplistically, this increase occurs because the microwave energy augments the effects of the resonant magnetic field and pushes more electrons into the higher energy "flipped spin" state shown in Figure 2. Normally, there is an equilibrium established between electrons absorbing energy and electrons relaxing back to ground state, so there is always a supply of electrons to absorb energy and generate a signal. As long as there are more electrons at lower energy, increasing power will increase the electron transitions and, consequently, the signal amplitude. However, the microwave energy interferes with normal relaxation processes, keeping more electrons in the higher energy state. At some power, an insufficient number of electrons remain in the lower state to maintain energy absorption, and the signal begins to decrease (Figure 7, left). This is called homogeneous broadening, and occurs in systems in which all paramagnetic centers are identical. However, in most food and biological systems, there are multiple radical and paramagnetic components, each of which has different saturation behavior. This mixture leads to inhomogeneous broadening, characterized by a leveling off with power as components that saturate are balanced by others that don't. In practice, the power saturation behavior must be determined for each sample, and all quantitative analyses should be conducted at powers below saturation.

While high powers can be a disadvantage quantitatively, they also provide a means of detecting and differentiating multiple paramagnetic species in complex samples. This is illustrated in Figure 7(right) where increasing the power saturates the initial signal and reveals a second underlying signal which either doesn't saturate or saturates at higher powers. The underlying species may be a second radical or multiple radicals. However, quite commonly in biological systems the second species is a metal that appeared only as drifting background at low powers, but becomes prominent at high powers.

Thus, it is important to know when multiple species are present and to run under power conditions appropriate to the experiment. For example, underlying metals can be lead to large errors when integrating free radical signals. Interference from metals can be minimized by running at powers as low as possible while maintaining good signal to noise. Usually this means powers less than 5-10 mW, and in some systems even lower. EPR signals in foods are often weak, and the temptation to run at 20 mW or higher to increase signal strength is great. However, multiple paramagnetic species are almost always present. Thus, high powers should be used with great caution and only after running the sample of interest at a full range of powers to determine saturation behavior.

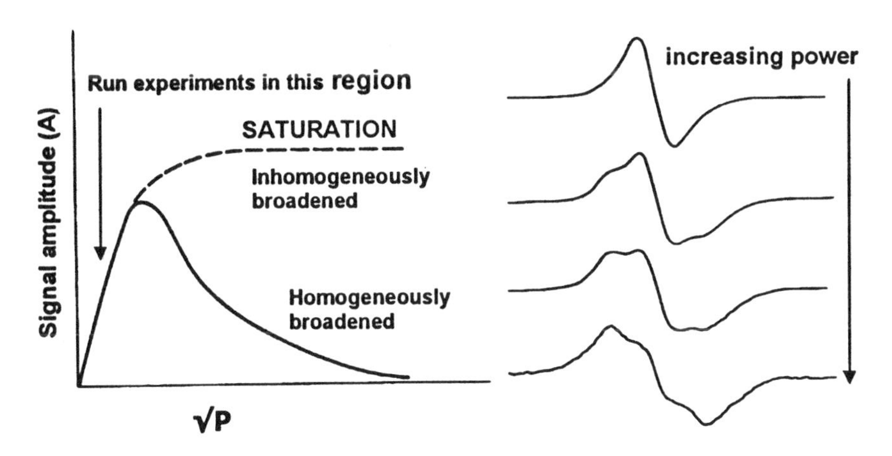

Figure 7. Effects of microwave power on EPR signal amplitude and lineshape. Left. EPR signal amplitude increases linearly with the square root of power up to a saturation point, beyond, which signals, remain constant or decrease. For most purposes, EPR experiments should be run at powers below saturation. Right: Multiple signals under the same curve but having different saturation characteristics can be differentiated by collecting spectra at various power levels. The radical dominant in the top spectrum saturates at higher powers, where the broad underlying spectrum becomes visible. Adapted from (3) with permission. Copyright Wiley-Interscience, 1972.

Modulation Amplitude Effects On Signal Amplitude And Shape

Modulation amplitude (H_M) is part of the phase sensitive detection used in EPR to improve the signal-to-noise ratio by enhancing the sample signal and filtering out microwave noise *(3-5)*. To accomplish this, a small-amplitude ac magnetic field is superimposed on the dc detector field (Figure 8). At or near resonance of the sample, the modulation alternately adds to and subtracts from a portion of the absorption at a set frequency, most commonly 100 kHz. This selectively amplifies the signal at 100 KHz, which is then rectified in the detector. The difference in the dc and ac currents through a complete frequency cycle generates an approximately first derivative EPR signal. It can be seen from the left diagram in Figure 8 that the *amplitude* of the modulation corresponds to the portion of the absorbance curve scanned. As H_M increases up to the line width (the absorption peak in the diagram), the signal intensity increases. However, higher modulation leads to broadening and distortion of the lines and decrease in the signal amplitude, A.

Modulation amplitude thus presents the paradox of multiple effects -- enhancing EPR signal strength while at the same time distorting and sometimes obscuring signal lines. As H_M increases up to the signal linewidth (peak of the absorption signal), the signal intensity increases (left, Figure 8). Hence, modulation plays a very important role in enhancing detection of weak signals, as are often found in food and biological materials. However, when H_M exceeds the intrinsic signal linewidth, the actual linewidth displayed approaches the modulation amplitude and the signal broadens (bottom right, Figure 8). The result is distortion of lineshape (top right, Figure 8) and loss of resolution, particularly of narrow lines. For example, a porphyrin with many narrow lines can be reduced to a narrow singlet at high modulations.

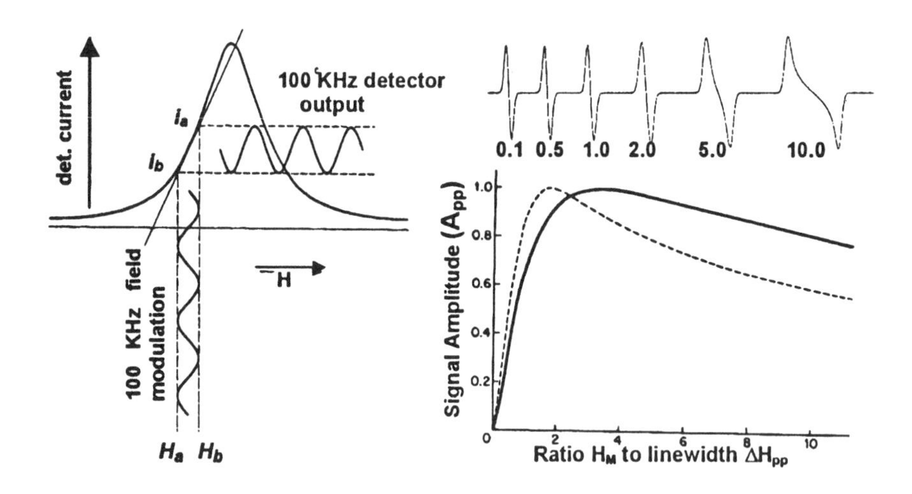

Figure 8. Effects of modulation amplitude (H_M) on EPR signal intensity and lineshape. Left: Modulating field applied to EPR signal alternately increases and decreases the signal strength. Right: Signal intensity increases with H_M up to H_M = 1/2 ΔH_{pp} (signal linewidth), then decreases. When H_M > linewidth, signals become stretched and distorted, and lines can be obscured. Figure components adapted with permission from (3-5). Copyright Wiley-Interscience, 1972; Academic Press, 1976; McGraw-Hill, 1972.

In practice, some compromise between signal detection and resolution must often be made, so the choice of H_M will depend on experimental goals. The following conditions listed provide some guidelines to appropriate modulation amplitudes for different applications *(4)*:

goal	approach
line resolution:	$H_M \leq 0.2\ (\Delta H_{pp})$
sensitivity (distortion tolerated)	increase H_M to obtain max. A_{pp}
sensitivity (minimum distortion):	increase H_M to obtain max. A_{pp}, then reduce H_M by a factor of 4 to 5

Temperature Effects In EPR Analyses

Temperature influences reactions, relaxation processes and molecular mobility, which in turn control radical lifetimes and signal intensities as well as signal lineshapes. At low temperatures, relaxation of electrons from their high-energy perpendicular orientation to the normal parallel orientation slows because there is less thermal energy, less Brownian motion, less molecular motion, and less molecular interaction. Consequently, radical lifetimes are longer and signal amplitudes increase. According to Curie's law, signal amplitude is inversely proportional to absolute temperature. As a general rule, therefore, it is an advantage to work at temperatures as low as feasible for all radicals whose signals are not motion-dependent (e.g. nitroxide spin labels and probes).

For many radicals, linewidths broaden as temperature increases. In some cases, lines broaden so much as temperature increases that spectra are not detectable at room temperature, and liquid nitrogen (77 K) or even liquid helium (4.2 K) temperatures are necessary to obtain sufficiently narrow lines. Oxygen and metal complexes, particularly hemes and cytochromes are good examples of this behavior (Figure 9). For some radicals, lifetimes are too short for detection at room temperature, or initial radicals undergo temperature-dependent transformations to other radical forms or products. Sulfur radicals (Figure 9) *(9)* and nitric oxide radical adducts of hemes *(12)* are examples of this. Low temperatures also reduce the interference of water, as will be discussed later. On the other hand, low temperature measurements have some disadvantages. Most radicals power saturate much more readily at lower temperatures; molecular immobilization can restrict tumbling, leading to line-broadening and sometimes loss of resolution; and freezing can break bonds, generating artifact radicals. This means that for low temperature experiments power usually must be reduced and extra precaution must be taken in conducting analyses and interpreting data. For each paramagnetic species there usually will be an optimum temperature that maximizes detection and minimizes negative effects and interferences. This temperature must be determined independently for each system, depending on the species to be detected, and then *that temperature must be controlled precisely and maintained throughout the EPR measurement.* Sometimes radicals can be detected only in a very narrow range of temperatures. Dewars with cryogenic fluids provide approximate temperatures and are satisfactory for screening purposes. However accurate analyses of radical species, spectral characteristics, or signal intensity (radical concentrations) requires temperature controllers that pump liquid nitrogen or helium vapors at defined flow rates to generate a range of temperatures and maintain specific temperatures. Thermocouples situated in the cavity (or better, in the sample tube when possible) should be used to verify the exact temperature of the samples.

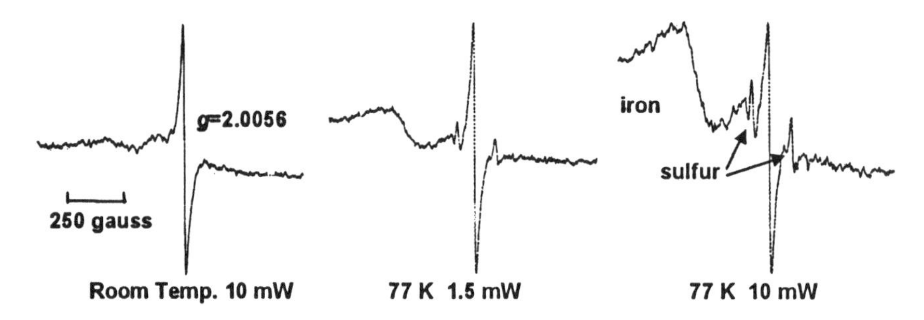

Figure 9. Effect of temperature on EPR signals of extruded cornmeal. Singlet peptide radical is dominant at room temperature, although a metal component is apparent in the background drift. Iron and sulfur signals appear at liquid nitrogen temperature and increase with power, while the peptide radical saturates. Modulation amplitude: 5 gauss, gain for all samples. (Schaich, unpublished data).

Water Effects In EPR Analyses

It can fairly be stated that water is the bane of the EPR experimentalist's existence! As a solvent, water increases molecular mobility and rates of reaction, reducing radical lifetimes. As a proton donor, water increases the rates of radical quenching. These two effects seem obvious. Less intuitive, however, is the reduction of EPR signal intensity by non-resonant absorption of microwave energy by water.

Water reduces the EPR cavity Q, which is a measure of microwave energy actually delivered to the sample *(3-5)*. Microwaves generated by the klystron have electric and magnetic components resonating perpendicular to each other,

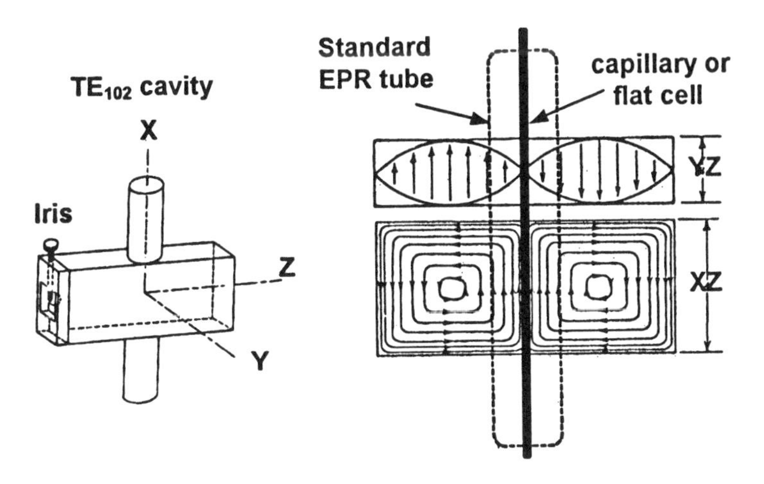

Figure 10. Diagram of rectangular EPR cavity with standing wave patterns from separated electrical (YZ) and magnetic (XY) components of the microwave energy. Standard EPR tubes (dotted lines) protrude into the electrical field; this interference is eliminated by using a flat cell or capillary tube (solid line). Adapted with permission from (3), copyright Wiley-Interscience, 1972.

as shown in Figure 10. Cavities are optimized to concentrate the magnetic component where it contacts the sample, while interaction with the electric component of the microwaves is minimized. Thus, during analysis, sample tubes pass through the zero point of the electric waves and dense concentration of magnetic waves. Water has a high dielectric constant, so when samples in normal tubes have water present, they protrude into the electric field and absorb energy. This decreases their absorption of energy from the magnetic field, and the signal intensity decreases (Figure 11A).

Several methods are used routinely to reduce or eliminate the effects of water, although none are without drawbacks:

1) *Dry the samples.* This removes the interfering water, but drying can break bonds and lead to artifactual signals, especially with lyophilization. Whatever the drying method, samples must be rigorously protected from oxygen before and after dehydration to prevent secondary formation of peroxyl radicals.

2) *Freeze the samples.* This reduces the dielectric constant of water, thus greatly limiting the interaction with the electric field. However, freezing can also generate new radicals because it breaks molecular bonds and it disrupts cells and tissues. Fast freezing minimizes ice crystal formation and tissue disruption, but the rapid expansion of material breaks EPR tubes. For dewar operation, stepwise reduction of temperature through a series of cryogenic fluids prevents tube breakage and minimizes tissue disruption. Most accurate analyses

require monitoring EPR signals as temperatures are gradually reduced using a temperature controller.

3) *Use a flat cell or capillary tube.* This limits water interference in two ways. The small cross-section minimizes interaction with the electric field, while the smaller sample size reduces the actual amount of sample and water present. This is the best method for room temperature analyses of either solids or liquids. The small sample size may be an advantage for precious materials, but it also limits detection sensitivity.

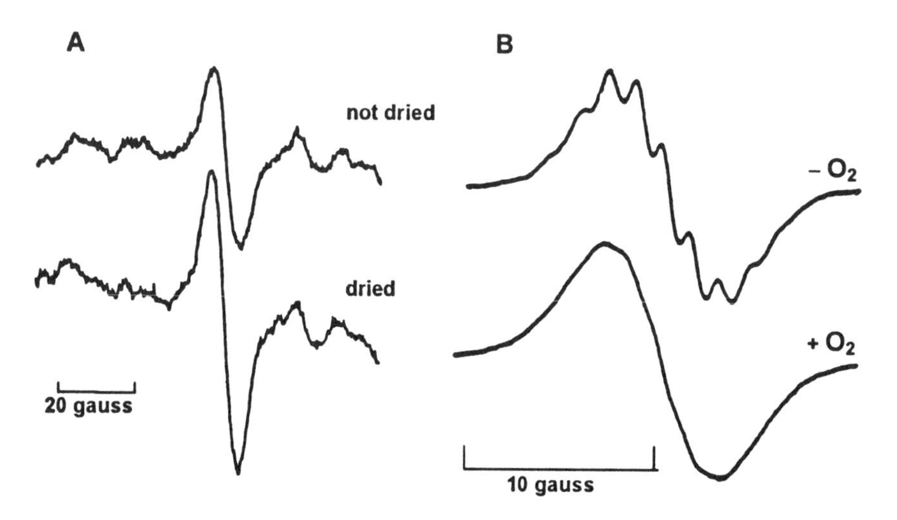

Figure 11. Modification of EPR signals by water and oxygen. A. Reduction of signal intensity in extruded cornmeal by water. Lower spectrum: sample dried under vacuum. Schaich, unpublished. B. Loss of hyperfine structure when spectral lines in Zn tetra-phenyl porphyrin are broadened by oxygen. Adapted with permission from (5), copyright Academic, 1976.

Oxygen Effects On EPR Signals

That oxygen is a biradical has several important consequences in EPR. Oxygen has a very broad spectrum of its own which can contribute to non-linear backgrounds. Oxygen decreases relaxation times of other radicals, leading to line broadening and loss of resolution (Figure 11B). At room temperature, the overall signal intensity is not affected, but at low temperatures where some signals saturate more readily, intensity can appear to increase. Oxygen reacts at diffusion controlled rates with most radicals, thereby converting initial species to peroxyl radicals. The process can be followed in solution by appearance of the typical g-values (*g*=2.01-2.02) and asymmetry of peroxyl radicals. However, in lyophilized materials, a different radical without the shape of peroxyl radicals is produced. It is difficult to distinguish between radicals initially present, radicals produced during lyophilization, and peroxyl radicals resulting from addition of oxygen to any of these radicals.

Oxygen effects can be observed at extremely low oxygen concentrations, as low as 10^{-10} M and oxygen solubility in water and most solvents is considerable. This makes scrupulous removal or exclusion of oxygen a challenging task that most often requires high vacuum lines, vacuum-sealed sample tubes, or glove boxes. Use of high purity nitrogen atmospheres (contain ppm levels of O_2) and sparging of solutions with inert gases are often not sufficient. Argon is superior to nitrogen because it is heavier (doesn't escape from the solution, sample, or flask as fast) and considerably less contaminated with oxygen.

Therefore, possible roles of oxygen must be evaluated for each experimental system in order to determine which effects can be tolerated and which must be eliminated or avoided. The critical issues are to control oxygen access to samples, and to keep oxygen content constant in all samples.

Detecting Short-lived Radicals

The discussion above has assumed static measurements of relatively stable radicals in a standard EPR tube, but it applies equally to all types of EPR measurements. Often the radicals of greatest interest are the ones with the shortest lifetimes, and these require special techniques to generate and maintain detectable steady-state concentrations of radicals *in situ* in the EPR cavity. Space limitations do not allow a discussion of these methods here. Further information is available in Swartz et al. *(3)*.

Summary and Conclusions

EPR is unique in being able to detect free radicals directly, and it is blind to non-paramagnetic species. Thus, EPR can be a powerful tool for studying free radical production and reactions in foods. In this paper, fundamental concepts necessary to understand EPR and to obtain spectra accurately and reproducibly have been reviewed. Three guiding principles for EPR experimentation are:

1) Never assume *a priori* that the EPR signal detected comes from the source expected or that lack of a signal means no radical was formed. EPR requires instantaneous radical concentrations of about 10^{-9} M for detection. Primary radicals with shorter lifetimes or generated at lower concentrations will not be detected in static measurements, and more stable secondary radicals can often become dominant.

2) Signal assignments require measurements. Key attributes of EPR

spectra (g-values and hyperfine splittings) must be measured to interpret the spectra accurately. This requires accurate calibration of microwave field scans and determination of microwave frequency with appropriate instrumentation and standards.

3) Instrumental conditions of power, modulation amplitude, and time constant (not discussed), as well as sample conditions of temperature and presence of water and oxygen must be carefully controlled to detect desired paramagnetic species and to provide maximum resolution and accurate quantitation of EPR signals.

The intense current interest in free radicals in foods will surely lead to increased EPR studies of a wide range of foods to monitor radical production and reactions during processing and storage. With EPR, important information can be gained about radical reactions that may be responsible for food qualities and deterioration. Adherence to the guidelines offered in this paper will ensure that the data obtained reflects the radicals and reactions being studied rather than secondary or artifact radicals, that the quantitation is accurate, and that the interpretations are sound.

References

1. Angier, N. *New York Times Magazine*, **1993**, *Apr. 25*, pp. 62-64.
2. Aruoma, O.I.; Halliwell, B., Eds. *Free Radicals and Food Additives*; Taylor and Francis: London, 1991.
3. Swartz, H. M.; Bolton, J. R.; Borg, D. C. *Biological Applications of Electron Spin Resonance*; Wiley-Interscience: New York, NY, 1972.
4. Wertz, J. E.; Bolton, J. R. *Electron Spin Resonance: Elementary Theory and Practical Applications;* McGraw-Hill: New York, NY, 1972.
5. Borg, D. C. In *Free Radicals in Biology,* Vol. 1; Pryor, W. A., Ed.; Academic: New York, NY, 1976, pp. 69-148.
6. Bersohn, M.; Baird, J.E. *An Introduction to Electron Paramagnetic Resonance*; W.A. Benjamin, Inc.: New York, NY, 1966.
7. *Electron Spin Resonance (ESR) Applications in Organic and Bio-Organic Materials*; Catoire, B., Ed.; Springer-Verlag: Berlin, 1992.
8. Poole, C. P., Jr. *Electron Spin Resonance: A Comprehensive Treatise on Experimental Techniques;* Interscience Publishers: New York, NY, 1967.
9. Schaich, K. M. *CRC Crit. Rev. Food Sci. Nutr,.* **1980,** *13*, 89-129.
10. Hadley, J. H., Jr.; Gordy, W. *Proc. Nat. Acad. Sci. U.S.A.,* **1974,** *71*, 3106-3110.
11. Symons, M. C. R. J. Chem. Soc. Perkins Trans. II, 1974, 1618-1620.
12. Schaich, K. M. NO• production during thermal processing of beef: evidence for protein oxidation. This volume.

Chapter 3

Free Radical Generation during Extrusion: A Critical Contributor to Texturization

K. M. Schaich

Department of Food Science, Rutgers University, 65 Dudley Road, New Brunswick, NJ 08901-8520

EPR has detected free radicals in extruded cornmeal, wheat flour, potato granules, and plantain flour. The *g*-values in all cases are in the range g=2.0055-2.0060, indicating nitrogen-centered radicals, most likely on proteins. Extensive EPR and chemical analyses have shown that the radicals are integrally involved in the chemical changes leading to texturization in these materials. High heat causes peptide scission, which yields N• radicals that are the major component of EPR signals. Under conditions of high shear, S-S bonds are also broken, and sulfur radicals then are evident in the EPR spectra. Oxidizing lipids transfer radicals to protein side chains, and this is reflected in spectral lineshapes. Addition of free radical scavengers during extrusion dramatically diminishes and alters both EPR spectra and extrudate textures. Chemical and microscopic evidence supports EPR data and shows that free radical-mediated protein-protein and protein-starch crosslinking are key determinants of extrudate physical structure.

Introduction

Despite extensive industrial use of extrusion, the molecular mechanisms of extrusion texturization remain poorly understood. In the Center for Advanced Food Technology at Rutgers University, we have been interested in determining specific reactions responsible for texturization during extrusion so that we can learn how to control these reactions to generate desired textures. During early work in this project, adventitious discovery of EPR (Electronic Paramagnetic Resonance) signals in cornmeal extrudates prompted exploratory investigations of free radical formation during extrusion. The EPR signals in cornmeal were broad singlets nearly identical to signals from lysozyme and other proteins irradiated or reacted with oxidizing lipids, and *g*-values were 2.0045-2.0059, indicating nitrogen-centered radicals (Figure 1A). Furthermore, signal intensities and lineshapes correlated with the degree of crosslinking in the extrudates (Figure 1B), suggesting a functional role for the radicals.

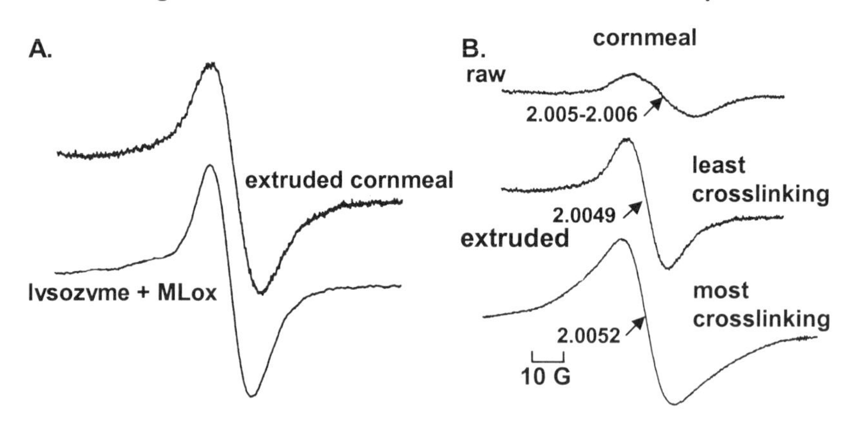

Figure 1. A. EPR spectrum of extruded cornmeal; spectrum of lysozyme reacted with oxidizing lipids included for comparison. Adapted from (1). Copyright American Assoc. of Cereal Chemists, 1999. B. EPR spectra from cornmeal extruded at 195 °C and different moistures correlated with degree of protein crosslinking in the extrudates. Relative signal intensities: cornmeal, 9; least crosslinking (30% moisture), 42; most crosslinking (20% moisture), 122.

This discovery led us to study free radical production systematically in extruded wheat flour to determine whether free radical production was related to extrusion conditions and chemical changes in wheat proteins, and to distinguish the relative contributions of heat, shear, and oxidizing lipids in formation of protein radicals.

Materials and Methods

Extrusion was conducted on a Werner and Pfleiderer ZSK30 twin screw extruder (Werner and Pfleiderer, Ramsey, NJ) equipped with two co-rotating self-wiping screws. Commercial hard red wheat flours (Bay State Milling Co., Minneapolis, MN) were extruded under puffing conditions according to conditions described by Schaich *(1)*, Rebello *(2)*, Koh et al *(3)*, and Wang *(4)*.

EPR analyses were conducted on a Varian E-12 EPR spectrometer interfaced to and controlled by a MassComp 5500 minicomputer and equipped with an X-band (9.5 GHz) microwave bridge, 100 kHz modulation, variable rate signal averaging, and Hewlett Packard 505C frequency counter *(1,5)*. Typical instrumental setting were 5-10 mW power, 5-10 G modulation amplitude, 10,000 gain, 0.03 sec time constant.

Results and Discussion

EPR signals from wheat flour extrudates were a mixture of nitrogen and sulfur centered radicals, with evidence of alkoxyl or peroxyl radicals as well (Figure 2). They varied quantitatively and qualitatively with extrusion conditions. The dominant signal in all extrudates was from nitrogen-centered radicals with broadened shoulders consistent with oxyl radicals. Weaker downfield peaks (left of the central line) from sulfur radicals appeared in some extrudates *(1)*.

EPR signal intensities were not random; they varied with extrusion conditions and showed strong correlations with high temperature and low moisture *(1)*. This indicates that the free radicals were not experimental artifacts but were integrally related to the extrusion process. Furthermore, EPR signals also correlated with SH/SS and HPLC changes and with protein loss *(2)*. Most importantly, extrudates with high EPR signals also showed the most dramatic alterations in protein mol wt distributions. HPLC peaks from gliadins, some low mol wt glutenins, and albumins and globulins were lost, and polymeric and low mol wt fractions increased *(2)*. These observations supported a critical role for free radicals in the chemical mechanisms of protein changes during extrusion.

To distinguish the relative contributions of heat, shear force, and lipid oxidation in production of free radicals during extrusion, raw flour was heated in an oven at 180 °C for varying periods, cooled, and analyzed by EPR at 77 K. EPR spectra were narrow singlets (g=2.0056) that increased with heating time (Figure 3). These are highly localized nitrogen radicals that arise from peptide scission. No sulfur radicals were evident even with long heating times and with addition of moisture. Either heat does not induce disulfide bond scission or the sulfur radicals were too short-lived for detection. The very weak nitrogen signals detected in starch derived from traces of nitrogenous contaminants. Starch radicals were never detected. At the present time, it is not known whether starch radicals are not formed or, more likely, they anneal too rapidly for detection *(1)*.

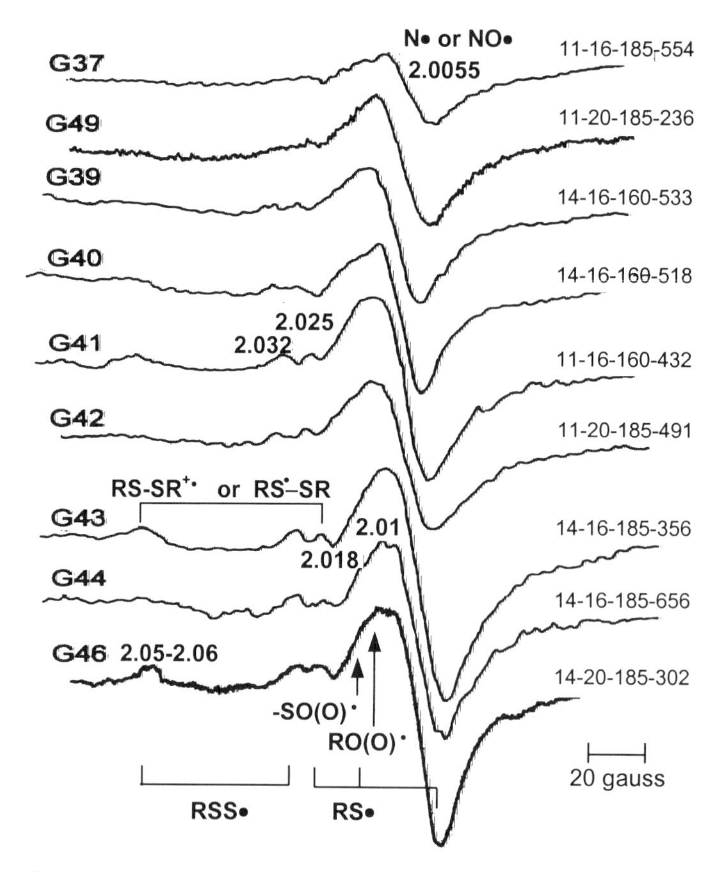

Figure 2. Low temperature EPR spectra from extruded wheat flour. g-values of major peaks and associated radical species are noted. Number sequences at right of spectra -- % protein, moisture, temperature, SME. G37-G46: extrusion run code numbers. Adapted from (1) with permission. Copyright American Assoc of Cereal Chemists, 1999.

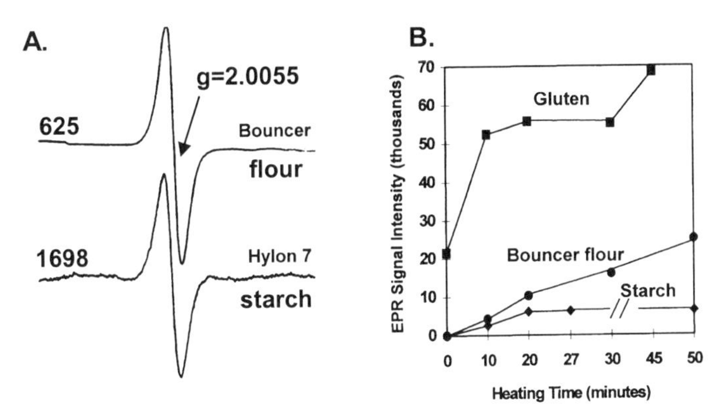

Figure 3. Heat production of nitrogen radicals. A. Room temp. EPR spectra from flour and Hylon 7 amylose, 180 °C for 10 min. Relative intensities noted at left. B. Nitrogen radical production as a function of heating time at 180 °C. Adapted from (1) with permission . Copyright American Assoc. of Cereal Chemists, 1999.

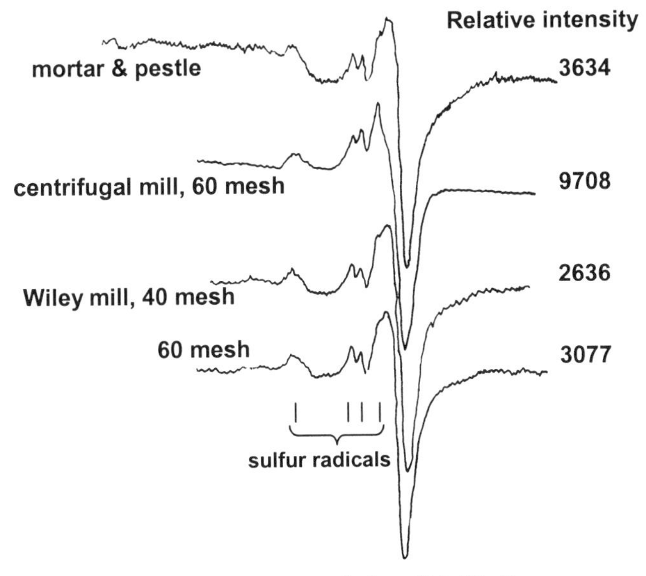

Figure 4. EPR signals of wheat flour extruded at 185 °C die temperature, 16% moisture, 225 g/min mass flow rate, and 500 rpm screw speed, then ground for analysis by various methods. Sulfur peaks in EPR signals show that shear stress (grinding) produces sulfur radicals via scission of S-S bonds.

Disulfide bonds are broken by shear force, however, as was evident in grinding experiments (Figure 4). Grinding is known to produce free radicals, so sample preparation for EPR analysis normally uses gentle grinding by hand in a mortar and pestle. EPR signals of extrudates ground in this manner are qualitatively identical to EPR signals from intact strands of extrudates (data not shown). The signal remained largely unchanged when extrudates were ground in a Wiley mill to 40 mesh (e.g. for chemical analysis), but at 60 mesh sulfur radicals increased. Samples ground to 60 mesh in a centrifugal mill showed marked shifts in the spectra, with diminished nitrogen radicals (center line), dominant sulfur radicals (shoulder from sulfur oxyl radicals and downfield peaks from thiyl and disulfide radicals), and much higher signal intensity. Chemical analyses of SH/SS contents showed an order of magnitude increase in available sulfhydryls

These results show that shear force breaks disulfide bonds and, further, provide a note of caution that materials in which SH/SS or other protein analyses are to be conducted must be handled gently to avoid artifacts and misleading results.

Raw flour and starch reacted with oxidizing methyl linoleate to determine the ability of lipids to induce radicals in flour proteins gave EPR signals that were broad singlets with significant wing structure but no sulfur radicals (Figure 5), consistent with previous evidence that oxidizing lipids can only react with surface residues on proteins and generally cannot break disulfide bonds (6).

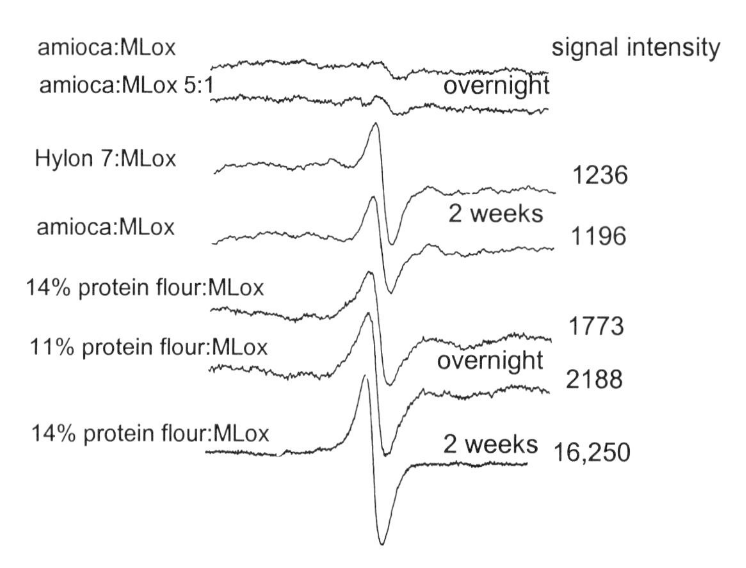

Figure 5. Free radicals induced in wheat flours and starch by oxidizing methyl linoleate, MLox (10:1 weight proportions). Adapted from (1) with permission, Copyright American Assoc. of Cereal Chemists, 1999.

The EPR lineshape is typical of composite spectra in which signals from many different but closely related radicals overlap. In this case, signals probably result from radicals on multiple side chains, particularly tryptophan, histidine, arginine, and lysine which are accessible to the lipids and have easily abstractable nitrogens *(6,7)*. The weak radical signals in starch again were nitrogen-centered, not carbon, arising from nitrogenous contaminants rather than the starch itself.

From these results, we propose that in EPR signals from wheat flour the dominant center line is from nitrogen-centered radicals produced during heat-induced peptide scission, down-field peaks and shoulders are from sulfur radicals induced by shear scission of disulfide bonds, and the line broadening and wings are from protein side chain nitrogen or oxyl radicals transferred from oxidizing lipids.

Integrating results to this point, we proposed an initial Working Hypothesis to explain protein changes and texturization extrusion. Under pressure, proteins denature; they become increasingly susceptible to fragmentation by heat and shear, and more amino acid side chains also become accessible to reaction with lipids. Fragmentation or reaction with lipids forms protein radicals, some of which recombine to generate a crosslinked network. Extensive crosslinking of large proteins yields an insoluble coarse network that provides the structural framework. Solubility, microscopic, chemical, and textural analyses suggest that starch acts primarily as the filler in the dispersed phase (fine network), along with some soluble proteins, protein-starch complexes, and protein fragments (Schaich, unpublished data; Wang, 2000). The overall texture in a given product is determined by the balance between the two fractions, as shown in Scheme 1.

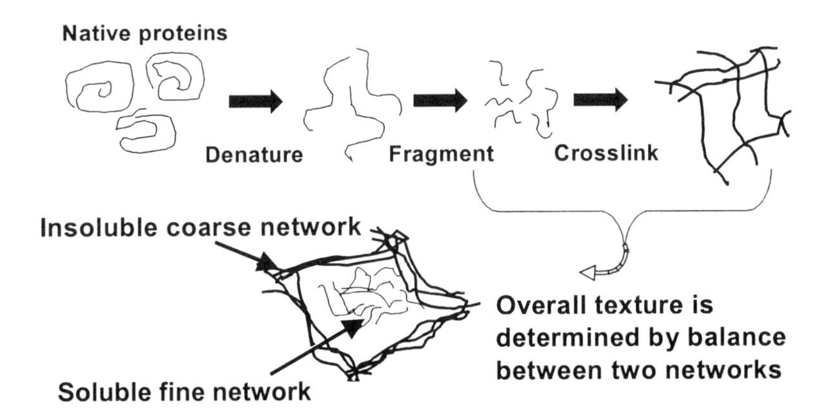

Scheme 1. Proposed reaction scheme explaining how protein changes during extrusion lead to texturization of final products.

To test this Working Hypothesis, 0%, 0.5%, 1% and 2% cysteine were added to wheat flour as a radical scavenger during extrusion (3). The original Bouncer flour (14% protein) was extruded at process moisture 16% (w/w), constant die temperature 185°C, 225 g/min mass flow rate, and screw speed 500 rpm -- conditions that provided optimum wheat flour expansion, flavor and textural quality in the earlier experiments.

Cysteine selectively quenched N• signals of the peptide backbone, leaving sulfur resonance behind in the process (Figure 6). The signals visually appeared to weaken as cysteine levels increased. The extrudates became increasingly light and electrostatic and did not pack. Hence, when extrudates were normalized to constant packing density, signal intensities remained surprisingly high. Computer analysis indicated that much of the signal was coming from the added cysteine rather than wheat protein, which is what would be expected if cysteine was acting to quench protein nitrogen-centered radicals:

$$RSH + P\text{-}N\bullet \longrightarrow P\text{-}NH + RS\bullet$$

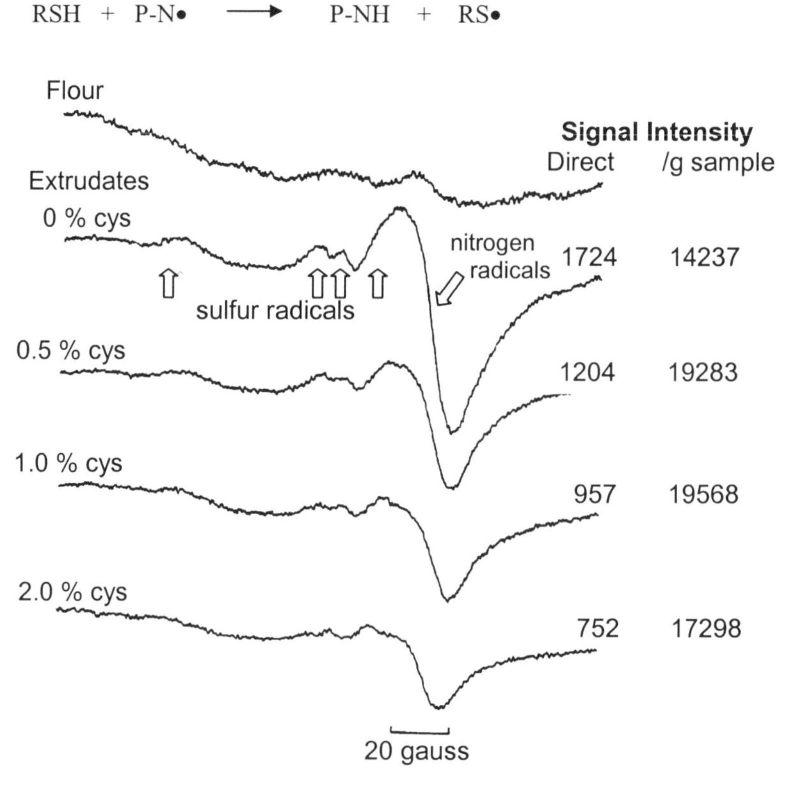

Figure 6. EPR signals of wheat four extruded with varying levels of cysteine added as a radical scavenger. Adapted from (3) with permission. Copyright American Assoc. of Cereal Chemists, 1996.

Cysteine dramatically increased the solubility of extrudate proteins in 1.5% SDS without the need for β-MCE to reduce S-S bonds (Figure 7). Increased low mol wt material and decreased polymeric material in SDS-PAGE analyses under reducing and non-reducing conditions showed that the solubilization resulted from radical quenching and prevention of crosslinking by radical recombination *(3)* (Figure 8). Proton donation from cysteine "fixed" random fragments of proteins and decreased the degree of polymerization so that previously insoluble proteins became soluble and molecular weights were reduced to a size that would enter the gels. Reducing gels merely released normal protein bands, so it appears that disulfide bonds did not contribute to new protein crosslinking, but rather acted in their normal positions to link larger aggregates of proteins crosslinked by radicals. There was no evidence for thiol-disulfide interchange in the extrudates. Thiol-disulfide interchange is a slow process that requires high moisture levels, whereas extrusion was run at low moisture and was complete within 2-3 minutes.

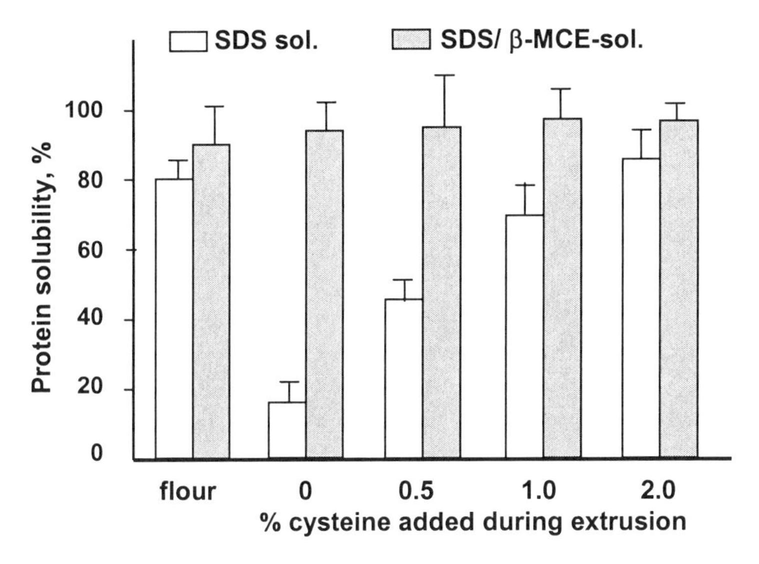

Figure 7. Protein solubility increases in wheat flour extrudates as a function of level of cysteine added as a free radical scavenger during extrusion. From Koh (3), used with permission. Copyright American Association of Cereal Chemists, 1996.

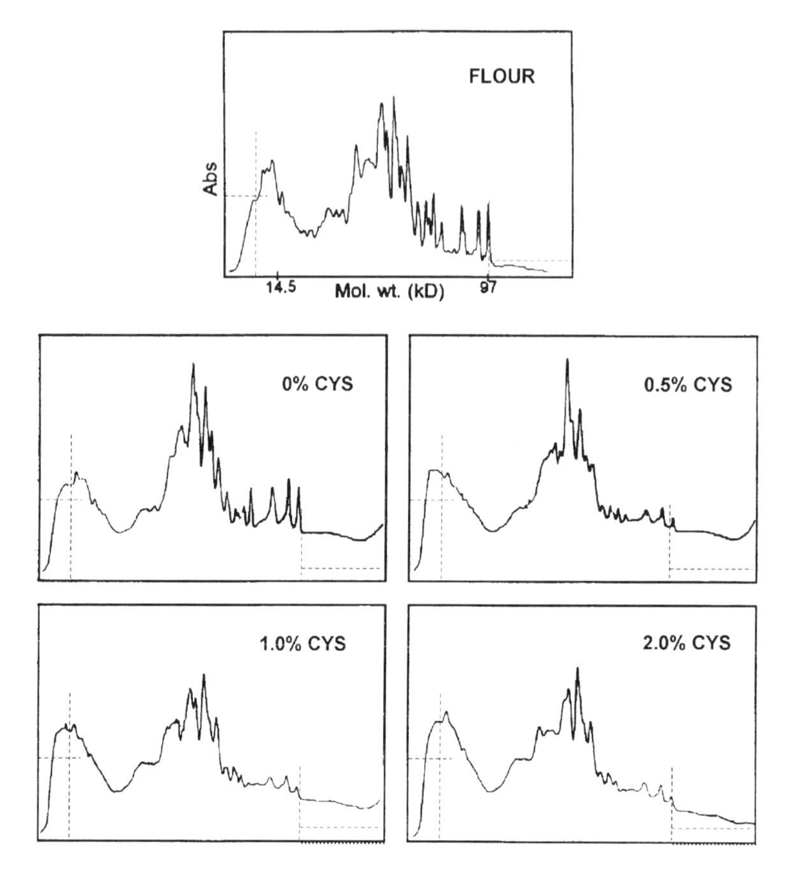

Figure 8. Densitometer scans of proteins extracted from wheat flour and extrudates separated by SDS-PAGE. Hatched lines provide reference to follow increase of low mol wt fragments and loss of high mol wt material with addition of cysteine. From (3), used with permission. Copyright American Assoc. of Cereal Chemists, 1996.

As in the initial extrusion study, extrudate properties correlated with radical production, or in this case scavenging, in the extrudates. As radical scavenging by cysteine increased, expansion ratio, cell size, and cell wall thickness all decreased (3) (Figure 9) because the crosslinked protein network that provided both structure and extensibility to the dough had been disrupted. Microscopy with differential staining of proteins and starch revealed mesh-like protein structures with starch dispersed within in control extrudates; with added cysteine the mesh structure was replaced by thin aligned strands of protein (Schaich, unpublished).

In investigating the nature of the increased polymeric material made

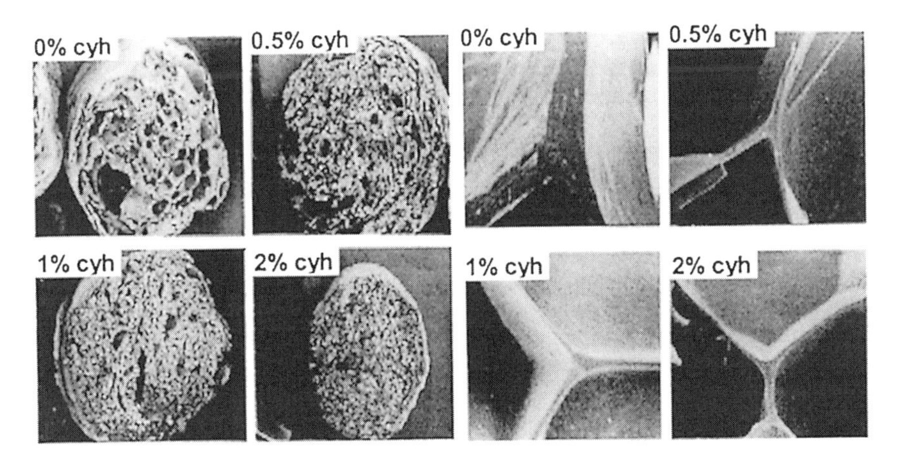

Figure 9. Radical quenching alters ultrastructure of extrudates. Expansion and cell size decrease (left) and cell walls thin (right) with cysteine. Adapted from (3) with permission. Copyright American Association of Cereal Chemists, 1996.

available by cysteine, periodic acid-Schiff base stain of PAGE gels detected protein-starch (P-S) complexes in addition to the protein-protein (P-P) known to be present. This was a very interesting discovery since starch radicals had not been observed, and it led to additional studies focussed on P-S complexes *(3)*.

To isolate proteins involved in protein-starch complexes, soluble proteins were removed from extrudates with SDS/β-mercaptoethanol; the insoluble residues were then treated with α-amylase to digest any starch present and release any proteins that had been bound. In separate analyses, the entire extrudate was treated with glucoamylase to remove starch and proteins were extracted. Protein was quantified by Kjeldahl analyses and protein glucosamine contents were measured as one form of protein-starch complex *(4)*.

Extrudates showed typical reduction in protein solubility relative to raw flour, up to about 15% depending on extrusion conditions. However, when the proteins released from insoluble residues by amylase were added to the soluble protein, recovery of protein was essentially quantitative (Figure 10). Protein glucosamine levels revealed more extensive protein-starch crosslinking in the residue fraction than in soluble proteins (Figure 11). Interestingly, glucosamine levels in both fractions decreased with cysteine addition, but the change in the soluble glucosamines was more than 50% (Figure 12). This indicates that two different mechanisms are involved in protein-starch complexation. Formation of soluble protein-starch complexes involves primarily free radical reactions and is inhibited by free radical scavengers, while complexation in the insoluble residues appears to result from both radical and non-radical reactions. Soluble protein-starch complexes showed statistically significant correlations to textural

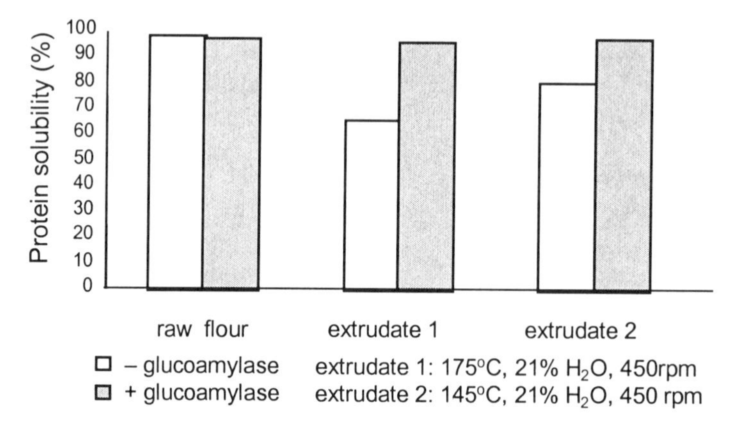

Figure 10. Glucoamylase digestion releases proteins complexed with starch and results in essentially complete solubilization of proteins in wheat flour extrudates. Other extrusion conditions (see Figure 11) show comparable behavior. Adapted from (4) with permission. Copyright American Assoc. of Cereal Chemists, 1996.

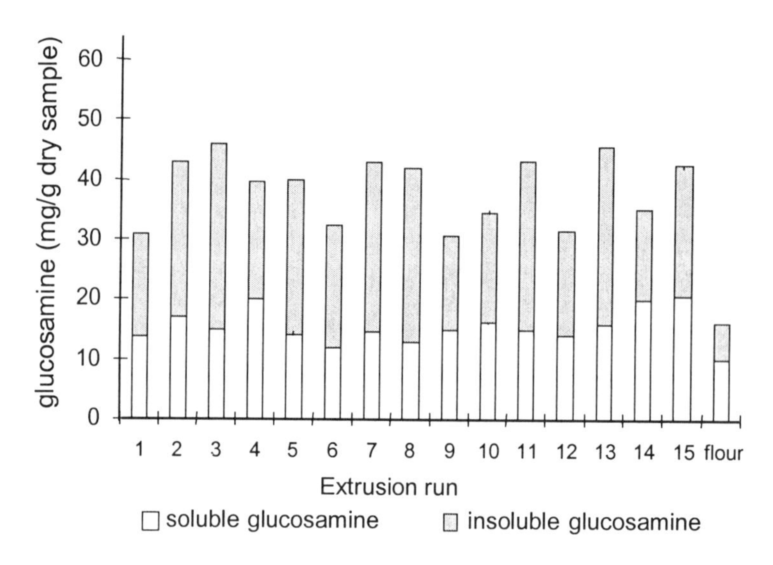

Figure 11. Protein-starch complexation in wheat flour extrudates as measured by glucosamine contents of SDS-soluble proteins and proteins released from insoluble residue by amylase. Extrusion runs are various combinations of 18, 21, and 24% moisture; 145, 160, and 175 °C die temperature; and 300, 375, and 425 rpm screw speed. Adapted from (4) with permission. Copyright American Assoc. of Cereal Chemists, 1996.

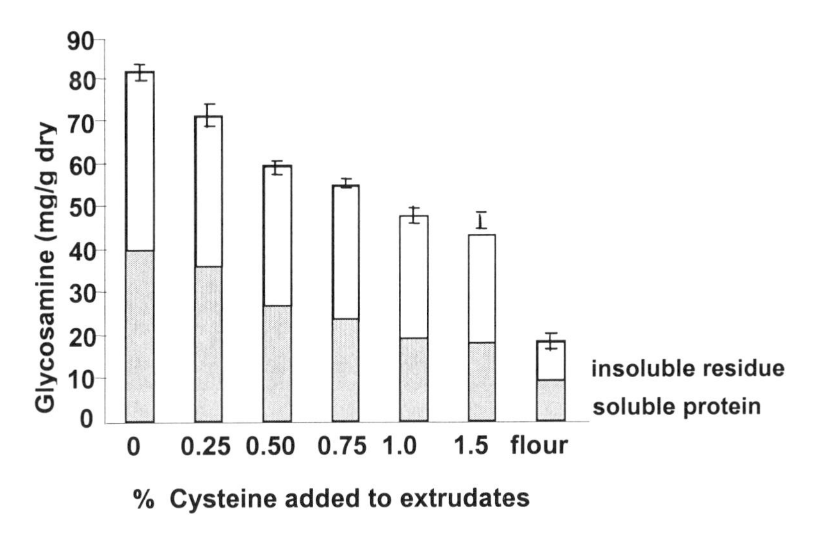

Figure 12. Effect of free radical scavenging by cysteine on distribution of glucosamine protein-starch linkages in soluble protein and insoluble residue fraction of wheat flour extrudates. Adapted from (4) with permission. Copyright Amer. Assoc. Cereal Chemists, 1996.

properties of extrudates, including bulk density, springiness, elastic modulus, fracturability, and hardness *(4)*. These observations suggest that free-radical mediated protein-starch complexation is important in forming the fine network dispersed in the structural protein matrix.

Summary and Conclusions

EPR and chemical evidence has been presented to show that free radicals formed in proteins play a critical and integral in chemical changes leading to texturization during extrusion. Radicals arise from heat-induced peptide scission, shear-induced S-S scission, and reaction of oxidizing lipids with amino acid side chains. The result is protein fragmentation and recombination of protein radical fragments with proteins and with starch to form high mol wt masses. Peptide crosslinking is the dominant process, with the role of disulfide bonds being interconnection of peptide-crosslinked protein masses. Quenching of the free radical processes leads to diminished expansion due to disruption of the structural protein network and protein-starch complexation. Results support a model that describes extrusion texturization as a mixture of a protein-based, largely insoluble structural network and a soluble network of starch and proteins dispersed within. Free-radical-mediated crosslinking of peptide and starch fragments forms soluble protein-starch complexes that significantly affect extrudate textures by modifying the fine network dispersed within the protein structural matrix.

References

1. Schaich, K. M.; Rebello, C. A. *Cereal Chem.* **1999**, *76*: 748-755.
2. Rebello, C. A.; Schaich, K. M. *Cereal Chem.* **1999**, *76*: 756-763.
3. Koh, B. K.; Karwe, M. V.; Schaich, K. M. *Cereal Chem.* **1996,** 73, 115-122.
4. Wang, S. *Ph.D. Thesis*, Rutgers University, New Brunswick, NJ, 2000.
5. Schaich, K. M. EPR methods for studying free radicals in foods, this volume, pp. xx.
6. Schaich, K.M.; Karel, M. *Lipids* **1976**, *11*, 392-400.
7. Schaich, K.M. *CRC Crit. Rev. Food Sci. Nutr.* **1980**, *13*, 189-244.

Chapter 4

CROSSPY: A Radical Intermediate of Melanoidin Formation in Roasted Coffee

T. Hofmann [1], W. Bors[2], and K. Stettmaier[2]

[1]Deutsche Forschungsanstalt für Lebensmittelchemie Lichten-bergstrasse 4, D-85748 Garching, Germany
[2]Institut für Strahlenbiologie, GSF Forschungszentrum für Umwelt und Gesundheit, 85746 Neuherberg, Germany

Although it is well-known for nearly half a century that free radicals are generated during roasting of coffee beans, neither their structures, nor their role as potential intermediates in melanoidin formation are as yet clear. EPR- and LC/MS-spectroscopic data as well as carefully planned synthetic experiments gave strong evidence that the structure of the radical detected in coffee melanoidins is the previously unknown protein-bound 1,4-bis-(5-amino-5-carboxy-1-pentyl)pyrazinium radical cation. Synthetic as well as quantitative experiments revealed that this radical species, which was named CROSSPY, is formed during roasting of the coffee beans from Maillard reactions of protein-bound lysine involving glyoxal and reductones, both formed during carbohydrate degradation. In addition, this CROSSPY was found to be an effective intermediate in radical assisted color development running predominantly via 2-hydroxy-1,4-bis-(5-amino-5-carboxy-1-pentyl)-1,4-dihydropyrazine as penultimate browning precursor.

Besides the unique aroma, the typical brown color developing during thermal food processing mainly originates from interactions between reducing carbohydrates and amino compounds, known as the Maillard reaction (*1,2*). E.g. in roasted coffee, breakfast cereals, bread crust, roasted meat or kiln-dried malt, this browning is highly desirable and is intimately associated in consumers minds with a delicious, high-grade product. This browning is mainly due to the so-called melanoidins, which are assumed to be water-soluble, brown colored, nitrogen-containing, polyfunctional macromolecules with masses up to 100 000 daltons *(1,2)*. Recent investigations relating the molecular weight distribution and the color potency of thermally treated mixtures of glucose and amino acids or proteins, respectively, revealed that carbohydrate-induced protein oligomerization generated melanoidins most effectively *(2)*. The knowledge on distinct structures and on the formation of these melanoidins is, however, as yet very fragmentary. To further improve the quality of processed foods such as roasted coffee, e.g. by controlling the non-enzymatic browning reaction more efficiently, a better understanding of the mechanisms of melanoidin formation is, however, required.

Studies on the non-enzymatic browning of carbohydrates and amino acids evidenced that, besides ionic condensation reactions *(3,4)*, also mechanisms involving free radical formation produce browning compounds in the Maillard reaction *(5-7)*. Because radicals have been detected in model melanoidins *(8,9)*, similar reactions might be involved in the formation of melanoidins during thermal food processing.

More than 40 years ago, free radicals were found to be formed upon roasting of green coffee beans accompanied by intense browning development *(10)*. The number of unpaired spins was estimated to be in the order of 10^{16}/g of roasted coffee powder *(10,11)*. Investigations on the influence of technological parameters on the concentration of that type of radical revealed that the radical formation is strongly accelerated with increasing time and temperature of the roasting procedure, and run in parallel with the color intensity of the roasted coffee *(10-14)*. The same type of radical could be detected in decaffeinated coffee as well as in imitation coffees *(13)*.

Investigations on the influence of the coffee brew preparation revealed the highest radical concentration of 10^{17}/g coffee grounds, and about 10^{14}/g in coffee brew *(11)*. These data clearly indicated that during preparation of the coffee brew the radicals are extracted only to some extend being well in line with a comparatively low concentration of unpaired spins in instant coffees *(12)*.

First systematic studies aimed at characterizing the precursors of these radicals in green coffee beans evidenced that the radical in coffee is most likely generated upon thermal reactions involving carbohydrates, and excluded phenols as natural source for these radicals *(14)*.

Although extensive investigations have been performed for more than 35 years, the chemical nature and the formation of the free radical in roasted coffee is still unknown. The objective of the present investigation was, therefore, (i) to characterize the chemical structure of the free radical formed during roasting of coffee beans, and (ii) to study whether this radical is somehow involved in melanoidin formation.

Experimental

Materials

Coffee brew was prepared from freshly ground, roasted coffee beans (Arabica, variety Caturra) by a percolation method using hot water at 95°C. The coffee/water ratio was 50 g/L. Phenols were removed from the coffee brew by treatment with poly(vinyl pyrrolidone) as reported in the literature *(14)*. The water used for the EPR experiments was stirred with an ion exchange resin (Amberlite MB-1; Merck, Darmstadt, Germany) for at least 24 h prior to use, to remove adventitious metal ions. The 1,4-diethylpyrazinium diquaternary salt was synthesized as recently reported by perethylation of pyrazine with triethyloxonium tetrafluoroborate (*7*).

Isolation of Melanoidins from Coffee Brew

A concentrated coffee brew (6 g/20 mL) was fractionated by gel filtration chromatography (5 × 75 cm) using Sephadex G-25 fine as the stationary phase and water (4 mL/min) as the eluent (*15*). Five fractions were collected, concentrated in a freeze-dryer and, then, analyzed by EPR spectroscopy.

Spectroscopic measurements

The EPR spectra were recorded at room temperature on a ESP 300 spectrometer (Bruker, Rheinstetten, Germany) using the experimental parameters reported recently (*7*). LC/MS spectroscopy was performed with a LCQ-MS (Finnigan MAT GmbH, Bremen, Germany) using electrospray ionization (ESI).

Reaction Mixtures of N_{α}-acetyl Amino Acids and Glucose

Binary mixtures of an amino compound (5 mmol) and glucose (5 mmol), respectively, were heated in phosphate buffer (4 mL; 0.5 mmol/L; pH 7.0) in closed vials at 95°C. After rapid cooling, an aliquot of each reaction mixture was analyzed by EPR spectroscopy.

Reaction Mixtures of Bovine Serum Albumin with Glucose or Glyoxal in the Absence/Presence of Ascorbic Acid

A mixture of glucose or glyoxal (1.25 mmol), respectively, and bovine serum albumin (0.05 mmol) was reacted in phosphate buffer (15 mL; 0.5 mmol/L; pH 7.0) at 95°C for 5 min. After cooling, high molecular melanoidins were isolated by ultracentrifugation with a cut-off of 100 000 Da (Centriplus, Amicon, Witten, Germany) as recently reported (*16*). The retentate was suspended in water (10 mL) and analyzed by EPR spectroscopy. The reaction mixture containing glyoxal was incubated with ascorbic acid (1 mmol) for 10 min at room temperature and, then, analyzed again by EPR spectroscopy.

Results and Discussion

To gain first insights into the chemical nature of the radical detected in roasted coffee, a coffee brew was freshly prepared by hot water extraction of roasted coffee beans and the brew was stirred with poly(vinyl pyrrolidone) in order to remove phenolic components. The suspension was filtered and then analyzed for paramagnetic behavior by using EPR spectroscopy. An intense broad signal of a radical could be observed fitting well with the radical reported in the literature *(10,13,14)*, and demonstrating that extraction of the phenols failed to remove the radical. These experiments confirmed data reported in the literature that phenols are not involved in generating the radical during coffee roasting *(14)*.

To study whether these radicals are somehow related to browning formation during coffee roasting, intense brown-colored melanoidins were isolated from a freshly prepared coffee brew by gel filtration chromatography using Sephadex G-25 fine as the stationary phase and water as the eluent (Figure 1).

The browning compounds of the coffee brew could be successfully separated into five fractions, amongst which fractions no. 1, 4 and 5 were intensely dark brown in color (Figure 1). Screening these fractions for paramagnetic behavior by EPR spectroscopy revealed the intense broad singlet displayed in Figure 2.

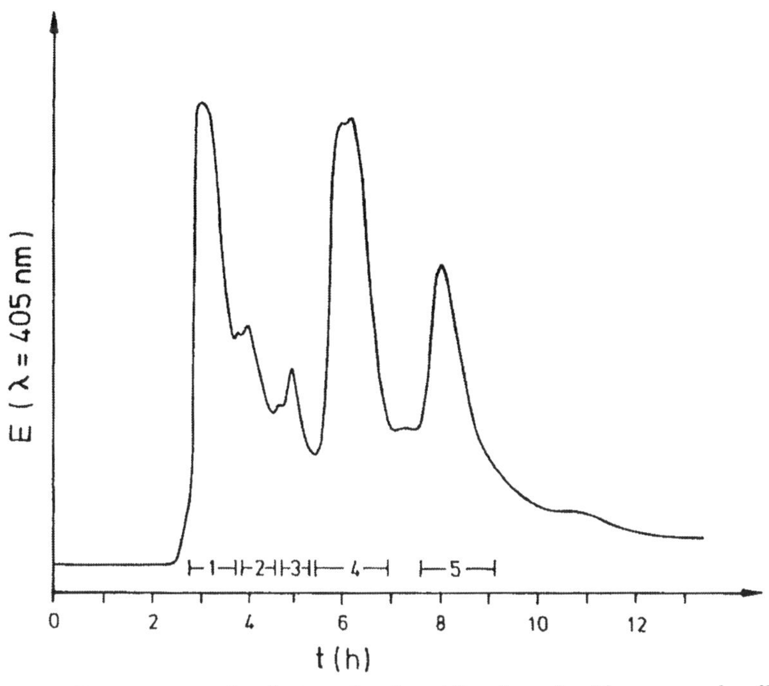

Figure 1. Separation and isolation of melanoidins from freshly prepared coffee brew by gel filtration chromatography.

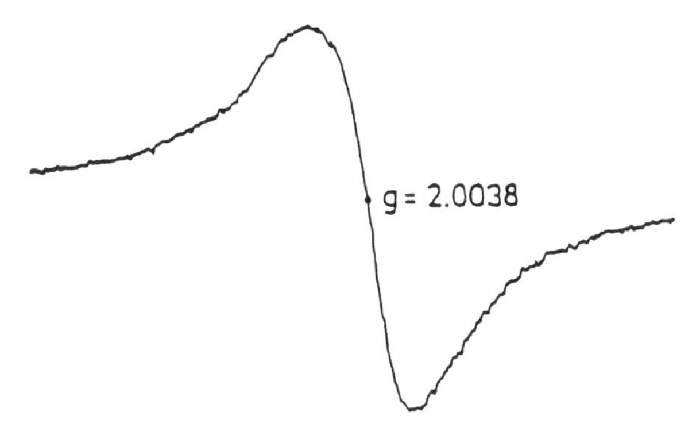

Figure 2. EPR signal of radical detected in melanoidin fraction no. 1, 4 and 5 isolated from freshly prepared coffee brew.

This type of radical exhibiting a g-value of 2.0038 was identical with the data reported in the literature *(13)*. The broad shape of the EPR-signal is typical for immobilized radicals showing restricted rotation and, therefore, implies that the radicals in coffee are somehow linked to the macromolecular melanoidins.

Structure Characterization of the Free Radical in Coffee Brew

To identify the natural precursors of this type of radical, several binary mixtures of mono-, di- and polysaccharides, amino acids and proteins have been thermally treated, brown colored melanoidins have been isolated by ultracentifugation and then analyzed by EPR spectroscopy for paramagnetic behavior (Hofmann, unpublished results). In a heated mixture of bovine serum albumin (BSA), which was chosen as a model food protein, and glucose, an intense radical with a g-value of 2.0038 was detected in the melanoidins *(15)*. The EPR signal of that radical, displayed in Figure 3, was identical with that detected in the coffee melanoidins (cf. Figure 2).

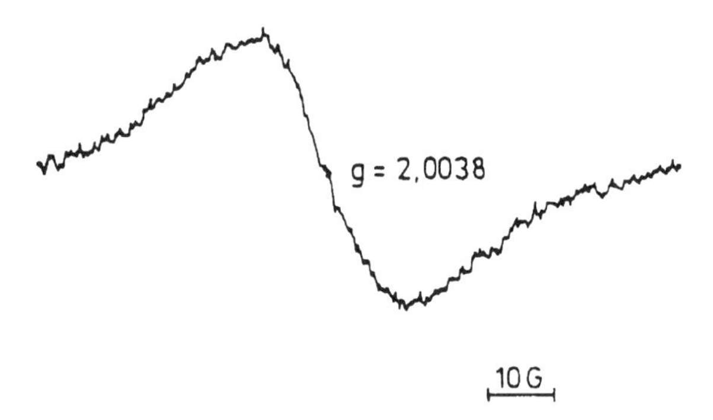

Figure 3. EPR signal of free radical detected in a thermally treated mixture of protein (BSA) and glucose.

To identify the amino acids in the protein, which are involved in radical formation, in a first experiment, we analyzed the amino acid composition in the coffee proteins before and after roasting of the coffee beans. The most drastic loss was observed for the amino acids lysine, arginine, cysteine and histidine, the amounts of which are summarized in Table I. On the basis of these results, it was assumed that the degradation of one of these amino acids is somehow involved in radical formation.

Table I. Most significant Changes in the Amounts of Amino Acids in Green Coffee Beans Occurring during Roasting

Amino acid	*Amino acids in coffee proteins [mol %]*	
	before roasting	after roasting
L-lysine	6.1	0.9
L-arginine	4.7	0.3
L-cysteine	1.6	0.3
L-histidine	2.2	1.9

In order to evaluate the ability of the side chains of these protein-bound amino acids in generating radicals, the corresponding N_α-acetyl-protected amino acids, which were chosen to model reactions of protein-bound amino acids, were heated in the presence of glucose and the reacted mixtures were analyzed for paramagnetic behavior by EPR spectroscopy (Table II).

Table II. Influence of Amino Acid Side Chains on Free Radical Formation When Reacted with Glucose

N_α-Acetyl-protected amino acid[a]	*rel. radical intensity [%]*
L-lysine	100
L-arginine	0
L-cysteine	0
L-histidine	0

[a] Binary mixtures of N_α-acetyl-protected amino acids and glucose (5 mmol each) were heated in phosphate buffer (4 mL; 0.5 mol/L; pH 7.0) for 15 min at 95°C.

As given in Table II, exclusively in the mixture containing the N_α-acetyl-L-lysine intense radical formation could be observed indicating the ε-amino group of the lysine side chain as an effective radical precursor. EPR spectroscopy revealed a spectrum which hyperfine structure is displayed in Figure 4. The splitting constants of 8.39, 2.92 and 5.40 G were in the range of those reported for the two equivalent nitrogens, four equivalent pyrazinium protons and four equivalent ε-protons of the ethyl groups of the 1,4-diethylpyrazinium radical cation reported in the literature *(5-7,17)*. Due to the striking similarity between the hyperfine structure of the EPR signal obtained in the glucose/N_α-acetyl-L-lysine mixture and a computer-simulated spectrum, the structure of the radical was assumed as the 1,4-bis-(5-acetylamino-5-carboxy-1-pentyl)pyrazinium radical cation.

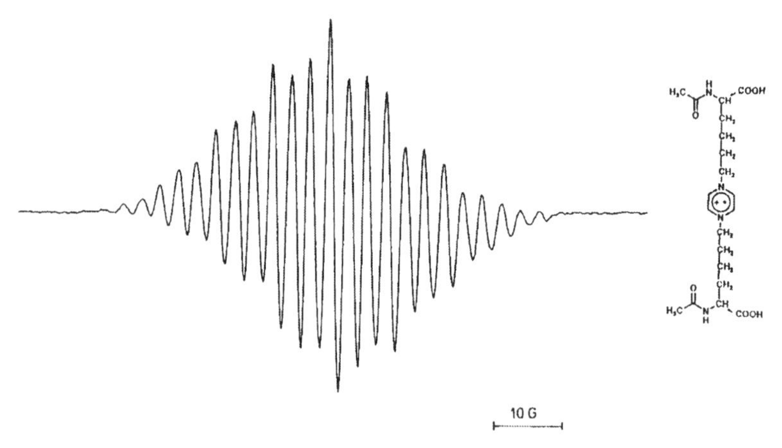

Figure 4. EPR signal of 1,4-bis-(5-acetylamino-5-carboxy-1-pentyl)pyrazinium radical cation detected in a heated mixture of N_α-acetyl-L-lysine and glucose

The proposed radical cation structure was further confirmed by means of LC-MS measurements. Besides the base peak at m/z 189 (100 %), corresponding to unreacted N_α-acetyl-L-lysine, an ion with m/z 424 (35%) was detected by LC-MS of the reacted glucose/N_α-acetyl-L-lysine mixture (unpublished results). Because the latter ion was expected for the proposed pyrazinium radical cation, it was further characterized by LC-MS/MS. As displayed in Figure 5, besides the molecule ion at m/z 424 a loss of 172 revealed the daughter ion m/z 252 (100 %) as the base peak, most likely resulting from the elimination of 5-acetylamino-5-carboxy-1-pentene from the 5-acetylamino-5-carboxy-1-pentyl side chain of the radical cation *(7)*. Taking all these data of the ESR and the LC-MS experiments into account, the structure of the free radical was identified as the 1,4-bis-(5-acetylamino-5-carboxy-1-pentyl)pyrazinium radical cation *(7)*.

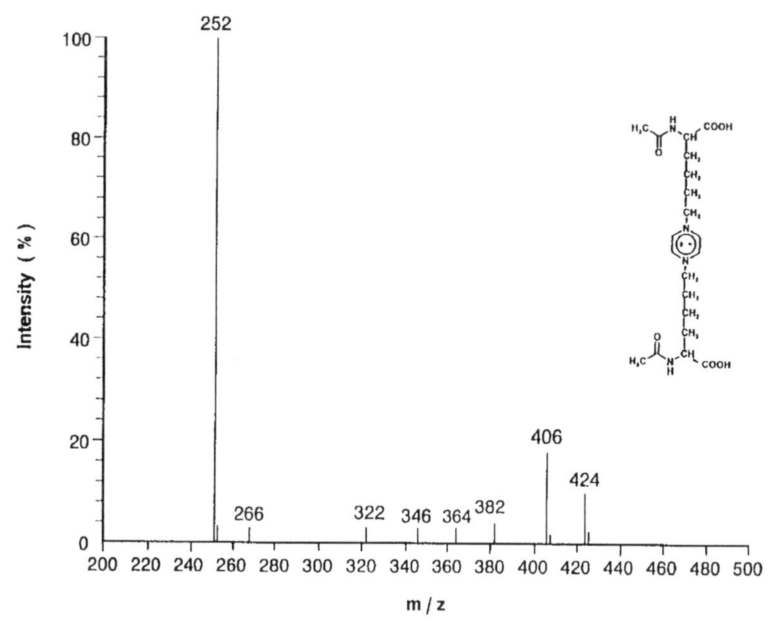

Figure 5. LC-MS/MS of the molecular ion (m/z=424) of 1,4-bis-(5-acetylamino-5-carboxy-1-pentyl) pyrazinium radical cation formed in a thermally treated mixture of N_α-acetyl-L-lysine and glucose.

The EPR spectroscopic data of the melanoidins isolated from coffee brew (cf. Figure 2) as well as from the BSA/glucose mixture (cf. Figure 3) revealed a very broad singlet of a radical. In contrast, the radical cation formed in the reaction of glucose and N_α-acetyl-L-lysine showed that hyperfine structure displayed in Figure 4. In order to check whether the broad signal of the melanoidins in Figures 1 and 2 represents also 1,4-bis-(5-amino-5-carboxy-1-pentyl)pyrazinium radical cations, however, in a protein-bound immobilized form instead, the following experiments were aimed at investigating the influence of peptide-bound L-lysine on the EPR spectrum. To achieve this, the N_α-9-fluorenylmethoxycarbonyl-L-alanyl-L-lysyl-L-leucyl-L-glycine was synthesized following the five-step reaction sequence displayed in Figure 6 *(18)*. Starting with N_α-FMOC-N_ε-BOC-protected L-lysyl-L-leucine (I), dicyclohexylcarbodiimide-catalyzed peptide-bond formation with L-alanine *tert*-butyl ester revealed N_α-FMOC-N_ε-BOC-L-lysyl-L-leucyl-L-glycine *tert*-butyl ester (II), from which the FMOC-protecting group was cleaved by exposure to morpholine yielding the N_ε-BOC-L-lysyl-L-leucyl-L-glycine *tert*-butyl ester (III). Peptide-bond coupling between the deprotected amino group of III and

FMOC-L-alanine afforded the N_α-FMOC-L-alanyl-N_ε-BOC-L-lysyl-L-leucyl-L-glycine *tert*-butyl ester (IV), which by exposure to trifluoracetic acid was cleaved to the target compound N_α-FMOC-L-alanyl-L-lysyl-L-leucyl-L-glycine (V).

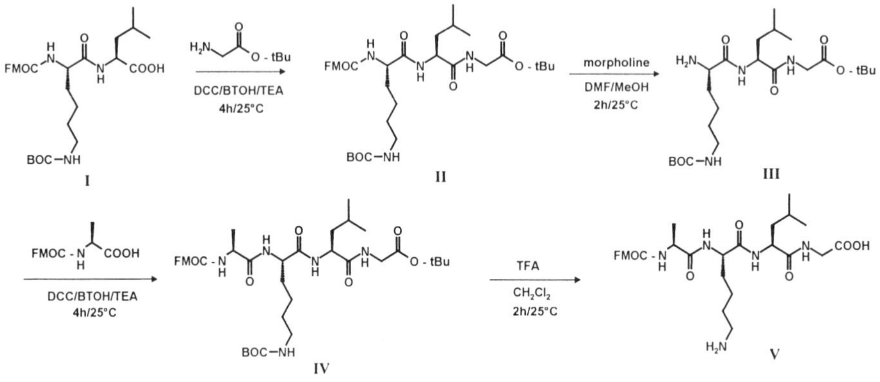

Figure 6. Synthetic sequence used for the preparation of N_α-FMOC-L-alanyl-L-lysyl-L-leucyl-L-glycine (V).

This N-terminal protected tetrapeptide was then thermally treated in the presence of glucose and the EPR spectrum, outlined in Figure 7, was compared with the spectroscopic data obtained from the BSA/glucose (cf. Figure 3) as well as the N_α-acetyl-L-lysine/glucose mixture (cf. Figure 4). The data clearly show that the incorporation of the lysine into a peptide leads to the formation of a radical showing a broad EPR signal and an identical g-value of 2.0038 as observed in the melanoidins formed from protein and glucose. These data clearly indicate that in the case of the tetrapeptide the formation of 1,4-disubstituted pyrazinium radical cations leads to intermolecular cross-linking and dimerization of the peptide.

These data demonstrate that in the reaction of the peptide, and the protein as well, crosslinking reactions lead to a strongly restricted rotation of the pyrazinium radical and to an immobilization of the radical, thus inhibiting the resolution of the EPR signal. We therefore proposed the broad singlet to represent protein-bound 1,4-bis-(5-amino-5-carboxy-1-pentyl) pyrazinium radical cations, which we named "CROSSPY".

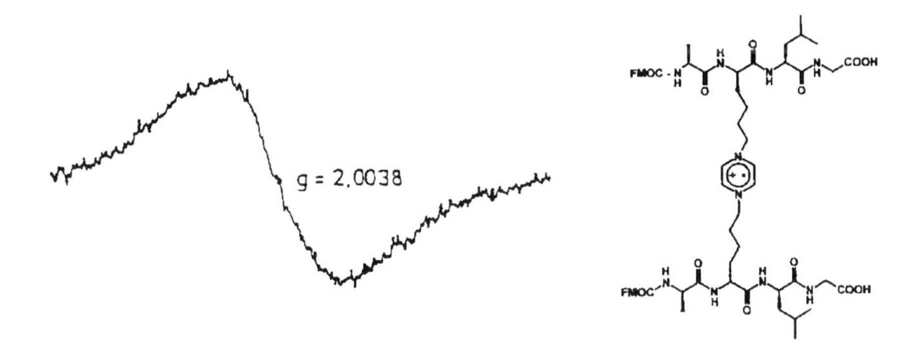

Figure 7. EPR spectrum of CROSSPY measured in a mixture of glucose heated in the presence of N_α-FMOC-L-alanyl-L-lysyl-L-leucyl-L-glycine.

Formation of CROSSPY

Recent investigations on radical formation in heated aqueous solutions of glucose and L-alanine revealed the carbohydrate fragmentation product glyoxal as an effective radical precursor *(7)*. It could be shown that after an induction period of 10 min radical formation was induced by a redox reaction of the glyoxal with reductones formed during carbohydrate degradation *(7)*. Because quantification experiments revealed high amounts of glyoxal (18500 µg/Kg) in roasted coffee (Hofmann, unpublished results), this C_2-compound was supposed to play also a key role in radical formation in coffee. In order to check whether, in addition, reductones are also involved in the formation of CROSSPY, first, a binary mixture of glucose and N_α-acetyl-L-lysine was thermally treated for 5 min and then analyzed by EPR spectroscopy (Table III).

As shown in Table III no radicals could be detected clearly demonstrating that the CROSSPY formation was still in the induction period. In order to check the influence of reductones on radical formation, this thermally pre-treated mixture was incubated in the presence of ascorbic acid at room temperature. Analysis of the mixture by EPR spectroscopy revealed that instantaneously after reductone addition the radical cation was generated (Table III). To investigate the effectivity of carbohydrate-derived reductones in CROSSPY formation, in comparative experiments, acetylformoin as well as methylene reductinic acid, both well-known to be formed during thermal treatment of hexoses (*19*), were added to the thermally pre-treated mixture. Both the Maillard reaction products were found to rapidly induce radical formation, however, in somewhat lower effectivity when compared to ascorbic acid (Table III).

Table III. Influence of Reductones on the Generation of Radical Cations in the Induction Period of a Thermally Treated Glucose/ N_{α}-acetyl-L-lysine Mixture

reductone[a]	*Rel. radical intensity [%]*
(without additive)	0
Ascorbic acid	100
Acetylformoin	92
Methylene reductinic acid	84

[a] A mixture of N_{α}-acetyl-L-lysine (2 mmol) and glucose (2 mmol) in phosphate buffer (2 mL; 0.5 mol/L; pH 7.0) was heated for 5 min at 95°C. After cooling, the reductone (0.5 mmol) was added, the mixture was maintained for 10 min at r.t. and was, then, analyzed by EPR spectroscopy.

In order to confirm the C_2-carbohydrate fragment glyoxal and proteins in generating CROSSPY upon redox reactions with reductones, we, then, heated BSA in the presence of glyoxal and analyzed the reacted mixture for paramagnetic behavior by EPR spectroscopy (Figure 8A). In comparison to the thermally treated mixture of BSA and glucose (Figure 3), no radical formation could be observed in the system containing glyoxal (Figure 8A). But addition of ascorbic acid to the thermally pre-treated solution instantaneously generated the immobilized radical as displayed in Figure 8B *(16)*. The broad signal exhibited a g-value of 2.0038 being identical with the CROSSPY radical detected in the melanoidins of the BSA/glucose mixture (Figure 3) as well as in coffee melanoidins (Figure 2). These data unequivocally confirm that reductones are able to initiate CROSSPY formation from glyoxal and lysine side chains of proteins. In addition to the broad signal, a doublet with a splitting constant of 1.8 G was detected in the BSA/glucose/ascorbic acid mixture (Figure 8B) and identified as the ascorbate radical *(20)*, thereby, demonstrating that the redox reaction is achieved by a single-electron transfer mechanism *(16)*.

Taking all these experimental data into consideration, the following reaction pathway was proposed for CROSSPY formation by carbohydrate-induced protein oligomerization (Figure 9): Glyoxal formed by an amine-assisted oxidation of carbohydrates in a very early stage of the Maillard reaction reacts with the ε-amino group of protein-bound lysine to form the corresponding glyoxal imine (I). Such reactive intermediates could be detected in aqueous solutions of N_{α}-acetyl-L-lysine and glyoxal by LC-MS (Hofmann, unpublished results). Upon redox reaction with reductones formed by carbohydrate degradation, these imines are then rapidly reduced into the protein-bound N-(5-amino-5-carboxy-1-pentyl)aminoacetaldehyde (II), which, upon dimerization, forms the 1,4-bis-(5-amino-5-carboxy-1-pentyl)-1,4-dihydropyrazine (III). Such dihydropyrazines are well-accepted in the literature to undergo subsequent oxidation, thereby, giving rise to the radical cation IV *(16)*.

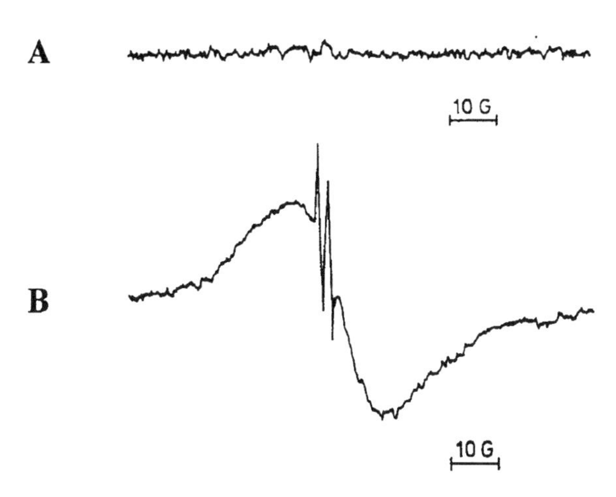

Figure 8. EPR signal of a heated mixture of (A) BSA and glyoxal, and (B) BSA and glyoxal incubated with ascorbic acid.

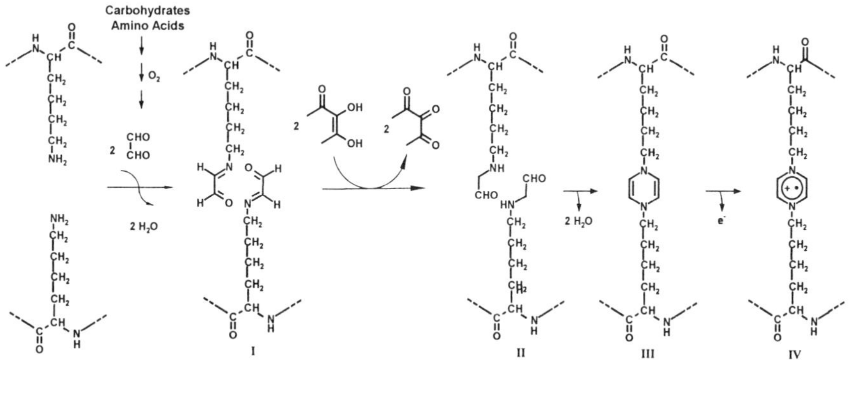

Figure 9. Reaction pathway leading to protein crosslinking by reductone-induced CROSSPY formation from lysine side chains and glyoxal.

CROSSPY as a Key Intermediate in Browning Development

On the basis of the observation that CROSSPY containing solutions undergo rapid browning, we assumed that these radical cations might be important intermediates in browning development and melanoidin formation during coffee roasting. In order to check this hypothesis, in a preliminary experiment, the time course of browning development was monitored in a heated mixture of glucose and N_{α}-acetyl-L-lysine by measuring the absorption at 420 nm and, in parallel, the CROSSPY formation was followed by EPR spectroscopy (Figure 10).

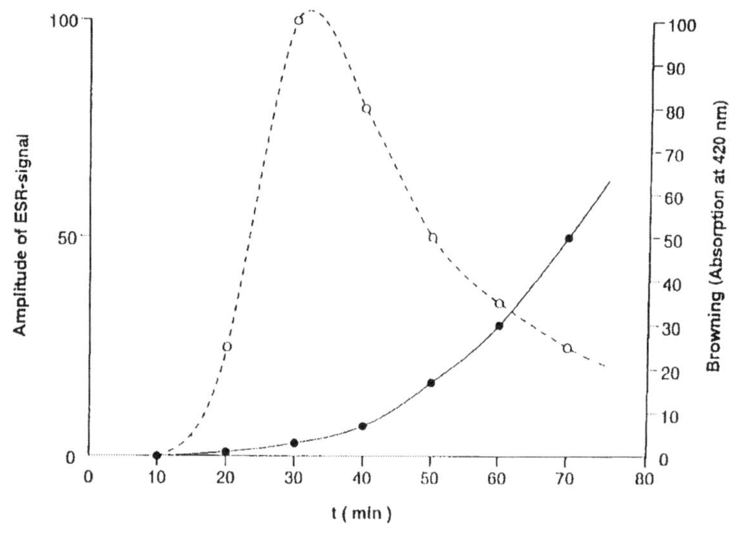

Figure 10. Time course of browning development and CROSSPY formation in a thermally treated mixture of glucose and N_{α}-acetyl-L-lysine.

As displayed in Figure 10, neither radicals, nor browning products could be detected within the first ten minutes of thermal treatment. But after that induction period, the concentration of CROSSPY increased rapidly. After running through a maximum at about 30 min the radical concentration decreased again, whereas the formation of browning products was strongly accelerated. On the basis of the tight relationship observed between CROSSPY degradation and browning formation, CROSSPY was assumed as an important intermediate in melanoidin formation *(16)*.

It is believed in the literature that such radical cations are quite unstable and undergo further disproportionation to form a 1,4-bis-(5-amino-5-carboxy-1-pentyl)-1,4-dihydropyrazine and the corresponding 1,4-bis-(5-amino-5-carboxy-1-pentyl)pyrazinium diquaternary salt, abbreviated as diquat (Figure 11). The dihydropyrazines are well-known to be rapidly oxidized, thus, regenerating the radical cation. Although the diquats are known to be very unstable in polar solvents *(21)*, investigations on how these diquats react further on have as yet not been performed systematically. Investigations on the reactions of these diquats were, therefore, assumed to give important informations on their role in non-enzymatic browning.

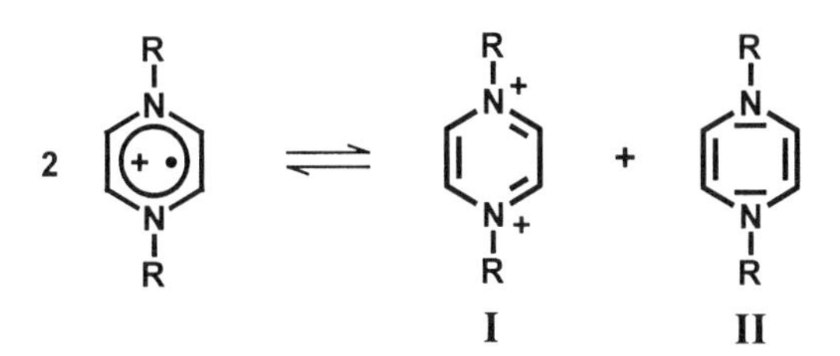

Figure 11. Disproportionation of CROSSPY leading to diquaternary 1,4-bis-(5-amino-5-carboxy-1-pentyl)pyrazinium ions (diquats, I) and 1,4-bis-(5-amino-5-carboxy-1-pentyl)-1,4-dihydropyrazines (II).

In order to identify reaction products of such diquats and to study their effectivity in color formation, we synthesized the 1,4-diethyldiquat as a suitable model substance for the 1,4-bis-(5-amino-5-carboxy-1-pentyl)diquat *(7)*. The diquat was extraordinarily labile in water and turned intensely dark colored instantaneously upon dissolving *(7,21)*. In order to identify primary reaction products formed, the reaction mixture was analyzed by LC-MS operating in the electrospray ionization (EI) mode. As displayed in an excerpt of the LC-MS spectrum in Figure 12, a base peak at m/z 155 (100 %) and an ion at m/z 138 (35 %) were monitored by LC-MS. The ion at m/z 138 is well in line with the formation of the 1,4-diethylpyrazinium radical cation, which was further confirmed by ESR spectroscopy (7). In order to identify the reaction product with m/z 155, LC-MS/MS measurements were performed showing a loss of 28 to the ion m/z 127, most likely resulting from the elimination of ethylene from an ethyl side chain. Because in a comparative experiment with oxygen-18 labeled water an analogous compound was detected with m/z 157 (Hofmann, unpublished results), the structure of that compound was proposed as 2-hydroxy-1,4-diethyl-1,4-dihydropyrazine formed from the diquat by addition of one molecule of water. In addition, an ion at m/z = 171 was detected by LC-MS, most likely corresponding to a dihydroxy-1,4-diethyl-1,4-dihydropyrazine *(7)*. These data clearly indicated that, upon dissolving the diquat in water, the monohydroxy-1,4-dihydropyrazine is formed as the major reaction product, followed by the 1,4-diethylpyrazinium cation radical and the bishydroxy-1,4-dihydropyrazine as minor compounds.

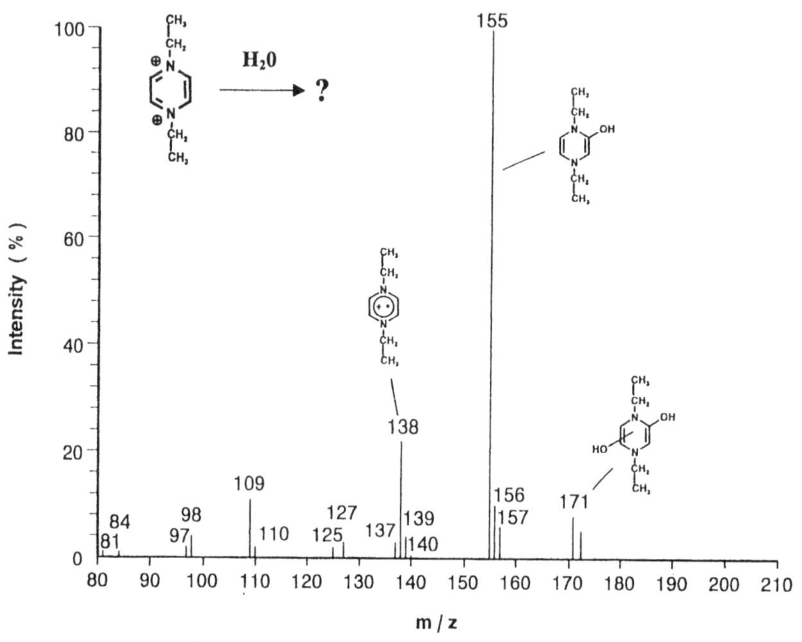

Figure 12. LC-MS of an aqueous solution of synthetic 1,4-diethyldiquat.

To gain further insights into the role of these hydroxydihydropyrazines in browning formation, the synthetic diquat was thermally treated in aqueous solution, and the time course of browning development was followed by measuring the absorbance at 420 nm (Figure 13). In addition, the formation of hydroxydihydropyrazine and radical cation was monitored by LC/MS and ESR spectroscopy. As outlined in Figure 13, the highest concentration of hydroxy-1,4-dihydropyrazine and radical cation was observed immediately upon dissolving of the diquat. During the first 10 min of thermal treatment the concentrations of hydroxydihydropyrazines and radical cations decreased rapidly to about 50 or 40 %, respectively, whereas prolonging of reaction time to 60 min led, however, only to a slight further decrease in their concentrations *(7)*. Monitoring the visible absorption in the heated diquat solution within a period of 60 min, revealed a drastic increase in color development during the first 10 min

(7). The color development then slowly approximated to a maximum value within the following 50 min. The tight relationship between the degradation of the hydroxydihydropyrazine and the formation of browning products suggested the hydroxydihydropyrazines as penultimate precursors in color development *(7)*.

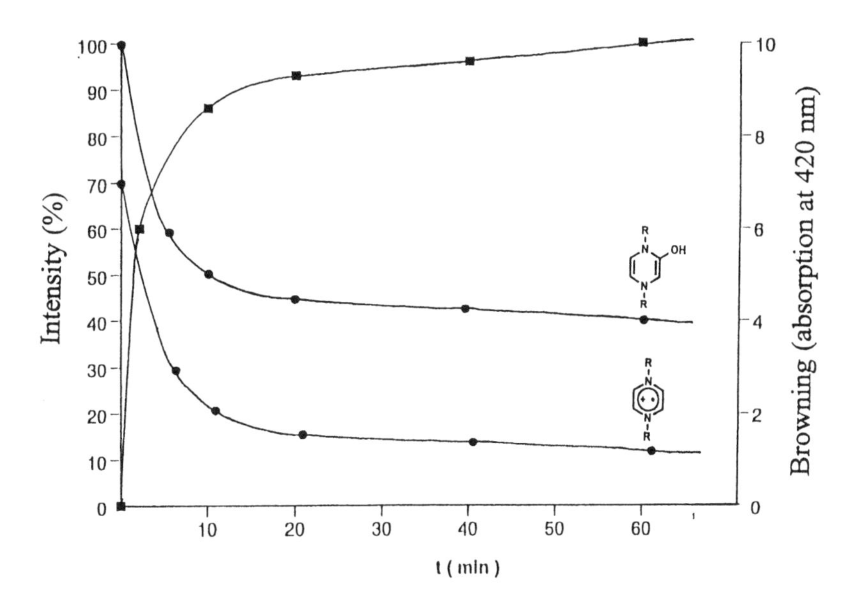

Figure 13. Time course of the formation of 2-hydroxy-1,4-diethyl-1,4-dihydropyrazine, 1,4-diethylpyrazinium radical cation (●) as well as browning products (■) during thermal treatment of an aqueous solution of synthetic 1,4-diethyldiquat.

On the basis of these experimental results a mechanism was proposed in Figure 14 showing the formation of browning compounds via CROSSPY as a radical intermediate. CROSSPY (I) liberated its diquarternary pyrazinium salt (II) and the 1,4-dihydropyrazine (III) upon disproportionation. Whereas the dihydropyrazine rapidly regenerates the CROSSPY by oxidation, the diquats were shown to subsequently form 2-hydroxy-1,4-dihydropyazines (IV) upon hydration as the primary reaction product *(7)*. Redox reaction between the diquat (II) and the hydroxydihydropyrazine (IV) was recently shown to regenerate CROSSPY and to form bishydroxylated dihydropyrazine (VI), most likely upon hydration of an intermediate hydroxylated diquat (V). Both the hydroxy- (IV) and the bishydroxydihydropyrazines (VI) possess strong nucleophilic character and were, therefore, proposed as penultimate monomers in an oligomerization reaction with electrophilic intermediates such as, e.g. the diquats, leading to browning compounds *(7)*.

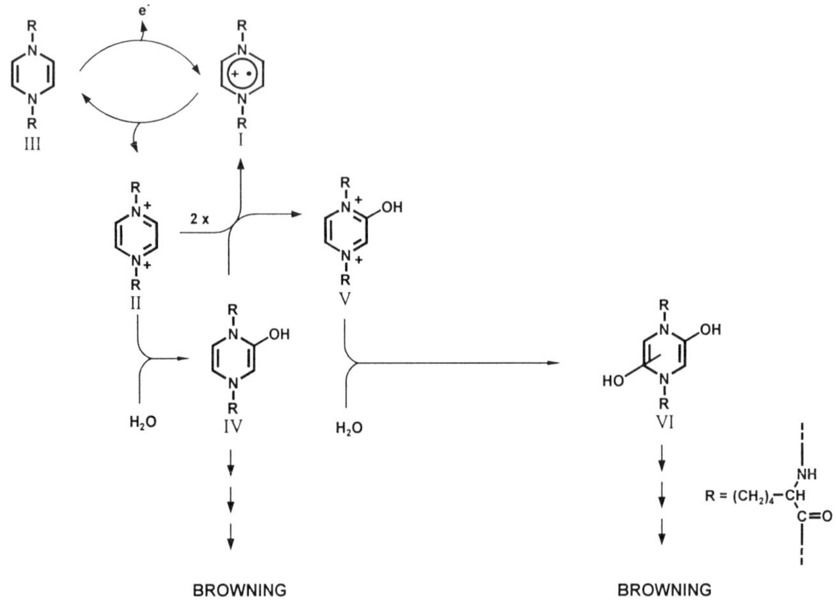

Figure 14. Reaction pathways leading to non-enzymatic browning via CROSSPY (I) and hydroxylated dihydropyrazines (IV, VI) as the key intermediates.

Preliminary investigations by LC-MS spectroscopy confirmed the formation of colored, high-molecular oligomers, but the exact chemical structures of these browning compounds have to be proven in future studies by using 1D- and 2D-NMR spectroscopic measurements.

Conclusions

The radical cation CROSSPY was identified as a previously unknown type of cross-linking amino acid involved in melanoidin formation during roasting of coffee beans. Also for other browned foods such as wheat bread crust or roasted meat a tight relationship between the formation of CROSSPY and browning development could be observed (Hofmann, unpublished results), thus, demonstrating that color formation involving CROSSPY is a general reaction principle during thermal food processing. This is also well in line with the fact that CROSSPY is not present in foods which did not show any browning upon thermal treartment such as, e.g. milk or microwave-cooked foods. Such detailed information on the chemical species involved in non-enzymatic browning might prove useful in finding new ways in accelerating the desired color formation during thermal food processing, such as giving microwave foods an attractive brown color, and also in preventing undesired colorization, e.g. during long-time storage of foods.

References

1. Ames, J. M. In *Progress in Food-Proteins - Biochemistry*; Hudson, B.J.F., Ed.; Elsevier Applied Science, London, 1992, pp 99-153.
2. Hofmann, T. *J. Agric. Food Chem.* **1998**, *46*, 3891-3895.
3. Hofmann, T. In *Chemistry and Physiology of Food Colorants*; Ames, J.M.; Hofmann, T.; Eds.; ACS symposium series, 2000, in press.
4. Ledl, F.; Severin, Th. *Z. Lebensm. Unters. Forsch.* **1978**, *167*, 410-413.
5. Namiki, M.; Hayashi, T.; Kawakishi S. *Agric. Biol. Chem.* **1973**, *37*, 2935-2936.
6. Hayashi, T.; Ohta, Y.; Namiki, M. *J. Agric. Food Chem.* **1977**, *25*, 1282-1287.
7. Hofmann, T.; Bors, W.; Stettmaier, K. *J. Agric. Food. Chem.* **1999**, *47*, 379-390.
8. Mitsuda, H.; Yasumoto, K.; Yokoyama, K. *Agric. Biol. Chem.* **1965**, *29*, 751-756.
9. Wu, C.H.; Russel, G.; Powrie, W.D. *J. Food Sci.* **1987**, *52*, 813-816.
10. O'Meara, J.P.; Truby, E.K.; Shaw, T.M. *Food. Res.* **1957**, *22*, 96-100.
11. Santanilla, J.D.; Fritsch, G.; Müller-Warmuth, W. *Z. Lebensm. Unters. Forsch.* **1981**, *172*, 81-86.
12. Troup, G.J.; Hutton, D.R.; Dobie, J.; Pilbrow, J.R.; Hunter, C.R.; Smith, B.R.; Bryant, B.J. *Medical Journal of Australia* **1988a**, *148*, 537-538.
13. Troup, G.J.; Wilson, G.L.; Hutton, D.R.; Hunter, C.R. *Medical Journal of Australia* **1988b**, *149*, 147-148.
14. Gonis, J.; Hewitt, D.G.; Troup, G.; Hutton, D.R.; Hunter, C.R. *Free. Rad. Res.* **1995**, *23*, 393-399.
15. Packert, A. PhD Thesis, 1993, University of Hamburg
16. Hofmann, T.; Bors, W.; Stettmaier, K. *J. Agric. Food. Chem.* **1999**, *47*, 391-396.
17. Namiki, M.; Hayashi, T. *J. Agric. Food Chem.* **1975**, *23*, 487-491.

18. Vinale, F.; Fogliano, V.; Schieberle, P.; Hofmann, T. *J. Agric. Food Chem.* **1999**, *47*, 5084-5092.
19. Hofmann, T. *J. Agric. Food Chem.* **1998**, *46*, 3918-3928.
20. Bors, W.; Michel, Ch.; Stettmaier, K. *J. Magn. Reson. Anal.* **1997**, *3*, 149-154.
21. Curphey, T.J.; Prasad, K.S. *J. Org. Chem.* **1972**, 37, 2259-2265.

Chapter 5

Radical Induced Formation of D-Glucosone from Amadori Compounds

R. Liedke and K. Eichner

Institut für Lebensmittelchemie der Universität Münster, Corrensstrasse, 45, D-48149 Münster, Germany

Amadori rearrangement products (ARP) are the first stable intermediates in the Maillard reaction between reducing sugars and amino acids. The oxidative decomposition of ARPs leads to the formation of osones like D-glucosone. The radical induced formation of D-glucosone in model systems containing ARPs is dependent on the chosen reaction parameters like pH, temperature, and water activity. Our investigations showed that, in general, the formation of D-glucosone is increased in the presence of transition metals; whereas, the yield of the D-glucosone is greatly reduced by elimination of the metal ions (addition of EDTA). Under these conditions the non-oxidative enolization reactions leading to the deoxyosones gain importance. The assumed radical pathway could be confirmed by EPR and spin trapping of radicals with DMPO. Currently, oxidative pathways in the Maillard reaction are being discussed for their relevance in medicine and health (autoxidative glycosylation of tissue protein, diabetes, aging).

Radical reaction mechanisms during the early Maillard reaction were first detected by Namiki et al. (*1, 2*). He identified *N,N'*-dialkylpyrazine-cation radicals that originated from the primary Schiff base formed by reaction between glucose and amino acids. The glycolaldehyde alkylimine formed by a reverse aldol reaction of the Schiff base leads to a dialkylpyrazinium radical cation after self-condensation. The formation of dialkylpyrazinium radical cations, which could be detected by EPR spectrometry, represents an alternative pathway of the Maillard reaction; it starts at the very beginning of the reaction, well before the formation of Amadori rearrangement products and depends on the pH value; it starts around pH 7 and increases up to pH 11.

In a similar way, EPR spectroscopy of orange-brown melanoidins, which were isolated from heated aqueous solutions of bovine serum albumin and glycolaldehyde, revealed the protein-bound 1,4-bis(5-amino-5-carboxy-1-pentyl)pyrazinium radical cation (CROSSPY) as a previously unknown type of cross-linking of proteins in vivo and during food processing (*3*). CROSSPY could be found in wheat bread crust, roasted cocoa, as well as coffee beans.

On the other side, glyoxal may be formed by oxidation of the above mentioned glycolaldehyde alkylimine; the glyoxal can react with the amino groups of amino acids (for example, the ε-amino group of protein-bound lysine) to form carboxymethyl-lysine (CML).

Furthermore, there are different pathways for the oxidative decomposition of Amadori products. Ahmed et al. (*4*) described the formation of *N*-ε-carboxymethyl-lysine (CML) by oxidation of ε-fructose-lysine which is promoted in the presence of iron and by increasing the pH value. CML can be formed during food processing as well as in vivo. On the other side, the enaminol form of Amadori compounds can be oxidized through a radical mechanism, especially in the presence of heavy metal ions like copper (*5*); as shown in Figure 1, glucosone is formed in this way, alternatively to the formation of the 3- and 1-desoxyosones via 1,2- and 2,3 enolization of Amadori compounds, respectively.

The reactive oxygen species (ROS) formed during autoxidation of Amadori compounds (superoxide- and hydroxyl-radicals) may cause oxidative damage of proteins in vivo (*6*) and may be responsible for the "oxidative stress". But glucosone is also formed during thermal food processing where the Maillard reaction takes place. Furthermore, glucose and other α-hydroxyaldehydes or α-hydroxyketones may undergo radical induced oxidation processes leading to very reactive glycosones and ROS which may damage proteins (*6, 7*). According to Wells-Knecht et al. (*8*) also glyoxal is formed during autoxidation of glucose; as already mentioned, it may cause crosslinking of proteins and formation of CML. Moreover, glucosone is formed during irradiation of glucose and fructose (*9, 10*).

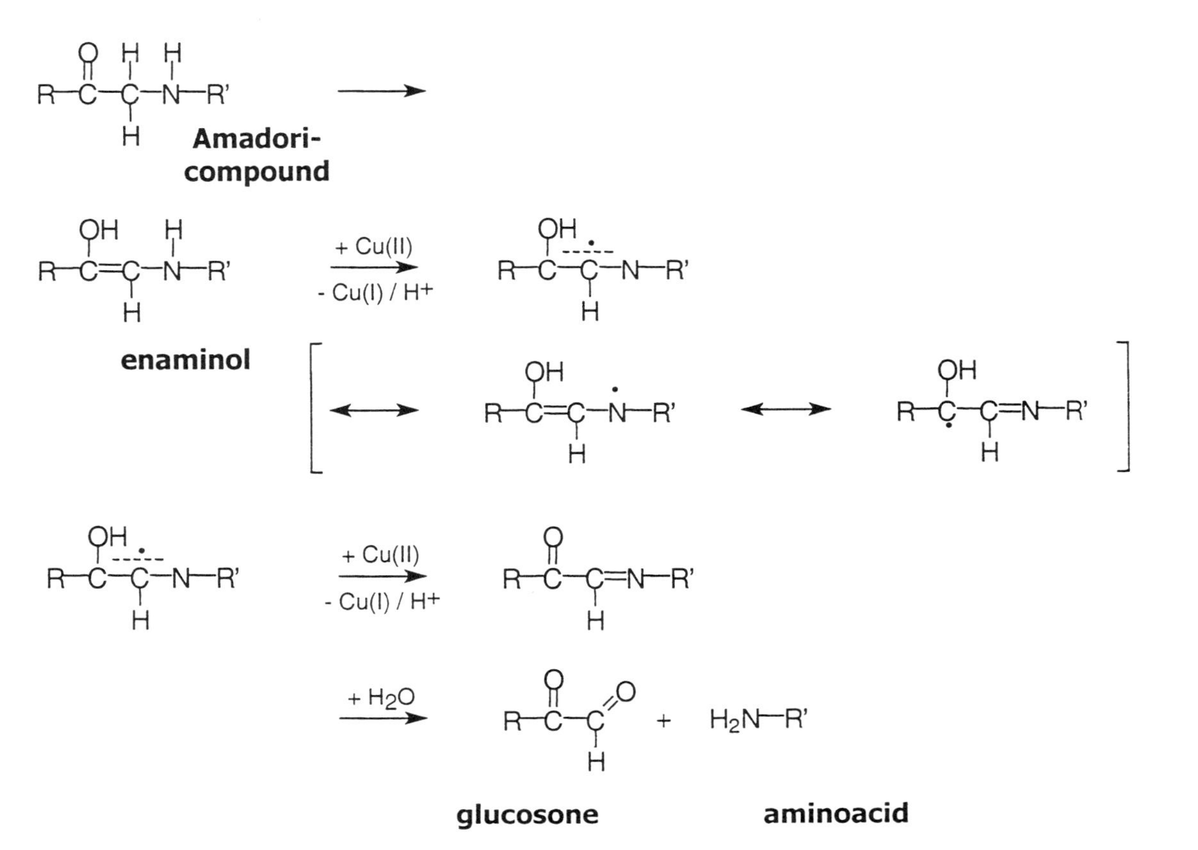

Figure 1. Formation of glucosone by radical induced decomposition of Amadori Compounds

Radical Induced Decomposition of Amadori Compounds

EPR measurements after Spin Trapping.

In preliminary EPR measurements with aqueous model solutions of fructose-alanine (Fru-Ala) heated to 80 °C for 2 hrs only a very small amount of radicals could be detected. In order to prolong the lifetime of the radicals, spin trapping with 5,5-dimethyl-1-pyrroline-*N*-oxide (DMPO) was applied. By formation of adducts with DMPO the free radicals are stabilized because of delocalization of the free electron between the nitrogen and the oxygen atom.

For formation of the DMPO-adduct to 100 µl of a buffered (pH 7.0) aqueous model solution of Fru-Ala (1 mol/L) containing 50 µmol/L copper(II)-ions 10 µl of DMPO were added. After an incubation time of 15 min at room temperature part of the solution was filled into a glass-capillary which was put into the cavity of an EPR-spectrometer. The EPR measurements were performed with a Bruker EMS 104 – EPR Analyzer at 9.5 GHz (X-band); microwave power set to 12.5 mW.

In Figure 2 the EPR-spectrum of the adduct between DMPO and hydroxyl radicals is shown. Since the splitting constants for the nitrogen atom giving 3 signals and the hydrogen atom in ß-position giving 2 signals are identical (a_N = 14.7, a_H = 14.7), only 4 EPR signals (intensity relation 1:2:2:1) are obtained. In order to promote the oxidative decomposition of the Amadori compound, Cu-ions (50 µmol/L) were present. After addition of DMPO decomposition of the Fru-Ala could be pursued. Figure 3 shows the EPR spectrum of the model system after 15 min. The lines having the highest intensity (indicated with 0) can be correlated to the adduct between DMPO and the OH radical (DMPO-OH, splitting constants a_N = 14.7 G, a_H = 14.7 G, compare Figure 2). The lines in Figure 3 indicated with X (splitting constants a_N = 16.3, a_H = 23.2) can be associated to DMPO-adducts with an organic radical bearing a hydroxyl group in the α-position. For comparison Figure 4 shows the well known DMPO adduct of the ethanol radical with splitting constants of a_N = 16.1 and a_H = 23.1 (*11*) and the possible adduct of DMPO with the primary radical arising from the oxidative decomposition of Fru-Ala showing a similar structure and having nearly identical splitting constants (cf. Figure 3). The described results are clearly showing, that in the presence of heavy metal ions free radicals are induced during decomposition of Amadori compounds indicating that Amadori compounds may be oxidized to osones as shown in Figure 1.

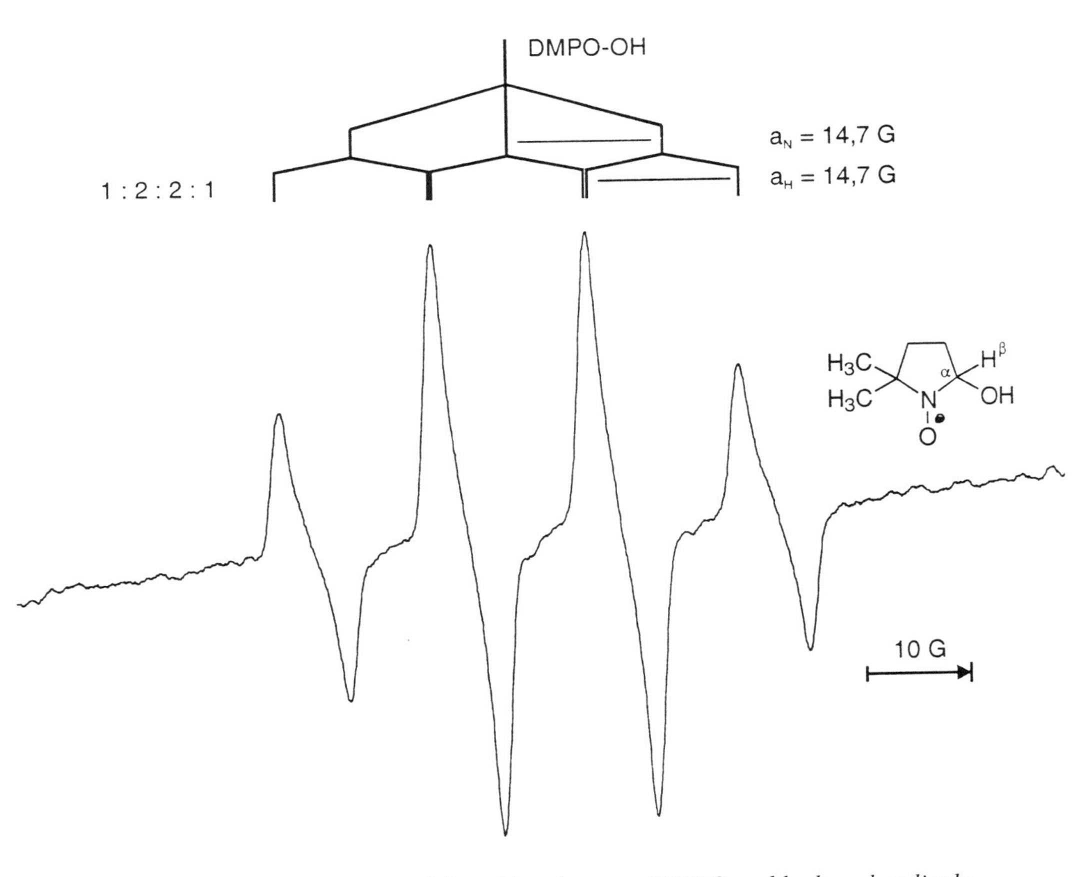

Figure 2. EPR-spectrum of the adduct between DMPO and hydroxyl radicals

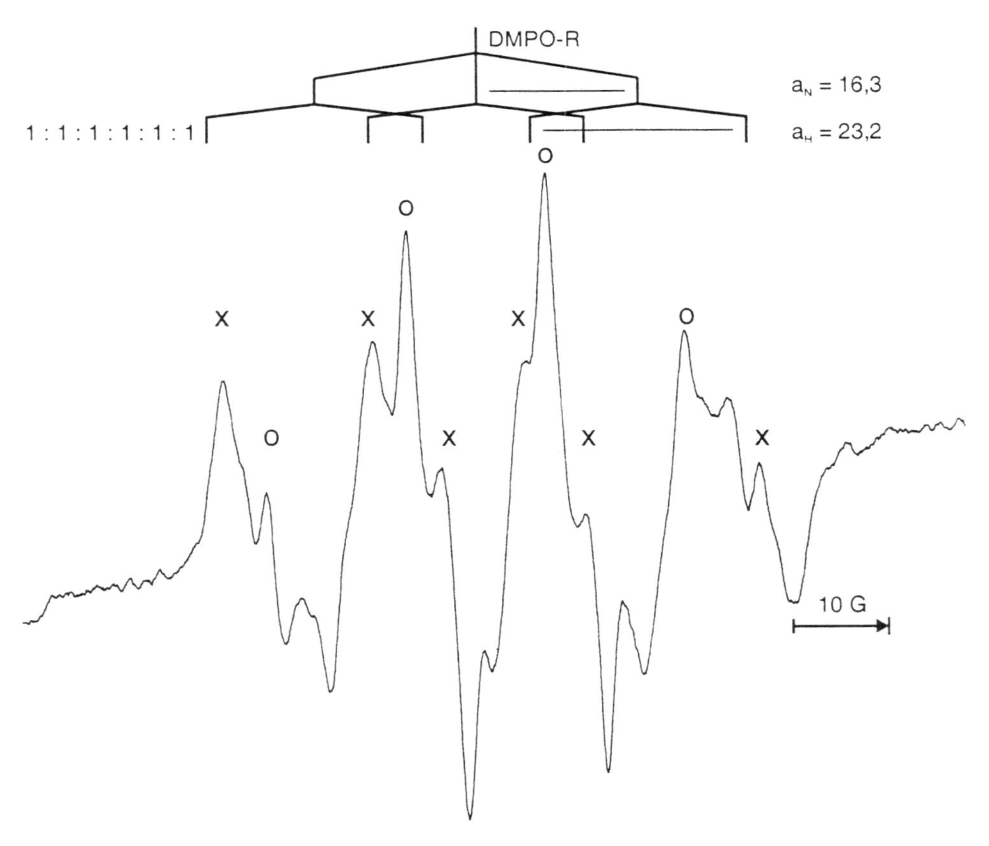

Figure 3. EPR-spectrum of the DMPO-adducts from a model solution containing fructose-alanine (Fru-Ala)

Scavenging of radicals by reaction with the 2,2-diphenyl-1-picrylhydrazyl radical (DPPH).

DPPH is a stable radical (absorption maximum: 529 nm) which can react with a second radical or a hydrogen atom to a stable adduct or to reduced DPPH (DPPH-H), respectively, showing no absorption maximum in the visible range. Therefore, the occurrence of free radicals can be determined by measuring the decrease of the absorption at 529 nm. This method is less suitable for determination of radicals formed during decomposition of Amadori compounds, because it is not connected with spin trapping; in this way only very unstable radicals formed in a very short period can be determined. Therefore, the DPPH-method is mainly applied for estimation of the hydrogen donating, radical scavenging properties of antioxidants (*12*).

Analysis of deoxyosones and glucosone formed by decomposition of Amadori compounds by gas chromatography

Aqueous model solutions of Fru-Ala (12 mmol/L) were adjusted to three different pH values (3.0 / 5.0 / 7.0) with citric acid. As a trapping agent for the deoxyosones and osones the solution contained 24 mmol/L o-phenylenediamine (OPDA). In this way different derivatives of quinoxalines are formed.

To determine the influence of heavy metal ions on the decomposition of Amadori compounds the solution contained 50 μmol/L copper(II)-ions. To determine the decomposition of Amadori compounds in the absence of heavy metal ions the solution contained 100 μmol/L EDTA as a chelating agent.

For following the decomposition of the Amadori compound 2.0 ml of the solution are heated at 80 °C or 98 °C in an airthight vessel for 6 hours. The solution is cooled and part of it is freeze-dried over night. The residue is then prepared for gas chromatography.

To the dried residue of the model solution 0.5 ml pyridine containing hydroxylammoniumchloride are added. For oximation of the sugars and the Amadori compounds the solution is heated in an airtight vessel at 70 °C for half an hour. After cooling 0.25 ml *N,O*-bis-(trimethylsilyl)-acetamide and 0,08 ml trimethyl-chlorsilane are added. Then the solution again is heated at 70 °C for half an hour. In this way all oximes and the quinoxalines formed by reaction of the deoxyosones and osones with OPDA are silylated and after that analyzed by gas chromatography (Figure 5).

Gas chromatography was performed with a Perkin Elmer Autosystem XL with FID (HP-5 column: 30 m / 0.25 mm / 0.25 μm; carrier: 1.2 ml/min

Figure 4. DMPO-adducts with a α-hydroxyethyl radical and the radical form of an Amadori product

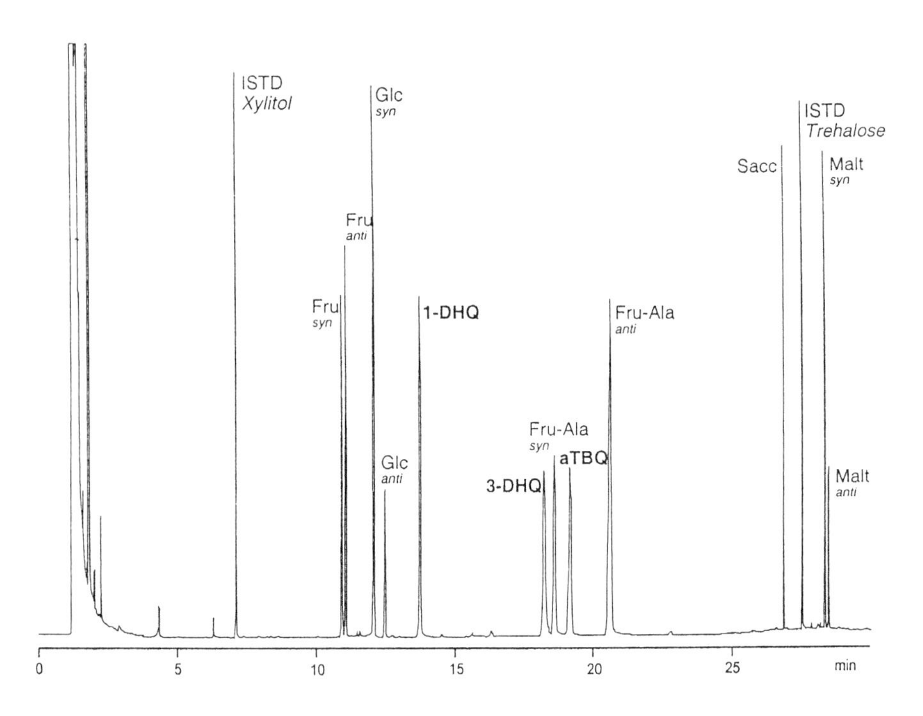

Figure 5. Separation of Amadori compounds and quinoxaline derivatives by gas chromatography

(1-DHQ = 1-deoxyosone-quinoxaline; 3-DHQ = 3-deoxyosone-quinoxaline; aTBQ = D-glucosone-quinoxaline = D-arabino-tetrahydroxybutyl-quinoxaline)

hydrogen. Temperature is raised from 140 °C with 6 °C/min to 200 °C, holding for 12 min, then with 20 °C/min to 300 °C, holding for 3 min.
For identification and calculation we used self-prepared Amadori compounds and quinoxalines which were characterized by mass spectrometry. Calculation was performed using xylitol and trehalose as internal standards.

Results

Formation of deoxyosones and glucosone during decomposition of Amadori compounds.

The 1- and 3-deoxyosone as well as the glucosone formed during decomposition of Amadori compounds are very reactive intermediates of the Maillard reaction, yielding a series of heterocyclic compounds and melanoidins. In order to get informations about the real amounts of these intermediates formed under different reaction conditions, these compounds were trapped by adding *o*-phenylenediamine (OPDA) to the reaction mixture as it was shown by Ledl (*13*) for detection of the 1-deoxyosone in Maillard reaction mixtures. In this way different derivatives of quinoxalines are formed. These compounds as well as reducing sugars and the original Amadori compounds were analyzed by gas chromatography after oximation and trimethylsilylation (*14*), as it is demonstrated in Figure 5. It shows two peaks for the reducing sugars and the Amadori compounds (syn- and anti-form of the oximes) and one peak for non reducing sugars and the quinoxalines.

In Figure 6 the decomposition rate (mol %) of the Amadori compound fructose-alanine (Fru-Ala) at 80 ° and 98 °C in aqueous model solutions in the absence and presence of copper ions dependent on the pH value is demonstrated. These temperatures represent typical sections of the brewing process (mash process and wort cooking) where the Amadori compounds present in malt (*14*) are decomposed. Figure 6 clearly shows that decomposition of Amadori compounds increases by increasing the pH value; furthermore, copper ions strongly accelerate the rate of decomposition, because a higher amount of glucosone additionally is formed in the presence of heavy metal ions (*5*). Figure 7 shows that, as well as the decomposition of Amadori compounds, the formation of glucosone is favoured by increasing the pH value; at pH 3 almost no glucosone can be detected. Furthermore, it can be seen that in the presence of copper ions the amount of glucosone formed by oxidative decomposition of Amadori compounds is much higher than in the model solution containing EDTA. On the other side, as shown in Figure 8, in the presence of copper ions a much lower portion of the 1-deoxyosone is formed; again, its formation is

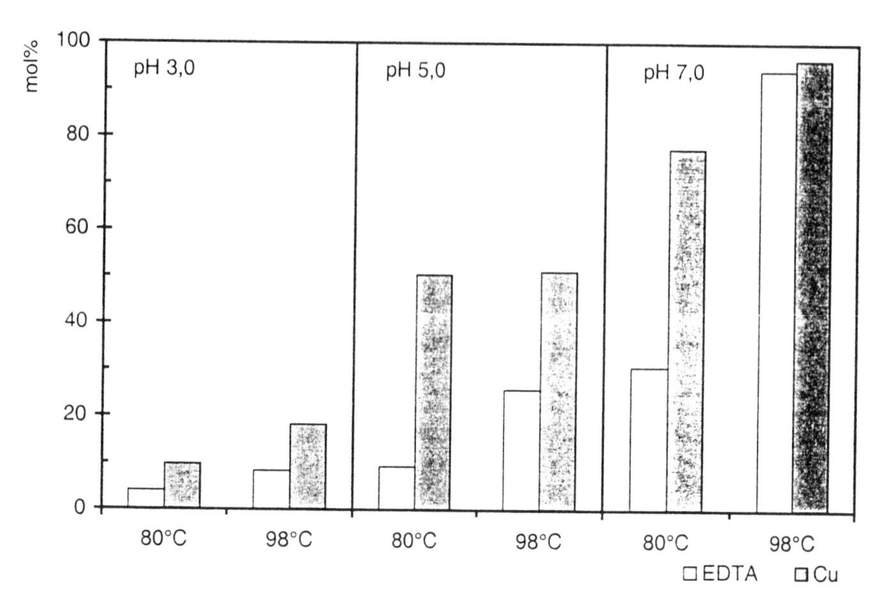

Figure 6. Decomposition of fructose-alanine dependent on the pH value in aqueous model systems in the absence and presence of Cu-ions (heating time: 6 hrs; Cu^{2+} : 50 µmol/L; EDTA : 100 µmol/L)

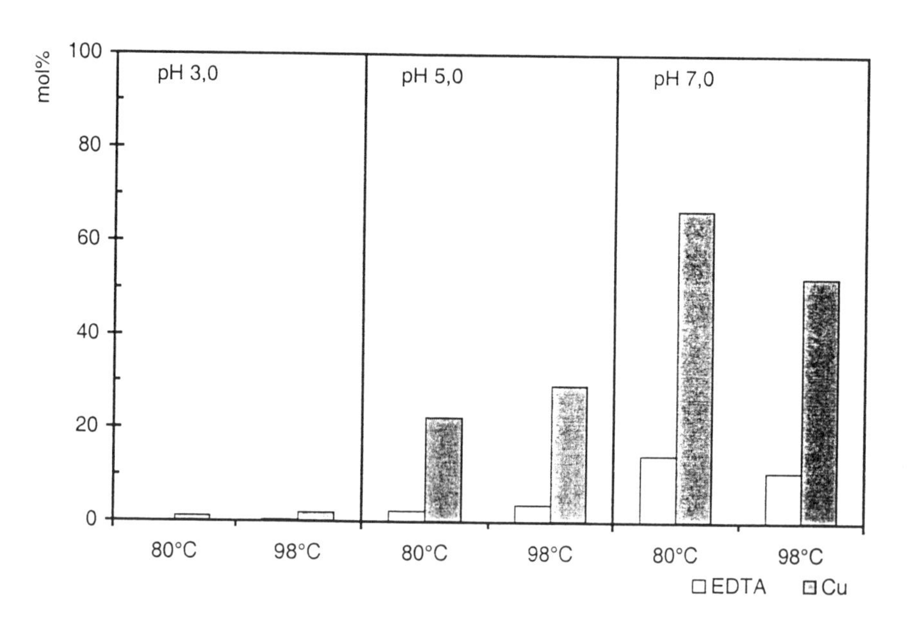

Figure 7. Formation of D-glucosone by decomposition of fructose-alanine dependent on the pH value in the absence and presence of Cu-ions (reaction conditions: cf. Figure 6)

favoured by increasing the pH value. Figure 9 demonstrates that the formation of the 3-deoxyosone has a maximum at pH 5 at 98 °C.

By measurement of the decrease of the stable DPPH radical in aqueous solutions of Fru-Ala heated for 2 hrs at 80 °C and different pH-values in the absence and presence of copper ions, it turned out that the amount of radicals formed by decompositon of the Amadori compound increases by increasing the pH value; in the presence of copper ions the radical concentration is further enhanced. Our results reveal that the concentration of free radicals in heated solutions of Amadori compounds has the same characteristic as the radical induced formation of glucosone (cf. Figure 7) supporting the radical mechanism taking place during the formation of the glucosone by decomposition of Amadori compounds.

Formation of deoxyosones and glucosone during the brewing process.

Using light barley malt the brewing process was simulated in a laboratory scale. Figure 10 shows the distribution of the 1- and 3-deoxyosone and the glucosone during the mash process and wort cooking. They again were determined by trapping with OPDA to form the quinoxaline derivatives which were determined by gas chromatography (cf. Figure 5). It can be seen that the portion of the glucosone generally is higher than the portions of the 1- and 3-deoxyosone, respectively. Moreover, contrary to the model solutions, a higher amount of 3-deoxyosone compared to the 1-deoxyosone is formed during the brewing process. These results may be explained by the fact, that a high surplus of glucose is present which, on the one side, may form glucosone by a radical induced oxidation, and, on the other side, may preferentially undergo a 1,2-enolization reaction at pH 5.8.

In Figure 11 the formation of the pertinent dicarbonyl intermediates during the brewing process dependent on the concentration of added copper ions is presented. It becomes clear that from a certain threshold concentration of copper ions (about 6 μmol/L) the formation of glucosone is strongly enhanced, whereas the concentrations of the 1- and 3-deoxyosone do not change.

Conclusions

The oxidative decomposition of Amadori compounds gains importance in the presence of heavy metal ions. It seems likely that during food processing also glucose undergoes oxidative decomposition forming D-glucosone.

Recently oxidative pathways within the Maillard reaction like radical induced crosslinking of proteins have been shown to have also physiological

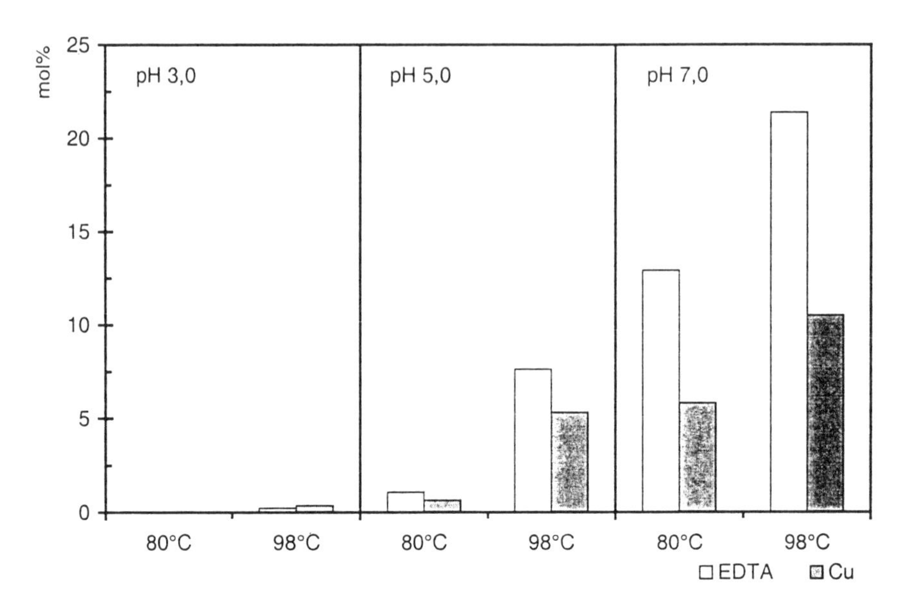

Figure 8. Formation of 1-deoxyosone by decomposition of fructose-alanine dependent on the pH value in the absence and presence of Cu-ions (reaction conditions: cf. Figure 6)

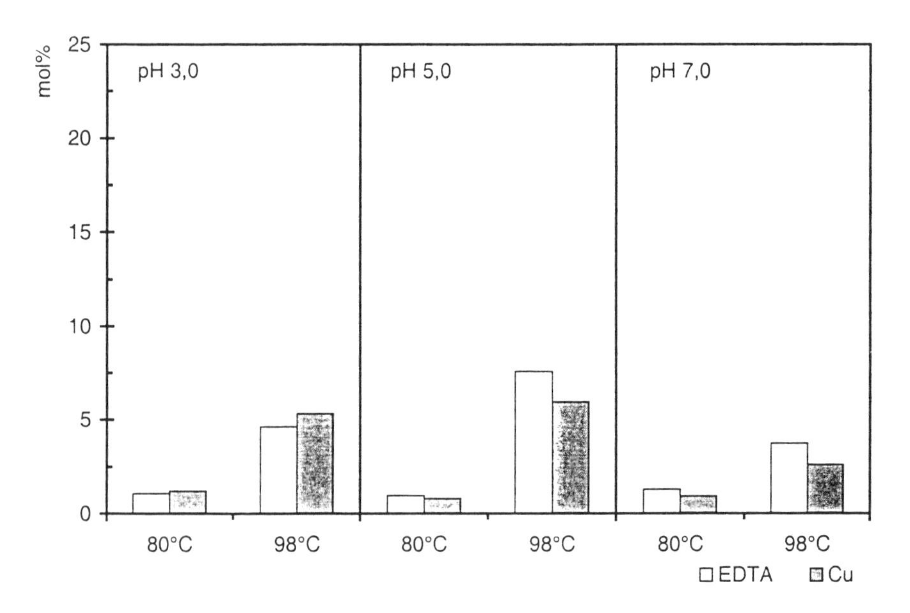

Figure 9. Formation of 3-deoxyosone by decomposition of fructose-alanine dependent on the pH value in the absence and presence of Cu-ions (reaction conditions: cf. Figure 6)

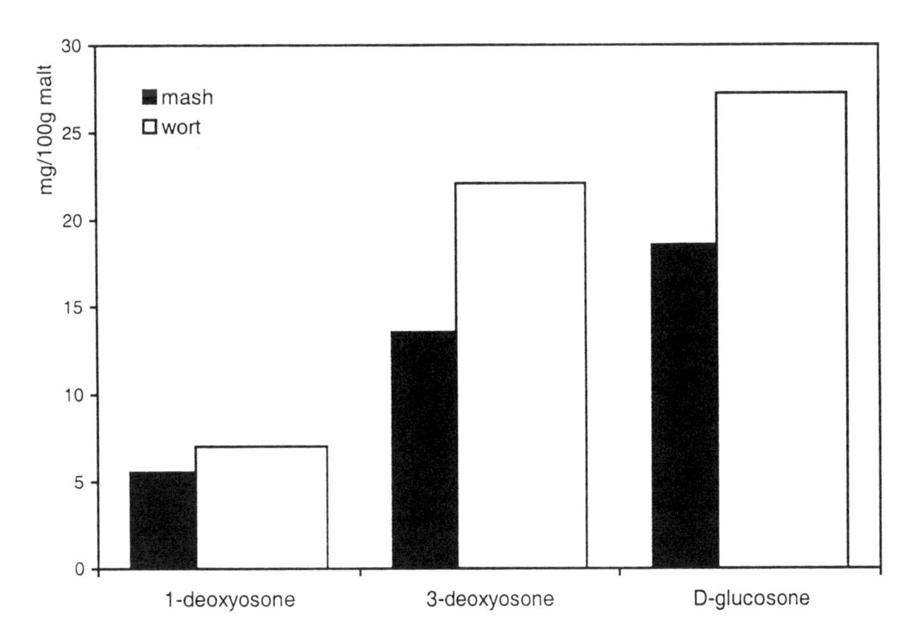

Figure 10. Formation of α-dicarbonyl intermediates during the brewing process

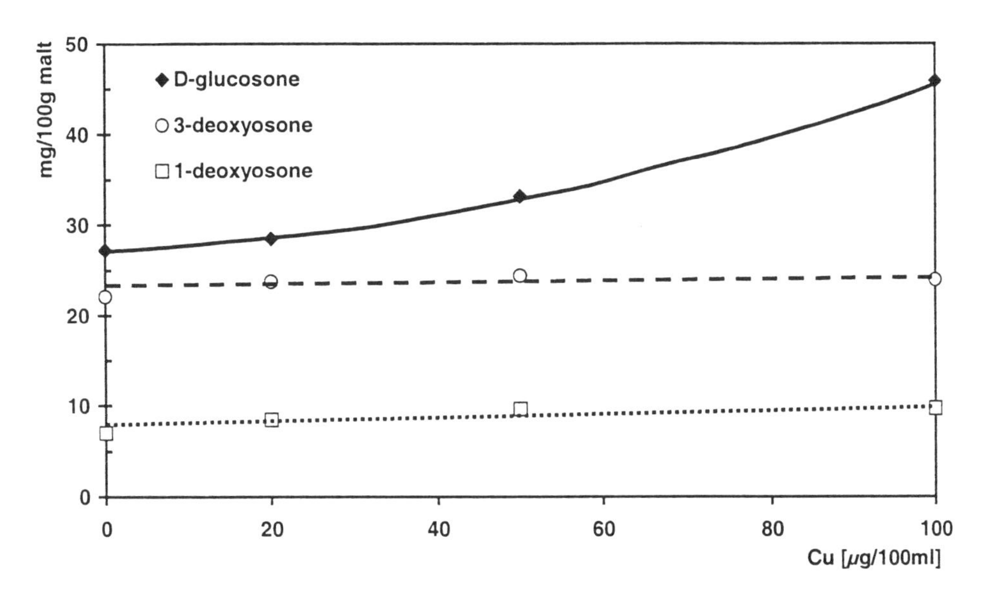

Figure 11. Formation of α-dicarbonyl intermediates during the brewing process dependent on the concentration of Cu-ions

relevance. Various cytotoxic roles for glucose have been proposed, including the slow nonenzymic glycosylation of proteins induced by glucose autoxidation causing conformation changes and functional alterations (*7, 8, 15*). Non-enzymic glycosylation and subsequent crosslinkig and browning reactions have been implicated in pathogenesis related to hyperglycaemia in diabetes and aging (*16*).

References

1. Namiki, M. *Advances in Food Research* **1988**, *32,* 115 – 184.
2 Namiki, M. *J. Agric. Food Chem.* **1975,** *23*, 487 – 491.
3. Hofmann, T.; Bors, W.; Stettmaier, K. *J.Agric. Food Chem.* **1999**, *47*, 391 – 396.
4. Ahmed, M.U.; Thorpe, S.R.; Baynes, J.W. *J. Biol. Chem.* **1986,** *261*, 4889 – 4894.
5. Kawakishi, S.; Tsunehiro, J.; Uchida, K. *Carbohydrate Research* **1991,** *211*, 167 – 171.
6. Yaylayan, V.A.; Huyghues-Despointes, A. *Crit. Rev. Food Sci. Nutr.* **1994**, *34*, 321 – 369.
7. Wolff, S.P.; Dean, R.T. *Biochem. J.* **1987**, *245*, 243 – 250.
8. Wells-Knecht, K.J.; Zyzak, D.V.; Litchfield, J.E.; Thorpe, S.R.; Baynes, J.W. *Biochemistry* **1995**, *34*, 3702 – 3709.
9. Den Drijver, L.; Holzapfel, C.W.; Van der Linde, H.J. *J. Agric. Food Chem.* **1986**, *34*, 758 – 762.
10. Niemand, J.G.; Den Drijver, L.; Pretorius, C.J., et al. *J. Agric. Food Chem.* **1983**, *31*, 1016 – 1020.
11. Zang, L.; Stone, K.; Pryor, W.A. *Free Radical Biology & Medicine* **1995**, *19*, 161 – 167.
12. Blois, M.S. *Nature* **1958**, *181*, 1199f.
13. Beck, L.; Ledl, F.; Severin, T. *Carbohydrate Research* **1988**, *177*, 240 – 243.
14. Wittmann, R.; Eichner, K. *Z. Lebensm. Unters. Forsch.* **1989**, *188*, 212 – 220.
15. Wells-Knecht, M.C.; Thorpe, S.R.; Baynes, J.W. *Biochemistry* **1995**, *34*, 15134 – 15141.
16. Harding, J.J. *Adv. Prot. Chem.* **1985**, *37*, 247 – 334.

Chapter 6

Factors Influencing Free Radical Generation in Food Emulsions

Eric A. Decker [1], D. Julian McClements[1], Wilailuk Chaiyasit[1], C. Nuchi[1], M. P. C. Silvestre[2], Jennifer R. Mancuso[1], Lawrence M. Tong[1], and Longyuan Mei[1]

[1]Department of Food Science, Chenoweth Lab, University of Massachusetts, Amherst, MA 01003
[2]Depto. De Alimentos, Fac. De Farmacia, Universidade Federal de Minas Gerais (UFMG), Belo Horizonte, MG Brazil

The oxidation of emulsified lipids is strongly influenced by the generation of free radicals from lipid peroxide-iron interactions at the emulsion droplet surface. Emulsion droplets stabilized by the anionic surfactants will attract iron and thereby exhibit both faster peroxide breakdown and lipid oxidation than noncharged and positively charged emulsion droplets. Chelators and sodium chloride decrease iron-emulsion droplet interactions thus decreasing peroxide breakdown and lipid oxidation. Similarly, protein-stabilized emulsions will attract prooxidant metals and oxidize faster at pH's above their pI's where the emulsion droplet is negatively charged. Decomposition of peroxides, tocopherol, methyl linoleate and salmon oil are all slower in emulsion droplets with thick interfacial membranes (e.g. Brij 700; 100 ethylenes in headgroup vs. Brij 76; 10 ethylenes in headgroup). These studies show that understanding the factors influencing interactions between lipids and prooxidants could lead to the development of new technologies to develop oxidatively stable lipid dispersions.

Introduction

Textbooks often refer to the oxidation of food lipids as autoxidation. The prefix "auto" is defined as "self-acting" thus the term, "autoxidation" has been used to describe the ability of lipids to oxidize by the self perpetuating generation of free radicals from unsaturated fatty acids in the presence of oxygen. Unfortunately, the term "autoxidation" can be misinterpreted as the ability of lipids to oxidize on their own. This is rarely the case in foods since there are normally prooxidants present or environmental conditions that cause and accelerate lipid oxidation.

Some prooxidants, including lipoxygenases and singlet oxygen generators, initiate lipid oxidation by directly interacting with unsaturated fatty acids to form lipid hydroperoxides (Figure 1). However, many prooxidants exert their effect by promoting the decomposition of hydrogen peroxide (originating from enzymes such as superoxide dismutase) and lipid hydroperoxides into free radicals. Free radicals originating from hydroperoxide breakdown then attack unsaturated fatty acids leading to the formation of lipid peroxides in the presence of oxygen. The decomposition of hydroperoxides into free radicals in foods is often accelerated by transition metals. This is normally the most important pathway for the acceleration of lipid oxidation in foods since metals and hydroperoxides are found in virtually all lipid containing food systems.

Transition metal-promoted hydroperoxide decomposition is important to the oxidative stability and quality of foods for several reasons. First, the abstraction of hydrogen from an unsaturated fatty acid results in the formation of a single alkyl radical. Following hydrogen abstraction, oxygen adds to the alkyl radical to form a peroxyl radical and subsequent abstraction of a hydrogen from another fatty acid or antioxidant to form a lipid hydroperoxide (Figure 1). These reactions by themselves do not result in an increase in free radical numbers. If these reactions were the only steps in the lipid oxidation reactions, the rapid exponential increase in oxidation that is commonly observed in lipids would not occur. Transition metal-promoted decomposition of lipid hydroperoxides results in the formation of additional radicals (e.g. alkoxyl and peroxyl) which exponentially increase oxidation rates as they start to attack other unsaturated fatty acids.

Free Radicals in Food: Chemistry, Nutrition and Health Effects (ACS Symposium Series)

1	SKU: mon0001640518 ISBN: 0841237417 - Books	L01-2-18-001-001-404	Good	$9.99

Subtotal:	$9.99
Shipping:	$3.99
Total Tax:	$0.94
Total:	$14.92

Notes:

Thanks for your order!

If you have any questions or concerns regarding this order, please contact us at serviceohio@hpb.com

HPB-Ohio
1835 Forms Dr
Carrollton, TX 75006
UNITED STATES
serviceohio@hpb.com

HPB-Ohio
1835 Forms Dr
Carrollton, TX 75006
UNITED STATES

To: Stephannie Spicer
2200 W UNIVERSITY BLVD APT 132
DURANT, OK 74701-3160
UNITED STATES

Marketplace:	Amazon US
Order Number:	4432715
Ship Method:	Standard
Customer Name:	Stephannie Spicer
Order Date:	9/18/2022
Marketplace Order #:	112-2434374-1962645
Email:	dmkmcx1813ljxlr@marketplace.amazon.com

Items:

Qty	Item	Locator	Condition	Price

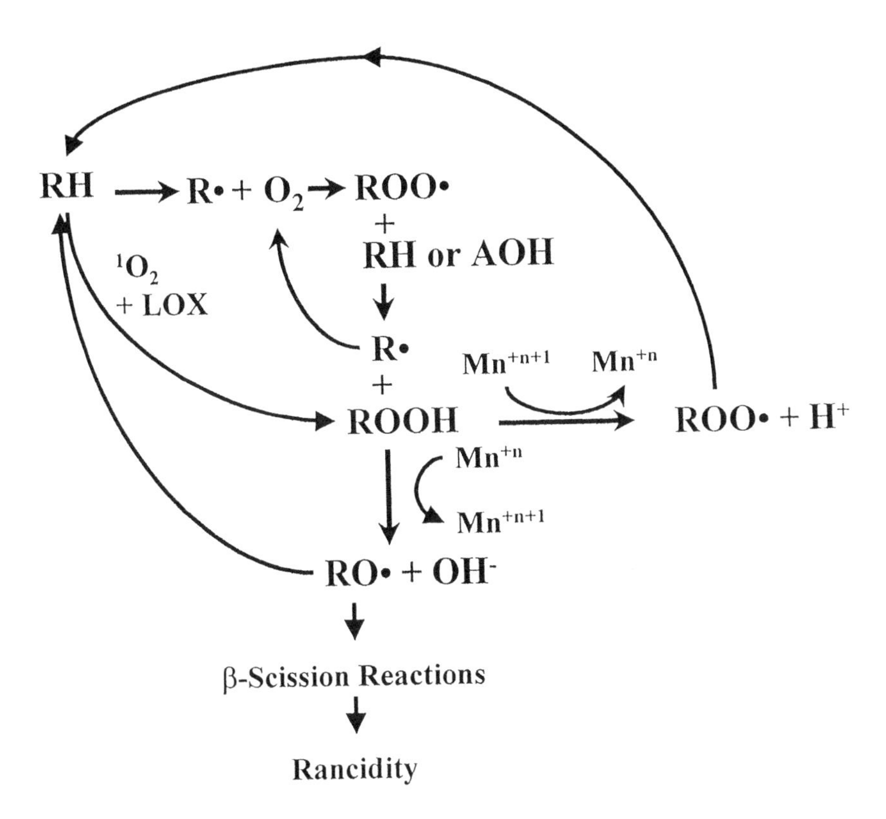

Figure 1. A schematic of the potential reaction pathways that impact the oxidative deterioration of foods. Mn^{+n+} and Mn^{+n+1} are transition metals in their reduced and oxidized states; RH, ROOH and AOH are an unsaturated fatty acid, lipid hydroperoxide and chain breaking antioxidant; and R•, RO•; ROO• are alkyl, alkoxyl and peroxyl radicals, and $^1O_2$ and LOX are singlet oxygen and lipoxygenase, respectively.

Secondly, lipid hydroperoxides by themselves have minimal impact on food quality. Lipoxygenases and singlet oxygen generators produce lipid hydroperoxides but do not directly accelerate the breakdown of unsaturated fatty acids into the compounds responsible for rancid aromas. It is the breakdown of lipid hydroperoxides that leads to β-scission reactions that decompose fatty acids into low molecular weight oxygenated compounds that are volatile enough to be perceived as rancidity. In lipids not subjected to high-energy radiation or elevated temperatures, metals are usually the major cause of hydroperoxide decomposition. An example of this can be seen in low density lipoprotein (LDL) where 15-lipoxygenase will form lipid peroxides in LDL but modification of apoprotein B, a process caused by the interaction of aldehydes originating from the breakdown of lipid peroxides with protein, only occurrs in the presence of copper that stimulates the breakdown of lipid peroxides (*1*). In addition, while lipoxygenases cause peroxidation of fatty acids and cholesterol esters, they have little impact on oxidation of tocopherols in LDL unless copper dependent peroxide decomposition also occurs (*2*).

Transition metals are common constituents of raw food materials, water, food ingredients and food packaging materials. While iron and other transition metals are common in all foods, their reactivity can vary greatly. Transition metals are normally bound to macromolecules to help control their reactivity in biological tissues. Free or low molecular weight complexes of metals are thought to be the major catalytically active metals in many foods and biological tissues. Reactive transition metals convert hydroperoxides into free radicals through the following redox cycling pathway (also shown in Figure 1):

1) $Mn^{n+} + ROOH \text{ or } HOOH \rightarrow Mn^{n+1} + RO\bullet \text{ or } HO\bullet + OH^-$

2) $Mn^{n+1} + ROOH \rightarrow Mn^{n+} + ROO\bullet + H^+$

Where Mn^{n+} and Mn^{n+1} are transition metals in their reduced and oxidized states; ROOH and HOOH are lipid and hydrogen peroxide; and RO•, HO• and ROO• are alkoxyl, hydroxyl and peroxyl radicals, respectively. The rate of hydroperoxide breakdown is dependent on the concentration, chemical state and type of metal. Iron and copper are the most common transition metals capable of participating in these reactions in foods with iron generally being more prevalent than copper. However, cuprous ions (Cu^{+1}) are 50-fold more reactive against hydrogen peroxide than ferrous ions (Fe^{2+}). The redox state of the metal is also important since Fe^{2+} decomposes hydrogen peroxide over 10^5 times faster than Fe^{3+} (*3*) and Fe^{2+} is 10^{17} and 10^{13} times more soluble than Fe^{3+} at pH 7 and 3, respectively (*4*) meaning that it is more apt to promote hydroperoxide decomposition in water-based foods. Hydroperoxide type also influences reaction rates with Fe^{2+} decomposing lipid hydroperoxides (10^3 M^{-1} sec^{-1}) at rates approximately an order of magnitude faster than hydrogen peroxide (76 M^{-1} sec^{-1}; 3).

Lipid hydroperoxides also exist in virtually all foods containing unsaturated fatty acids (*5*). In addition, hydrogen peroxide is found in foods via its use as a processing aid and its formation by numeous pathways in biological tissues such as skeletal muscle. Surfactants are another potential source of hydroperoxides, e.g. peroxides levels ranging from 12-35 μmole peroxide/g surfactant have been measured in Tweens and lechithin (*6*). Food lipids low in hydroperoxides would generally have concentrations in the range of 10-100 nmole/g lipid. These concentrations are often considered low in a food systems, however, they are an estimated 400-1000 times greater than the lipid peroxide concentrations found in blood plasma lipids (*7,8*) suggesting that significant lipid hydroperoxide formation occurs during the conversion of biological tissues into food. The hydroperoxide concentrations found in food lipids means that a food emulsion (mean droplet diameter of 0.4 μm) prepared with a typical oil (50 nmole lipid hydroperoxides/g oil) would contain 4 x 10^4 hydroperoxide molecules/droplet compareed to only 1 hydroperoxide for every 200-400 low density lipoprotein (LDL) particles *in vivo* (*8*).

The existence of both hydroperoxides and transition metals in foods suggests that the interactions of these compounds to produce free radicals is likely an important mechanism for the oxidative deterioration of many foods. The oxidative stability of many biological lipids is dependent on both hydroperoxide and metal concentrations with metals not being strong prooxidants in hydroperoxide-free lipids and hydroperoxides being relatively stable in the absence of metals (*5,7,8*). The role of metal-promoted hydroperoxide decomposition on food quality has been known for some time as food processors commonly use chelators to decrease the development of rancidity. The formation of free radicals by interactions between lipid hydroperoxides and transition metals is also important in *in vitro* models even when high purity double distilled water, salts and surfactants are used in experimental protocols. A recent study in our laboratory showed that oxidation of Tween 20-stabilized salmon oil emulsions was inhibited by both ethylenediaminetetraacetic acid (EDTA; 50 uM) and transferrin (31 μM) even though no exogenous metals were added to the system (*9*). The antioxidant activity of transferrin suggests that iron is the primary prooxidant since transferrin has a strong preference for the chelation of iron over other transition metals. This research highlights the fact that the ingredients used to prepare food emulsions (e.g. lipid, surfactant, buffer and water) would likely contain high enough metal and hydroperoxide concentrations to promote lipid oxidation.

Kinetic studies on lipid oxidation typically show a lag phase where lipid oxidation rates are slow followed by a rapid promotion stage. The length of the lag phase is often attributed to a combination of the fatty acid composition and the concentrations and activity of chain breaking antioxidants. However, the lag phase of lipid oxidation can also be dependent on the concentrations of lipid hydroperoxides and metals. In studies with LDL, the lag phase of Cu^{2+}-promoted oxidation decreases with increasing peroxide concentrations. These studies have been performed by the direct addition of pre-formed lipid peroxides or by formation of lipid hydroperoxides by lipoxygenases (*1,2*). Conversely, when lipid hydroperoxide concentrations in LDL are decreased by the addition of ebselen (*10,11*) or by utilization of glutathione and phospholipid hydroperoxide glutathione peroxidase (*12*), the lag phase of Cu^{2+}-promoted oxidation increased. Therefore, the time of the lag phase of lipid oxidation is likely due to a combination of lipid hydroperoxide, reactive tranisition metals and antioxidant concentrations as well as the degree of fatty acid unsaturation.

It has been known for sometime that metal promoted decomposition of hydroperoxides is involved in the oxidative deterioration of food lipids. However, the understanding of the factors that impact metal-hydroperoxide interactions is still poorly understood. Physical location of the reactants is one of the major factors influencing the ability of transition metals to decompose lipid hydroperoxides especially in foods that contain emulsified lipids. Emulsions contain several physically distinct environments including the lipid core, the droplet interfacial membrane and the aqueous phase (Figure 2). In emulsion systems, the molecular participants in lipid oxidation can concentrate into these different environments. For example, lipid hydroperoxides tend to concentrate at the emulsion surface due to the surface activity imparted into the fatty acid by oxygen. In oil-in-water emulsions, conditions under which transition metals and lipid hydroperoxides interactions increase generally increases lipid oxidation rates. Conversely, metal reactivity can be decreased by physically removing metals from the lipid and/or droplet surface, one of the mechanisms by which chelators and metal binding proteins inhibit lipid oxidation.

Relationship between metal-emulsions droplet interactions and lipid oxidation

Emulsion droplet charge can be an important factor in lipid oxidation rates in oil-in-water emulsions, as well as in other lipid dispersions. When corn oil-in-water emulsions are prepared with anionic (sodium dodecyl sulfate; SDS), cationic (dodecyltrimethylammonium bromide; DTAB) or nonionic (Brij 35) surfactants, iron-promoted lipid oxidation is in the order of SDS>Brij 35>DTAB (*13*). The prooxidant activity of iron in the SDS-stabilized emulsions increased at low pH (3.0) where iron solubility increases. The high

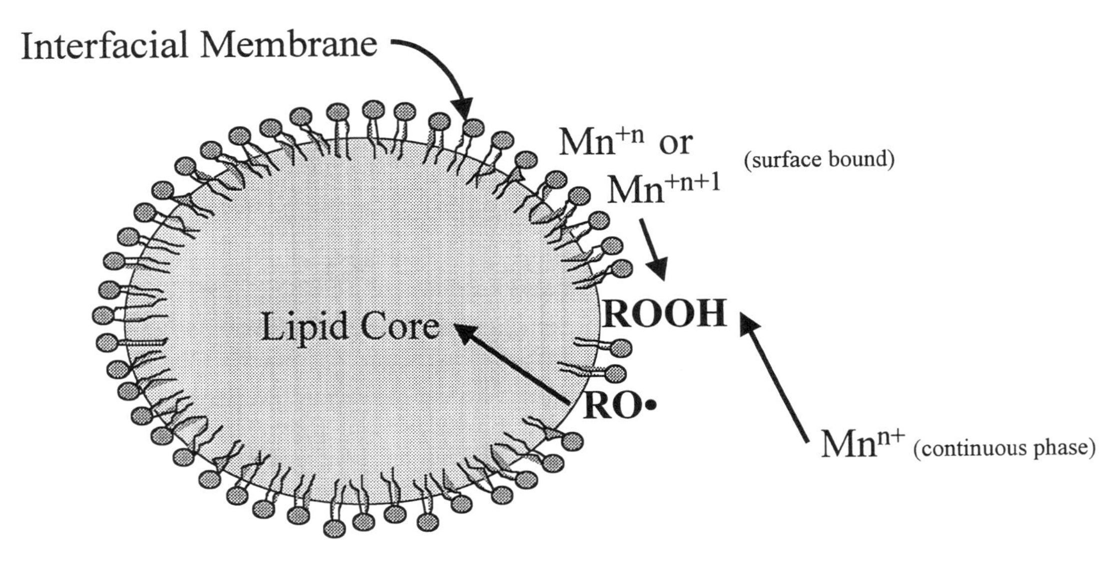

Figure 2. A schematic of the different physical environments of an oil-in-water emulsion and the potential location of the reactants in transition metal-promoted decomposition of lipid hydroperoxide. Mn^{+n+} and Mn^{+n+1} are transition metals in their reduced and oxidized states, respectively; ROOH is a lipid hydroperoxide and RO•, and ROO• are alkoxyl and peroxyl radicals, respectively. Components are not drawn to scale.

amount of interactions between lipid hydroperoxides and iron in SDS-stabilized emulsions at pH 3.0 can cause hydroperoxides to decompose almost as fast as they were formed (*13*). The ability of iron to associate with negatively charged emulsion droplets can be observed by measuring droplet charge (zeta potential) in the absence and presence of iron. At pH 3.0, both ferrous and ferric ions will associate with negatively charged SDS-stabilized droplets but not with positively charged DTAB-stabilized droplets or uncharged Brij 35-stabilized emulsion droplets (*14*). Iron interactions with emulsion droplets stabilized with nonionic surfactants will also occur as iron comes out of solution at pH 7.0 (*12*), an observation which may help explain why emulsions stabilized by nonioinc surfactants oxidize faster at pH 7.0 than 3.0 where iron is soluble and would partition into the continuous phase. However, it is unknown whether insoluble forms of iron can participate in lipid oxidation reactions.

The ability of iron to promote lipid hydroperoxide decomposition in emulsion droplets can be observed in hexadecane emulsions containing cumene hydroperoxide with peroxide decomposition rates being in the order of SDS>Brij>DTAB (*15*). The importance of emulsion droplet-metal interactions is especially true with the oxidized states of the iron. Ferric ions are only observed to decompose cumene hydroperoxide in anionic emulsion droplets suggesting that the ferric ions must be in direct contact with the emulsion droplet to be an active prooxidant. This is likely due to the 10^5 times slower reaction rates of ferric compared to ferrous ions making physical concentration of Fe^{3+} at the droplet surface an important factor in the ability of ferric ions to decompose hydroperoxides. Similarly, cupric ions must be associated with the surface of LDL to be able to promote hydroperoxide decomposition with cupric ions losing their prooxidant activity when copper binding sites are blocked by chemical modification of LDL (*16*).

Protein-stabilized emulsions are another system where the importance of emulsion droplet surface charge on iron-promoted lipid oxidation can be observed. In menhaden oil-in-water emulsions stabilized by whey protein isolate (pI = 5.1), iron-promoted lipid oxidation is faster at pH 7.0 where the emulsion droplets are anionic compared to pH 3.0 where the droplets are cationic. Removal of whey protein from the emulsion droplet surface by Tween 20 changes the droplets from positively charged to nonionic which increases lipid oxidation rates (*17*).

The importance of emulsion droplet charge on lipid oxidation rates also holds true in salmon oil-in-water emulsions in the absence of added metals where oxidation rates are again in the order of SDS $\geq$ Tween 20 $\geq$ DTAB. (*12*). The trend of increasing transition metal-lipid interactions resulting in increasing lipid oxidation rates also occurs in micelles and lipid bilayer model systems (*18-21*). For example, methyl linoleate is oxidized faster by transition metals in negatively charged micelles than in positively charged micelles (*18*) and Fe^{3+}-chelates promote more oxidation in negatively charged than uncharged phospholipid vesicles (*21*).

Further evidence that metal-emulsion droplet interactions are an important factor in lipid oxidation rates is provided by studies using chelators. Chelators inhibit lipid oxidation by altering metal redox potential, solubility or physical location. The importance of the ability of a chelator to impact lipid oxidation by altering the physical location of metals can be seen with chelators that inhibit lipid oxidation without preventing metal redox cycling. For example cuprous ions promote the oxidation of LDL by binding to apoprotein B and catalyzing lipid hydroperoxides decomposition through a redox cycling pathway. Histidine will inhibit copper-promoted LDL oxidation. Since histidine does not alter the redox activity of copper, its mechanism of action involves the physical removal of copper from the surface of LDL thus decreasing its ability to decompose lipid hydroperoxides (*23*).

Several common food additives also impact metal-emulsion droplet interactions. EDTA and phytate remove iron from the surface of emulsion droplets (as determined by zeta potential) and will decrease lipid oxidation in SDS-stabilized salmon oil-in-water emulsions (*14*). Iron that was precipitated onto the surface of nonionic emulsion droplets can be removed by EDTA or by increasing iron solubility by decreasing pH (*12*). Whey proteins also remove iron from SDS-stabilized emulsion droplets, a property that may contribute to their antioxidant activity (*24*). NaCl reduces the ability of iron to bind to anionic emulsion droplets by screening electrostatic interactions between charged species (*13*). Proteins such as transferrin, phosvitin, lactoferrin and serum albumin will form strong metal complexes. These proteins decrease the prooxidant activity of metals by physically minimizing metal-lipid interactions, altering the redox potential of metals and/or helping to maintain metals in their less active, oxidized states.

The processing of foods will also alter the physical location of metals in foods thus increasing their ability to promote lipid oxidation. Release of protein-bound iron by thermal processing has been implicated in the promotion of warmed-over flavor (*25*) and sodium-promoted release of protein-bound iron may be involved in the prooxidant activity of salt in meat products (*26*). Jacobsen et al. (*27*) reported that alterations in the physical location of iron in mayonnaise impacts oxidative stability. In their study, lipid oxidation in mayonnaise was accelerated by ascorbic acid, which was attributed to the ascorbate-dependent release of iron from phosvitin located at the emulsion droplet interface. The authors suggested that the prooxidant effect of ascorbate was due to its ability to both reduce iron to its more reactive ferrous form and release iron from phosvitin thus increasing the ability of iron to decompose existing lipid hydroperoxides.

Relationships between emulsion droplet membrane thickness and iron-promoted lipid oxidation

In addition to surface charge, it is possible that the thickness of the interfacial membrane of emulsion droplets may also impact lipid oxidation reactions by acting as a physical barrier between the reactants in lipid oxidation. For example, if emulsion droplets contained a thick interfacial membrane layer, it may be difficult for aqueous phase iron to interact with lipid hydroperoxides at the droplet surface or for free radicals originating from hydroperoxide decomposition to penetrate through the interfacial membrane to reach unsaturated fatty acids in the lipid core. This hypothesis was tested by preparing emulsions with Brij 76 and 700. These two nonionic surfactants have identical hydrophobic tail groups ($CH_3(CH_2)_{17}-$), but different polar head group lengths with Brij 700 containing 10 times more oxyethylene groups than Brij 76's. The rate of cumene hydroperoxide decomposition was not statistically different in emulsions droplets stabilized by Brij 700 than Brij 76 (Figure 3). In addition, in emulsions containing both cumene hydroperoxide and α-tocopherol, no differences in α-tocopherol oxidation rates were observed between the surfactants. In an emulsion containing cumene hydroperoxide and methyl linoleate, the longer hydrophillic head group of Brij 700 was able to decrease linoleate oxidation (as determine by headspace hexanal formation; Figure 5). These data suggest that the physical location of the different components was an important determinant in the ability of the hydrophilic head group to protect them from oxidation. The observation that the long hydrophilic head groups provide little against the oxidation of cumene hydroperoxide and α-tocopherol suggests that these components are partitioning in the same area (likely the emulsion droplet interfacial membrane) with the free radicals being produced from cumene hydroperoxide decomposition being able to readily oxidize α-tocopherol. The ability of Brij 700 to decrease methyl linoleate oxidation suggests that the thicker interfacial layer inhibited the ability of free radicals originating from cumene hydroperoxide to penetrate into the lipid core and oxidize the fatty acids.

Recent research in our lab has also shown that increasing surfactant hydrophobic tail group size can decrease lipid oxidation as can be seen in salmon oil-in-water emulsions stabilized by polyoxyethylene 10 lauryl ether (Brij-lauryl) or polyoxyethylene 10 stearyl ether (Brij-stearyl). Oxidation of salmon oil was greater in emulsions stabilized by Brij-lauryl than Brij-stearyl as determined by the formation of both lipid peroxides and headspace propanal. However, the ability of hydrophobic tail group size to alter oxidation rates is much less than alterations in oxidation rates observed in emulsions stabilized by surfactants with varying hydrophilic head group size or charge. This may not be due to the type of barrier provided by the hydrophobic tail group but instead the magnitude of the barriers formed since it is possible to produce oil-in-water

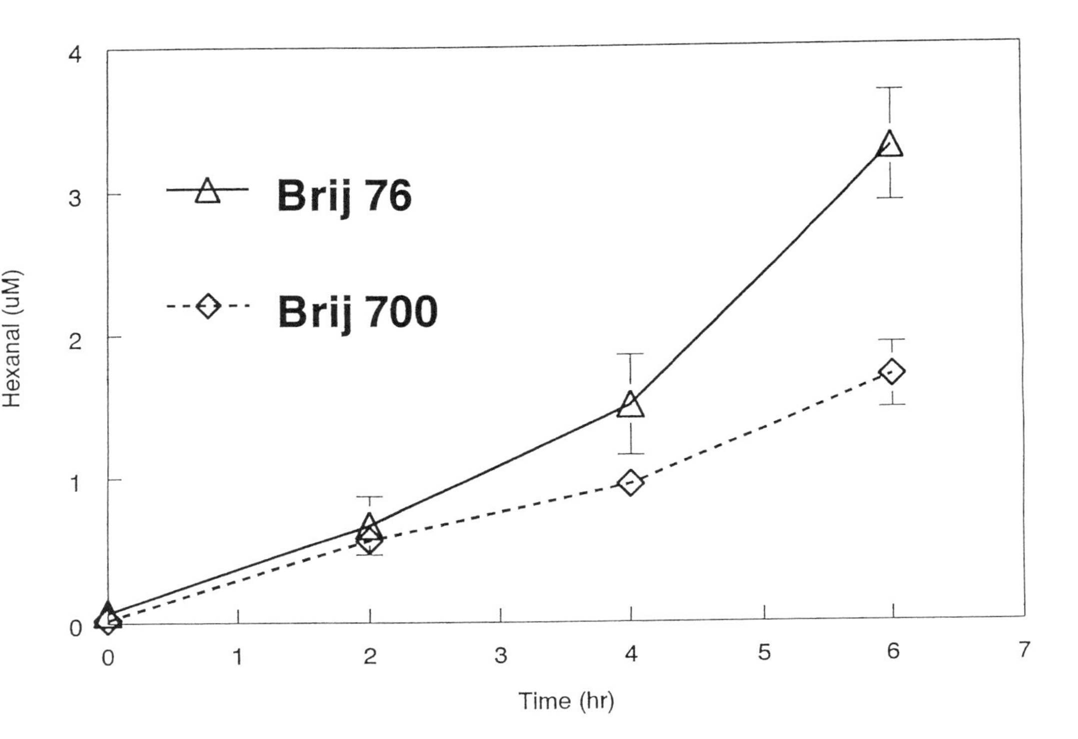

Figure 3. Decomposition of cumene hydroperoxide (initial concentrations ≈ 22 mmoles/L emulsion) in hexadecane emulsions stabilized by Brij 76 or Brij 700 at pH 3.0 and 55°C in the absence of added transition metal. Data points represent means (n=3) ± standard deviations. Some standard deviations lie within the data points.

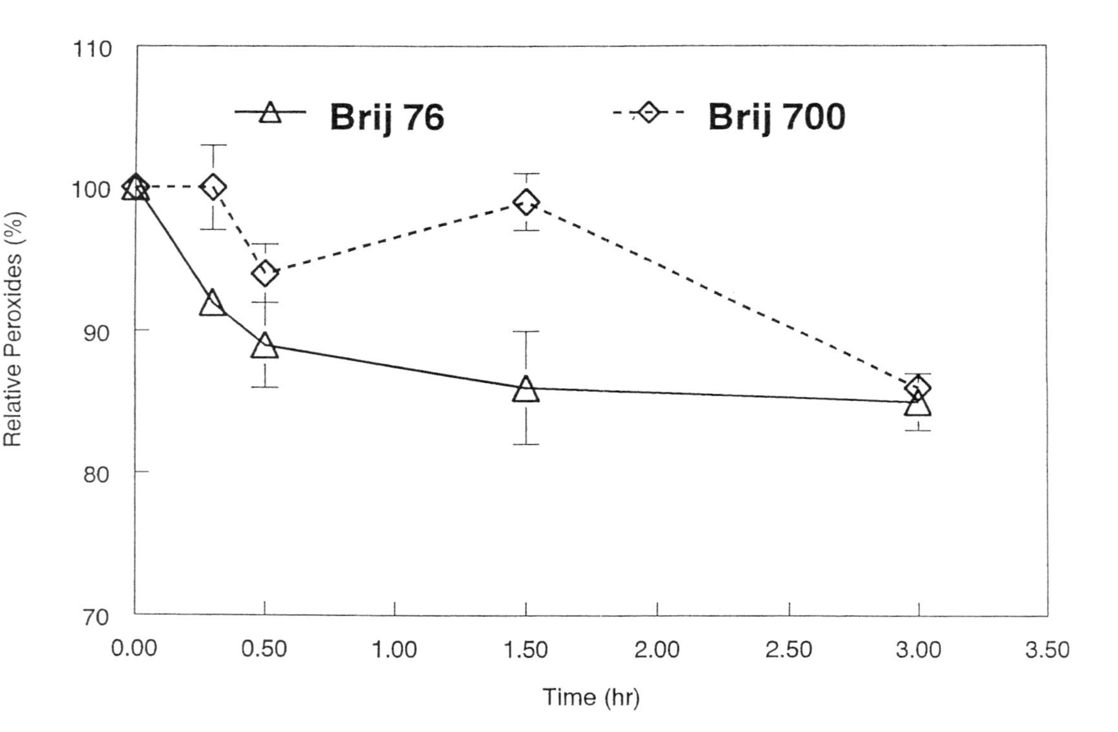

Figure 4. Changes in α-tocopherol concentration (initial concentrations $\approx$ 80 µmoles/ L emulsion) in hexadecane emulsions stabilized by Brij 76 or Brij 700 at pH 3.0 and 55°C in the absence of added transition metals. Emulsions contained 5.7 mmoles cumene hydroperoxide/ l emulsion. Data points represent means (n=3) ± standard deviations. Some standard deviations lie within the data points.

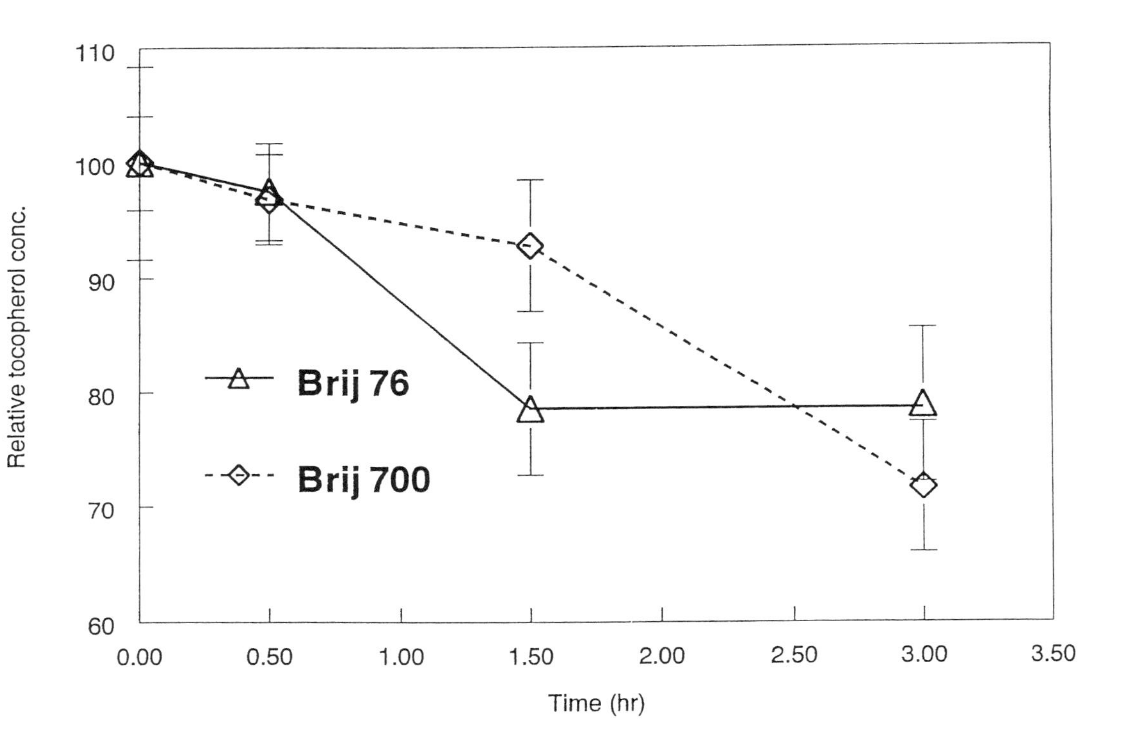

Figure 5. Formation of hexanal in hexadecane and methyl linoleate (0.1 mmoles/L emulsion) emulsions stabilized by Brij 76 or Brij 700 at pH 3.0 and 55°C in the absence of added transition metals. Emulsions contained 5.7 mmoles cumene hydroperoxide/ l emulsion. Data points represent means (n=3) ± standard deviations. Some standard deviations lie within the data points.

emulsions with much greater variations in interfacial membrane properties by altering hydrophillic head properties.

Conclusions:

Finding technologies to minimize transition metal-lipid hydroperoxide interactions may be an additional strategy for increasing the oxidative stability of food lipids. This could be accomplished by alterations in emulsion droplet charge, isolation of metals from the emulsion droplets by chelators and/or increasing emulsion droplet membrane thickness to provide a physical barrier. These strategies in combination with the utilization of traditional antioxidant additives could substantially increase the oxidative stability of lipids in food dispersions.

Acknowledgements

Some of the research presented in this review was partially supported by Grant 9901521 from the NRI Competitive Grants Program of the United States Department of Agriculture. Dr. Marialice P.C. Silvestre was supported by CNPq-Brazil.

References:

1. O'Leary, V.J.; Darley-Usmar, V.M.; Russell, L.J.; Stone, D. *Biochem. J.* **1992**, *282,* 631-634.
2. Lass, A.; Belkner, J.; Esterbauer, H.; Kuhn, H. *Biochem.J.***1996**, *314,* 577-585.
3. Dunford, H.B. *Free Rad. Biol. Med.* **1987**, *3,* 405-421.
4. Zumdahl, S.S., *Chemistry*, 2nd ed., D.C. Heath and Co.: MA, 1989.
5. Halliwell, B.; Gutteridge, J.M. *Meth. Enzymol.***1990,** *186,*1-85.
6. Mancuso, J.R.; McClements, D.J.; Decker, E.A. *J. Agric. Food Chem* **1999,** *47,* 4146-4149.
7. Girotti, A.W. *J. Lipid Res.***1998,** *39,* 1529-1542.
8. Patel, R.P.; Darley-Usmar, V.M. *Free Rad. Res.* **1999**, *30*: 1-9.
9. Mancuso, J.R.; McClements, D.J.; Decker, E.A. *J. Agric. Food Chem.* **1999,** *47,* 4112-4116.

10. Iwatsuki, I.; Niki, E.; Stone, D.; Darley-Usmar, V.M. *FEBS Letters* **1995**, *360,* 271-276.
11. Thomas, C.E.; Jackson, R.L. *J. Pharm. Exper. Ther.* **1991**, *256,* 1182-1188.
12. Thomas, J.P.; Kalyanaraman, B.; Girotti, A.W. *Arch. Biochem. Biophys.* **1994**, *315,* 244-254.
13. Mei, L.; McClements, D.J.; Wu, J.; Decker, E.A. *Food Chem.* **1998**, *61,* 307-312.
14. Mei, L.; Decker, E.A.; McClements, D.J. *J. Agric. Food Chem.* **1998**, *46,* 5072-5074.
15. Mancuso, J. R.; McClements, D.J.; Decker, E.A. *J. Agric. Food Chem.* **2000**, *48,* 213-219.
16. Wagner, P.; Heinecke, J.W. *Arterioscler. Thromb. Vasc. Biol.* **1997**, *11,* 3338-3346 .
17. Donnelly, J.L.; Decker, E.A.; McClements, D.J. *J. Food Sci.* **1998**, *63,* 997-1000.
18. Yoshida, Y.; Niki, E. *Arch. Biochem. Biophys.* **1992**, *295,* 107-114.
19. Fukuzawa, K.; Fujii, T. *Lipids* **1992**, *7,* 227-233.
20. Pryor, W.A.; Cornicelli, J.A.; Devall, L.J.; Tait, B.; Trivedi, B.K.; Witiak, D.T.; Wu, M.A. *J. Org. Chem.* **1993**. *58,* 3521-3532.
21. Fukuzawa, K.; Soumi, K.; Iemura, M.; Goto, S.; Tokumura, A. *Arch. Biochem. Biophys.* **1995**, *316,* 83-91.
22. Silvestre, M.P.C.; Chaiyasit, W.; Brannan, R.G.; McClements, D.J.; Decker, E.A. *J. Agric. Food Chem,* **2000**, *48*, 2057-2061.
23. Retsky, K.L.; Chen, K.; Zeind; J.; Frei, B. *Free Rad. Biol. Med.* **1999**, *26,* 90-98.
24. Tong, L.M.; Sasaki, S.; McClements, D.J.; Decker, E.A. *J. Agric. Food Chem.* **2000**, *48*, 1473-1478.
25. Han, D.; McMillin, K.W.; Godber, J.S.; Bidner, T.D.; Younathan, M.T.; Marshall, D.L.; Hart, L.T. *J. Food Sci.* **1993**, *58,* 697-700.
26. Kanner, J.; Harel, H.; Jaffe, R. *J. Agric. Food Chem.* **1991**, *39,* 1017-1021.
27. Jacobsen, C.; Adler-Nissen, J.; Meyer, A.S. *J. Agric. Food Chem.* **1999**, *47,* 4917-4926.

Chapter 7

Spin Label Study of Water Binding and Protein Mobility in Lysozyme

K. M. Schaich

Department of Food Science, Rutgers University, 65 Dudley Road, New Brunswick, NJ 08901-8520

To better understand, at the molecular level, protein behavior in reduced-water environments, lysozyme was labeled with TEMPO-iodoacetamide (TEMPO-IA), a small amphiphilic nitroxide spin label, and equilibrated over phosphorus pentoxide or saturated salt solutions to a full range (0-1.0) of water activities. Changes in solvent environment detected by the spin label at 0.11, 0.32, and 0.75 parallel the sigmoidal moisture sorption isotherm transitions. Changes in TEMPO spectra vs protein hydration support water binding in clusters at discrete locations rather than in uniform layers on the protein surface. Motion of the TEMPO-IA was very slow ($\tau \sim 10^{-8}$ sec) and increased progressively with hydration as predicted by the glass transition theory. Paradoxically, the order parameter increased with hydration. Apparently, with hydration dry protein, which had been fixed in random orientation, became more mobile and able to associate and align, thus increasing the order and orientation of the spin label. Overall, results support both sorption isotherm and glass transition models of hydration and molecular motion, and demonstrates unique promise for use of spin labels to obtain information about hydration localization and effects on solvent environment and molecular mobility in food systems.

Introduction

Most papers in this forum are considering free radicals actually formed in foods. However, there are cases in which stable free radicals can be *added* to foods as spin probes or spin labels to act as reporters reflecting the molecular environment within the food. Spin labels and probes are stable nitroxide radicals whose EPR spectra are sensitive to their environments *(1)*. When dispersed in systems free as *spin probes*, these stable nitroxides will give information about the solvent environment and mobility within the matrix *(2)*. When attached to a macromolecule as a *spin probe,* the nitroxide radical reflects hydration and motion of the molecule. In this study, the spin probe TEMPO (2,2,6,6-tetramethylpiperidine-N-oxyl radical) modified with an iodoacetamide group to provide a linker group and become a spin label (TEMPO-IA) was attached to lysozyme to gain information about water binding and mobility on this protein.

Information Available From Spin Labels

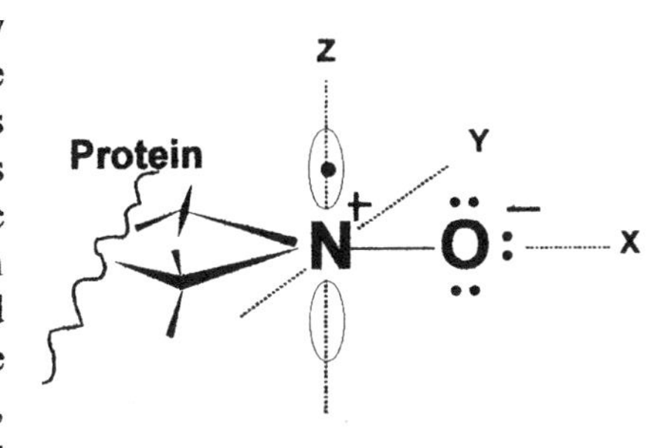

Nitroxide radicals are uniquely able to give this information because the molecule is asymmetric on its three axes (at right) *(1)*. Thus, as this radical interacts in a magnetic field, it gives a different spectrum depending on which axis is oriented parallel to the magnetic field in the spectrometer. In a single crystal, each orientation can be distinguished quite clearly *(1-3)*, as shown in Figure 1. The X-axis looks end-on at the oxygen; its orientation-specific signal shows a broadening and downfield shift due to the oxygen. From the y-axis side view, the nitroxide is planar and the asymmetry is minimized, and this is reflected in the narrow lines and signal centered at the nitrogen g-value. The z-axis looks down at the top of the nitroxide, where the rest of the molecule now induces considerable asymmetry in interaction with the magnetic field. As a consequence, in the EPR signal the lines broaden and the splittings (distance between lines) nearly double. In contrast, in solution the nitroxide is freely tumbling (i.e. it is rotating very fast) so all these different orientations average each other out, and the spectrum observed is just from the N-O itself -- very narrow, with typical hyperfine splittings of about 15 G typical of nitrogen radicals.

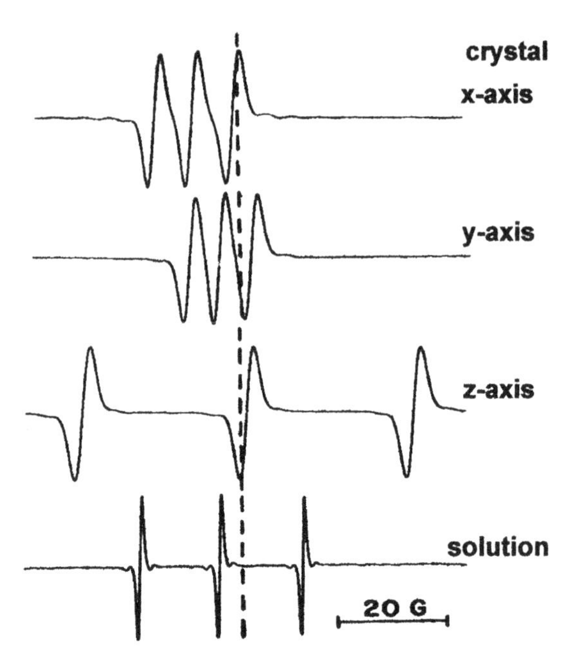

Figure 1. Spectra of nitroxide radicals change with the axis oriented parallel to the magnetic field. Adapted from with permission from (2). Copyright American Chem. Soc, 1969.

The EPR spectra shown in Figure 1 are from isolated crystals where there is no motional interference from other molecules or structures. When nitroxides are linked to molecules or dispersed in systems, these same axial symmetries come into play, now reflecting the environment in which they are located. Where nitroxides are free as in solution, the signal has the sharp narrow lines typical of the freely tumbling nitroxide (Figure 2a). When the nitroxide is fixed in space, a highly asymmetric signal results, with the two outer lines weaker than the center line (Figure 2b). The line furthest upfield (right of center) is most affected, and sometimes can become almost indistinguishable. Mixed environments in which some of the nitroxides are immobilized and some are more mobile can also be detected (Figure 2c).

There are two key features that provide information in spin label spectra. One is the hyperfine splitting, or distances between the three lines in the solution spectra, designated a_N *(1-3)*. Line splittings vary with the solvent (or environment), being smallest in hydrophobic aprotic solvents and increasing with polarity as the electron density shifts from the oxygen to the nitrogen, as shown for di-*t*-butyl nitroxide in Table I *(3)*.

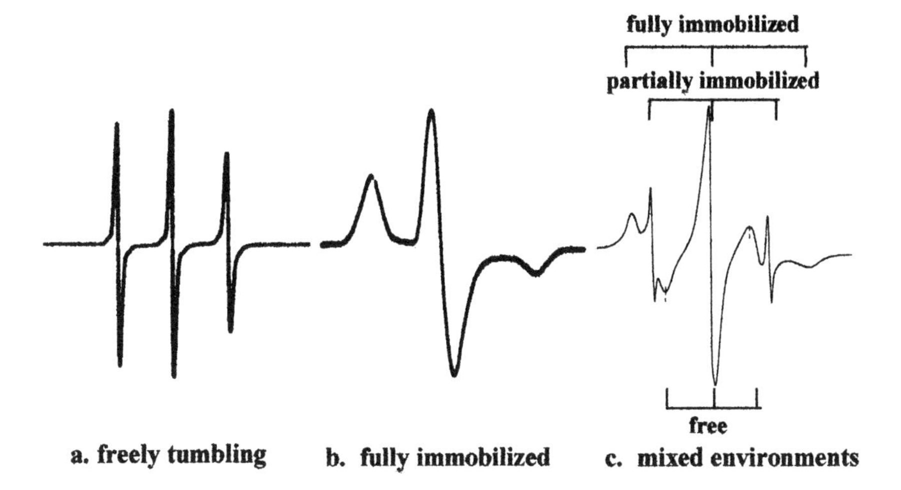

Figure 2. Types of spectra from nitroxide radicals attached to a macromolecule or dispersed in a viscous medium. Adapted from (1) with permission. Copyright Academic Press, 1976.

Table I. Variation of hyperfine splittings and g-values of a di-t-butyl nitroxide spin label in different solvents.[a]

solvent	a_N (gauss)	g-value	
water	17.16	2.0056	$>\overset{+\bullet}{N}-O^-$
ethanol: H_2O 1:1	16.69	2.0057	↑
methanol	16.21	2.0058	
ethanol	16.06	2.0059	
isopropanol	15.94	2.0059	
octanol	15.89	2.0059	
decanol	15.87	2.0059	
acetonitrile	15.68	2.0060	
acetone	15.52	2.0061	
Me propionate	15.45	2.0061	
hexane	15.10	2.0061	$>N-O\bullet$

[a] Data selected from *(3)*. Copyright Academic Press, 1978.

The second key feature is the anisotropy *(1,4)*, or degree of distortion, reflected in a) an increase in the distance between the two outer lines, b) an increase in the line width of the center signal because now it actually has several overlapping lines from multiple orientations in it, and c) a decrease in the intensity of the downfield line relative to the center line (Figure 3.). Note especially $a_{//}$ and $a_{\perp}$, $h_{(0)}$, $h_{(+1)}$, and $\Delta H_{(0)}$, the key features used to gain definitive quantitative information about nitroxide environment from the spectra.

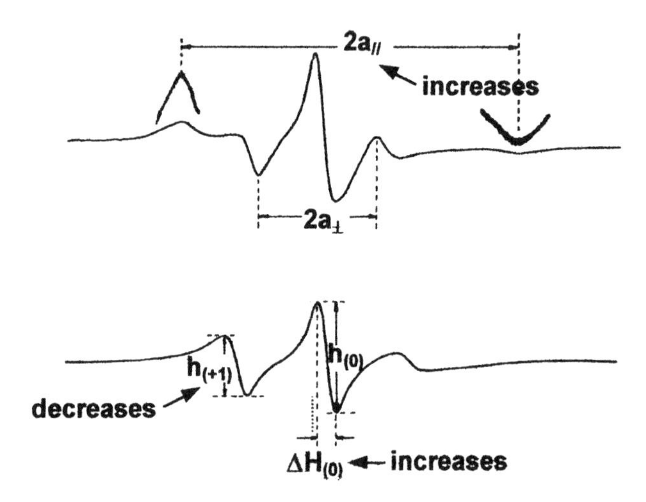

Figure 3. Anisotropy in EPR spectra reflects immobilization of the spin label. Changes in splitting constants $a_{//}$ and $a_{\perp}$, line intensities $h_{(0)}$ and $h_{(+1)}$, and the line width $\Delta H_{(0)}$ of the center line provide a means of measuring the degree of immobilization. Adapted from (5) with permission. Copyright Elsevier Science Publishers, 1990.

Using Spin Labels to Obtain Information about Hydration Environments of Proteins

Spin labels were used in this study to follow the extent and effects of water binding to a dry model protein, lysozyme, as water activity was increased. Where water is bound and what it is doing in food systems plays a critical role in textural characteristics and chemical stability of foods. The relationship of water to food characteristics has classically been described in moisture sorption

isotherms (Figure 4), where the actual moisture content is plotted against water activity (how much water in food is like pure water) of a food *(6)*. Three regions have been described in which water initially binds to polar sites in dry molecules then builds up to a monolayer (I), then multilayers and capillaries (II), and finally fluid water (III) *(7)*. In the hydration (adsorption) process, molecules change from rigid to plastic and mobile, to fully hydrated or dissolved, and the
process is reversed in dehydration. A food at a given water activity can have two different moisture contents, depending on whether that state is reached by adsorbing or desorbing water. This conceptual description was developed to explain hydration data in foods, but not everyone agrees that it is an accurate picture, and definitive physical evidence for the individual hydration states and corresponding physical properties is still lacking.

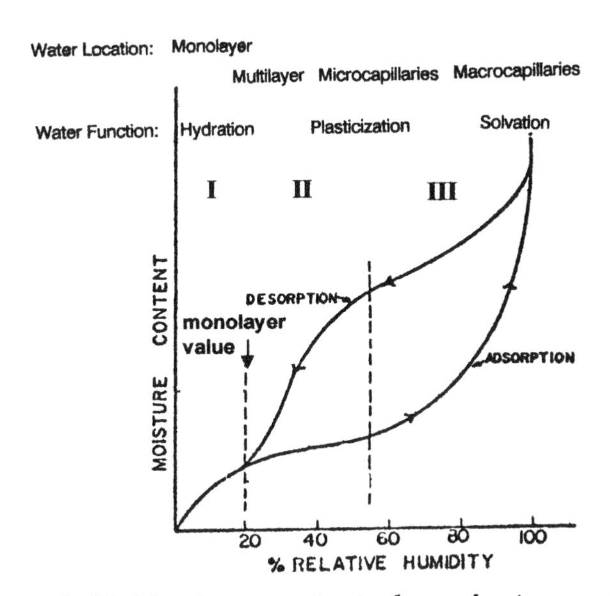

Figure 4. Model moisture sorption isotherm showing proposed function and location of water at different water activities in foods.

An alternative approach to understanding hydration in foods is the glass transition, a phenomenon in which a solid amorphous phase undergoes a discontinuous change from rigid glass to flexible rubber as a function of moisture and temperature *(7-11)*. Viewed in terms of molecular motion, the "glassy" state is characterized by extremely high viscosity rather absolute fixation. The viscosity can be decreased and molecular motion and migration increased either by raising the temperature or the moisture content of the system. There are a series of temperature-moisture combinations, called glass transitions, below which molecular motion is quite limited and the system is effectively rigid (e.g. a food is crisp and hard) and above which the material becomes flexible (Figure 5). As moisture increases above T_g, a food becomes rubbery, then increasingly sticky and soggy, and eventually the structure collapses as molecules are solubilized.

Glass transitions and moisture sorption isotherms are not incompatible concepts. Isotherms measure hydration while glass transitions reflect hydration behavior of a food system. Glass transitions modify the impact of moisture. The points at which plots of T_g curves and moisture sorption isotherms intersect indicate conditions under which molecules become mobile and thus foods will be unstable and have limited shelf life (Figure 5). Materials with high T_g's require high moisture before reactions occur and properties change. Materials with low T_g's will be reactive even when dry.

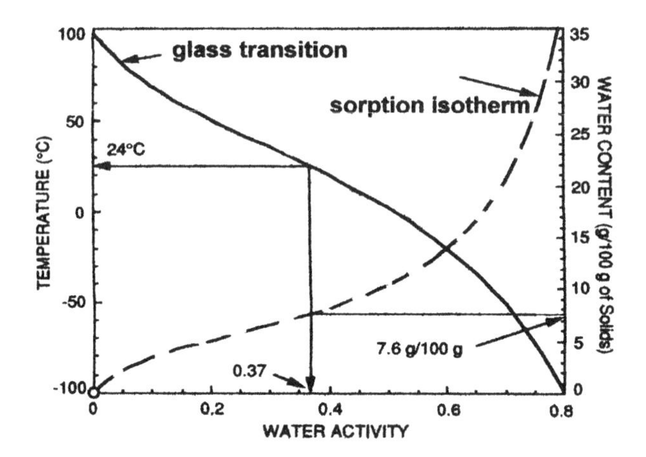

Figure 5. Interrelationship of glass transitions and moisture sorption isotherms of foods. At every condition to the left of the glass transition curve, this food will be in its glassy state regardless of water activity or moisture content. To the right of the glass transition, all molecules have some degree of motion, even in dry or low moisture foods. Adapted from (11) with permission. Copyright Instit. of Food Technologists, 1996.

Much still needs to be learned about the interrelationships between moisture sorption, T_g's , and food properties and stability. Given their sensitivity to both molecular motion and solvent, spin labels offer a unique method for monitoring phase changes at the molecular rather than property level. Thus, they were used to obtain information about the hydration level and pattern, the local reaction environments, and the molecular mobility in lysozyme as a protein model system. Lysozyme was selected because it is well-characterized as a protein, and labeling techniques for it are well-established *(12,13)*. *His*-15 is near the terminal end of lysozyme, easily labeled, and in the folded state; this site also is on the protein surface so it should be an early site for hydration. The spin label TEMPO-IA (the iodoacetamide of 2,2,6,6-tetramethylpiperidine-N-oxyl radical) was used because, as nearly the smallest of the probes, it will introduce the least motional disturbance in the protein. Because it is small and polar, the nitroxide reporting group will be situated close to the protein surface, and thus is very sensitive to the polarity and motion in the immediate environment of the *his* to which it is attached.

Materials and Methods

Lysozyme (hen egg white, 3x recrystallized) was obtained from Sigma Chemical Corp., St Louis, Mo. It was labeled at *his*-15 with TEMPO-IA (Molecular Probes, Inc., Eugene, OR) according to established procedures *(12,13)*, then lyophilized. Labeling was verified by NMR.

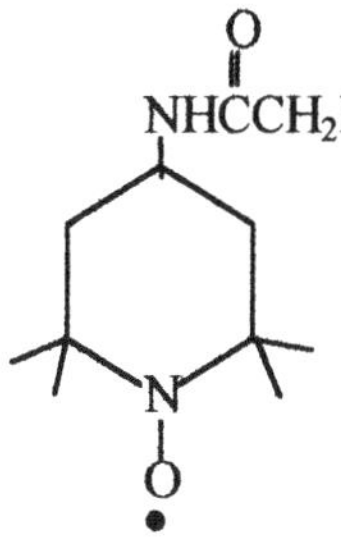

Samples of the labeled, lyophilized lysozyme in small Petri dishes were equilibrated for two weeks over the following saturated salt solutions in sealed jars to generate systems with different water activities (relative humidities): phosphorus pentoxide, 0; lithium chloride, 11; potassium acetate, 22; magnesium chloride, 32; potassium carbonate, 40; potassium iodide, 69; sodium chloride, 75; ammonium sulfate, 81; potassium chloride, 84; potassium nitrate, 94 *(14)*. Rehydration of the lyophilized lysozyme was slow, and preliminary experiments showed that two weeks were necessary for samples to reach moisture equilibrium (constant weight).

O
NHCCH$_2$I
N
O
•
TEMPO-IA

Samples were loaded into suprasil quartz EPR tissue flat cells, and EPR spectra were recorded under non-saturating conditions on a Varian E-12 EPR spectrometer interfaced to and controlled by a MassComp 5500 minicomputer and equipped with an X-band (9.5 GHz) microwave bridge, 100 kHz modulation, and variable rate signal averaging. Microwave frequencies were measured by a Hewlett Packard 505C frequency counter; magnetic field scans were routinely calibrated using a solution of Fremy's salt (peroxylamine disulphonate). Spectra were generated by repeatedly collecting and signal averaging 30 second scans over at least 5 minutes, and were presented as first derivatives of the absorption at resonance. A DPPH crystal mounted provided an internal field marker for normalizing successive scans. Spectral *g*-values were determined relative to a DPPH standard by established methods *(15)*. Data was collected, stored, and analyzed by computer. Typical EPR settings were 10 mW power, 2 G modulation amplitude, 10,000 gain, and 0.03 sec time constant.

To quantify changes in motion of the spin label, four characteristics were measured or calculated from the spectra. $\Delta H_{(0)}$ is the line width of the center line; $2a_N$ is the distance between the outer lines. Both are quite sensitive to solvent and immobilization. The order parameter (S) is the degree of immobilization relative to a fully rigid nitroxide (1 is fully rigid and 0 is fully free). The calculation of S varies with the system and especially with degree of anisotropy. Several methods for calculating S were tested, and the following method developed by Gaffney for anisotropic systems was found to reflect the spectra most accurately *(16)*. The equation includes a polarity correction to account for linewidth differences due to solvent change, in this case hydration, rather than immobilization:

$$S = \frac{a_{//} - [a_{\perp} + C]}{a_{//} + 2[a_{\perp} + C]} * 1.66 \qquad (1)$$

where

$$C = 4.06 - 0.053\,(a_{//} - a_{\perp}) \text{ MHz.} \qquad (2)$$

The rotational correlation time, τ, is a measure of how fast (in sec) the nitroxide is moving. Typical values for τ range from 10^{-11} to 10^{-6} sec (2,4). Because the spectra showed anisotropy and limitation in probe motion, the following equation was used to estimate τ *(5)*:

$$\tau = \text{const}\ \Delta H_{(0)} \sqrt{\frac{h_{(0)}}{h_{(+1)}}} \qquad (3)$$

using const = 2.55×10^{-9} for this spin label.

Results and Discussion

EPR spectra showed the expected progression from immobilized nitroxides at low a_w's to increased mobility at high a_w's (Figure 6). Spectra from dry and low a_w samples were broad with substantial anisotropy and no evidence of multiple species. Spectral widths decreased with a_w, but the anisotropy remained even as hydration increased, indicating the label was not free to rotate

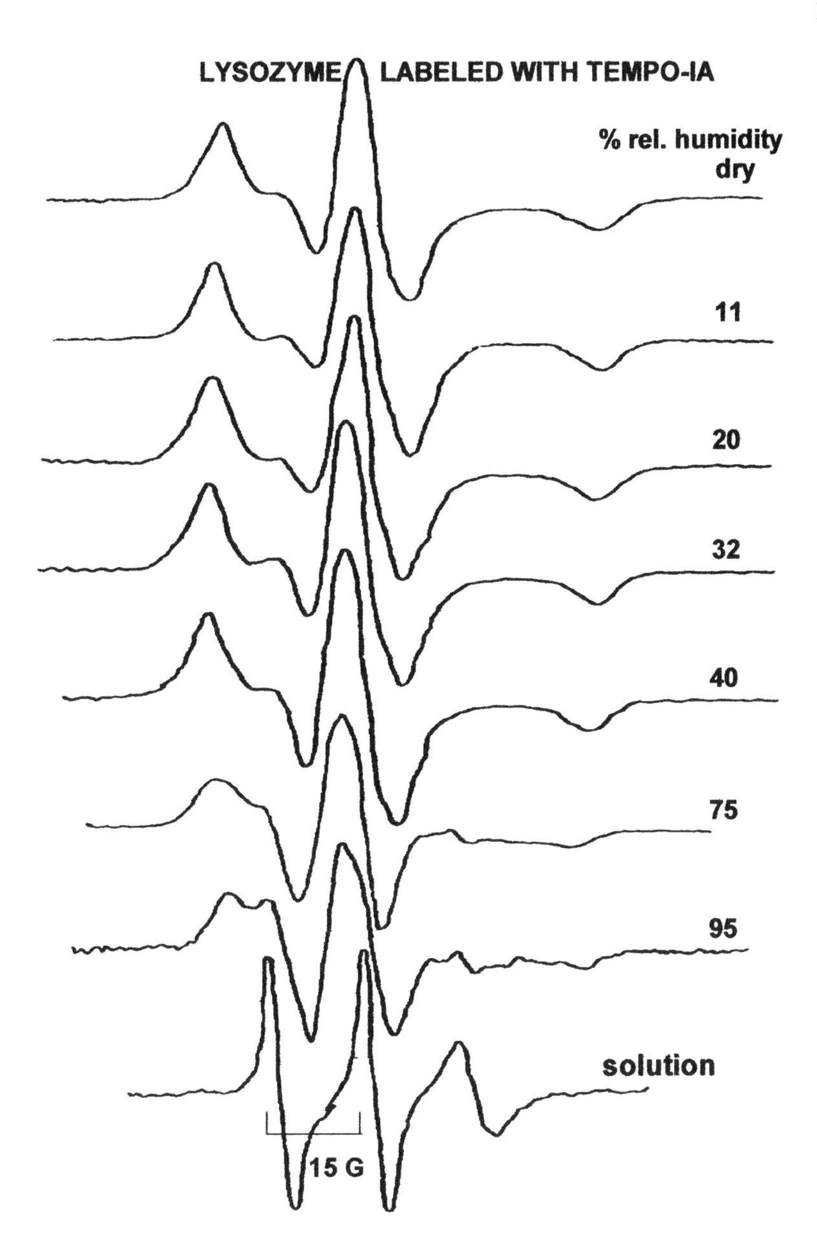

Figure 6. As water activity increases, EPR spectra show shifts from anisotropic immobilization to some motion of lysozyme spin labeled with TEMPO-IA at His-15. Multiple species are evident in spectra at a_w 0.75 and 0.95.

on the surface of the protein. At a_w 0.75 and 0.95, both immobile and mobile species were evident. The second signal was particularly evident in a_w 0.95 samples; the increased linewidth of the center line, $H_{(0)}$, resulted from overlapping center lines from the two components.

Analysis of the four spectral parameters quantitated the label and protein response to the increased moisture (Figure 7). Spectral widths ($2a_N$) and rotational correlation times decreased with increased moisture, although the values for a_w 0.75 samples were out of line for several measures, possibly due to inconsistent or incomplete hydration of the lysozyme particles. Closer examination of the data revealed some very interesting patterns. Signal width, the measure most sensitive to change in solvent, had inflection points at a_w 0.11, 0.32, and 0.75, which closely correspond to the inflection points of the sorption isotherm for lysozyme. This is very interesting since hydration data (Table II) indicates that the average hydration of lysozyme is less than 2 molecules of water per amino acids residue at a_w 0.95, certainly not enough for a spin label to experience as fluid water. Thus, to account for the spin label data, the water must be binding in clusters around polar surface sites, including *his-15*, rather than accumulating as a monolayer over the entire protein.

The rotational correlation times had weak inflections at a_w 0.40 and 0.90, which correlate well with changes in system behaviors as a function of a_w (Figure 4). Multiple species were not visually evident in the spectra until a_w 0.75 but the change in center linewidth indicates that unresolvable levels were already present at a_w 0.40. This is in mid-intermediate moisture region where, according to theory, micro- and macrocapillaries of water begin to be present, and spin labels located in these regions would be free to rotate. Overall, these three measures all support the functional descriptions of moisture sorption isotherms, although they suggest that attribute changes may actually occur at higher a_w's than usually ascribed.

This spin label data is consistent with previous studies of lysozyme showing that physical properties such as Young's modulus and activation energy of the mechanical glass transition *(17)*, tryptophan phosphorescence *(18)*, dielectric constant *(19)*, and hydrogen exchangeability *(20)* all exhibited sigmoidal or other nonlinear behavior as a function of water activity. The first $2a_N$ inflection point matches monolayer values of a_w 0.11 previously reported for lyophilized lysozyme *(21)*. Rupley *(22)* found that water clusters begin to form at 0.1 g water /g lysozyme and that monolayer coverage was not complete until 0.40 g water/g protein, near the second $2a_N$ inflection point. Populations of tightly bound, weakly bound, and "multilayer" or trapped water in lysozyme were identified by O^{17} and D_2 NMR *(23)* also at water activities corresponding to the $2a_N$ inflection points. TEMPO-labeled lysozyme, therefore, appears to behave the same as the unlabeled enzyme, so the label should accurately reflect the local solvent environment around *His*-15.

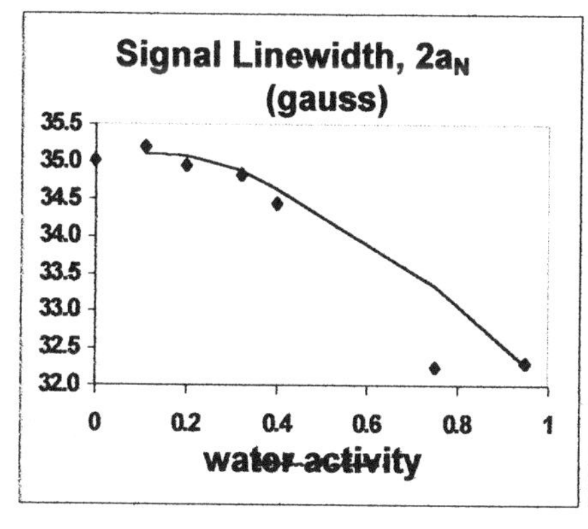

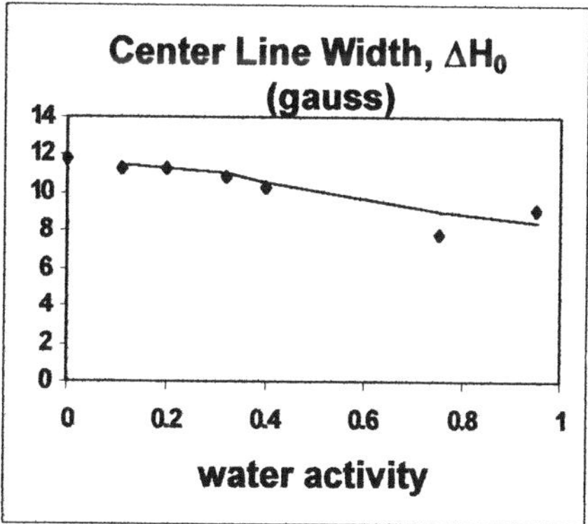

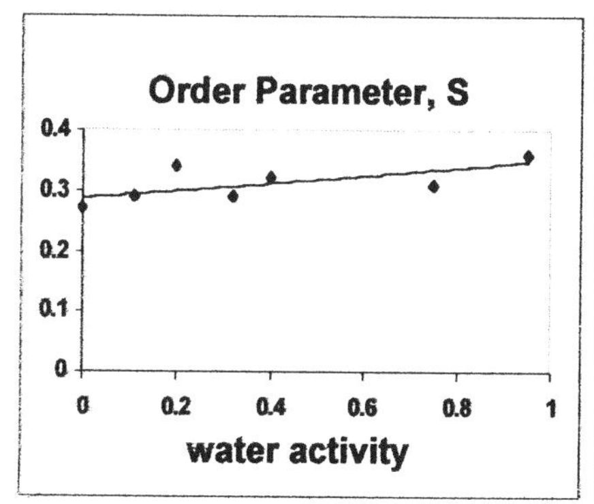

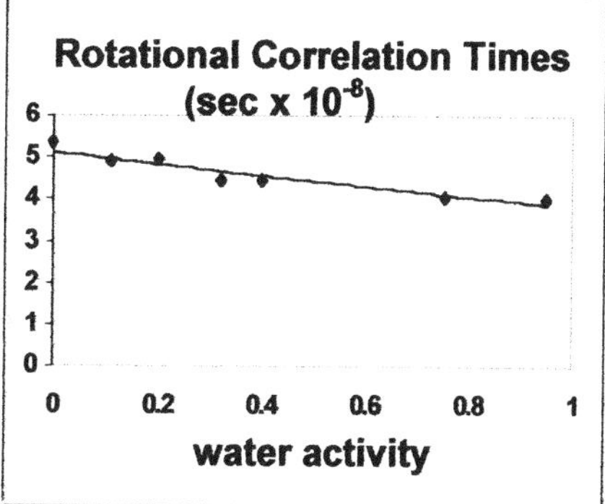

Figure 7. Changes in spectral parameters of spin-labeled lysozyme as water activity and system hydration increase. Lines between points are best fit curves.

Table 2. Hydration of Lyophilized Lysozyme[a]

a_w	g H_2O / g protein	moles H_2O[b] / moles protein	moles H_2O[b] / aa residue
11	0.06	49	0.38
20	0.08	65	0.50
32	0.11	89	0.69
40	0.14	110	0.85
75	0.23	185	1.43
95	>0.30	243	1.89

[a] Seymour Gilbert, Rutgers University, Center for Advanced Food Technology, unpublished, used with permission

[b] Based on mol wt 14.6 kD mol wt and 129 residues

There was support also for the glass transition theory in the measures of molecular order and motion. Morozov et al. determined the glass transition of lysozyme to range from approximately 140 to 220 K, for a_w 0.93 to dry, respectively -- well below room temperature *(24)*. Thus, full rigidity should not be present and the systems studied should reflect increasing mobilization of molecules already in the rubbery phase. This is indeed what was observed.

The S values were less than 0.5, indicating considerable motion even in the dry and low moisture lysozyme. In contrast to the solvent-sensitive parameters, S and τ showed smooth progressive change with increasing a_w. This suggests that each incremental addition of water contributed to molecular flexibility in the label region even if it is insufficient to change the solvent environment around the label. That S increased with moisture was not expected. S was calculated by several different equations specific for isotropic or anisotropic motion in several motional ranges, and they all gave the same trend. Apparently, in the amorphous lyophilized solid, lysozyme molecules are fixed in space and cannot pack together. However, with rehydration the protein chains become free to align, associate, and pack together, thereby increasing molecular ordering. Rotational correlation times were in the slow motion range (10^{-8} sec) and showed a smooth decrease with hydration as protein chains moved faster. It is known that spin labels reflect their own motion separate from the protein, motion of a localized segment of the protein, and action of the entire protein chain moving en masse *(25)*. Thus, the slight inflection of τ at a_w 0.40 may mark a shift point below which the motion arises predominantly from the label alone at the hydrated surface, and above which motion of the *his-15* chain and local protein region contribute increasingly as water penetrates the protein interior. Hydration was not sufficient in any of the rehydrated lysozymes to support full mobilization of the protein chains, and this was reflected in the τ values.

Summary and conclusions

In lyophilized lysozyme labeled with TEMPO-IA and rehydrated to a range of water activities, hydration remained low, even at high a_w's, and consequently molecular rotation was slow, 10^{-8} sec. Solvent environment changes detected by the spin label at 0.11, 0.32, and 0.75 parallel the sigmoidal moisture sorption isotherm transitions. Changes in TEMPO spectra vs protein hydration levels support water binding in clusters at discrete locations rather than in uniform layers on the protein surface.

Motion of the TEMPO-IA labeled lysozyme in rehydrated model systems was very slow, near the limit for EPR detection, and increased progressively with hydration as predicted by glass transition theory. Paradoxically, the probe order parameter decreased with moisture. In dry amorphous systems, protein chains were fixed in random orientations, but with hydration chains became more mobile and able to associate and align. Hence, molecular order and orientation detected by the label increased.

Some final qualification of the spectral interpretations must be offered. Interpretations of spin label data must be made with caution, particularly in systems with mixed probe orientations such as in this study. Order parameters are normally calculated relative to known probe orientations and assume fast motion. In unoriented samples with broad lines, absolute S values can only be calculates by extensive computer simulation *(4)*. However, S can be approximated using equations appropriate for the type of motion evident in the spectra, and these values are useful to show relative trends and changes between systems. Similarly, the 10^{-8} sec rotation times observed in the lysozyme systems are slow, very near the slow motion limit for standard EPR. Various mathematical treatments have been developed to accurately describe slow anisotropic motion, but are only approximations. Determination of transitions in molecular mobility of the probe and protein is complicated when mixed populations of spin labels are present because solvent effects ($\uparrow 2a_N$) counter-balance mobilization effects ($\downarrow 2a_N$) of water and overlapping lines obscure true linewidths. While conventional EPR measurements of spin-labeled molecules are clearly useful for range-finding, definitive studies should use spectral simulations *(4)* and advanced EPR methods such as saturation transfer *(26)* that are sensitive to slower motions (10^{-7} to 10^{-3} sec) to detect small changes in mobility and obtain more accurate, detailed information about spin labeled proteins at low moisture contents.

Acknowledging these qualifications, this study nevertheless demonstrates unique promise for use of spin labels to obtain information about hydration localization and effects on solvent environment and molecular mobility in food systems. Results provide support for both sorption isotherm and glass transition models of hydration and molecular motion. More detailed information about hydration and mobility in the lysozyme and other proteins upon rehydration may be obtainable by testing spin labels with different polarities and molecular sizes bound at various protein sites. It will be necessary to use computer simulations of spectra and develop saturation transfer EPR methods to study the slow molecular motions that are probably widespread in food systems.

References

1. I.C.P. Smith. In *Free Radicals in Biology, Vol 1*, Pryor, W.A., Ed. Academic Press: New York, NY, 1976, p. 149.
2. Griffith, O. H.; Waggoner, A.S. *Accounts Chem. Res.* **1969,** *2,* 17-24.
3. Jost, P. C.; Griffith, O. H. *Meth. Enzymol.* **1978**, *XLIX,* 369-407.
4. Schrier, S.; Polnasek, C.F.; Smith, I. C. P. *Biochim. Biophys. Acta* **1978,** *515,* 375-436.
5. Kitagawa, S.; Kametani, F.; Tsuchiya, K.; Sakurai, H. *Biochim. Biophys. Acta* **1990,** *1027*, 123-129.
6. Rockland, L. B.; Stewart, G. F. *Water Activity: Influences on Food Quality;* Academic Press: New York, NY, 1981.
7. Pomeranz, Y. *Functional Properties of Food Components.* Academic Press: New York, NY, 1991, Chapt 1: Water, pp. 3-23.
8. *Water Relationships in Foods*; Levin, H.; Slade, L., Eds. Plenum Press: New York, NY, 1991.
9. *The Glassy State in Foods*; Blanshard, J.V. M.; Lillford, P. J., Eds.; Notting-ham University Press: Loughborough, UK, 1993.
10. Roos, Y. H. *Food Technol.* **1995**, *49*, 97-102.
11. Glass transitions in low moisture and frozen foods: effects on shelf life and quality, IFT Scientific Status Summary; *Food Technol.* **1996**, *50*, 95-108.
12. Wien, R.W.; Morissett,J.D.; McConnell, H.M. *Biochemistry* **1972**, *11*, 3707-3716.
13. Ogawa, S., McConnell, H.M., and Horwitz, A.F. *Proc. Natl. Acad. Sci. U.S.A.* **1968.** *601*, 401-405
14. Rahman, S. *Food Properties Handbook*; CRC Press, Boca Raton, FL, **1995**, pp. 4-11.
15. Schaich, K.M. EPR methods for studying free radicals in foods; this volume. pp. xx.
16. Gaffney, B. J. In *Spin Labeling: Theory and Applications*; Berliner, L. J., Ed.; Academic Press: NY, 1976, p.567-571.
17. Morozov, V. N.; Morozova, T. Y.; Kacjalova, G. S., Myachin, E. T. *Int. J. Biol. Macromol.* **1988**, *10*, 329-336.
18. Shah, N. K.; Ludescher, R.D. *Photochem. Photobiol.* **1993,** *58*, 169-174.
19. Bone, S.; Pethig, R. Dielectric studies of the binding of water to lysozyme; J. Mol. Biol. **1982,** *157*, 571-575.
20. Poole, P. L.; Finney, J. L. *Int. J. Biol. Macromol.* **1983,** *5*, 308-310.
21. Schaich, K.M. and Karel, M. *J. Food Sci.* **1975**, *40,* 456-459.
22. Rupley, J. A.; Yang, P.-H.; Tollin, G. In *Water in Polymers*, ACS Symposium Series 127; Rowland, S. R., Ed.; Amer. Chem. Soc.: Washington, DC, 1980, pp. 111-132.

23. Lioutas, T. S., Baianu, I. C.; Steinberg, M. P. *Arch. Biochem. Biophys.* **1986,** *247*, 68-75.
24. Morozov, V. N.; Gevorkian, S.G. *Biopolymers* **1985,** *24*, 1785-1799.
25. Berliner, L. J. Methods Enzymol. **1978,** XLIX Enzyme Structures G, 419-479.
26. Hyde, J. S.; Dalton, L. R. In *Spin Labeling: Theory and Applications II*, Berliner, L.J., Ed.; Academic Press, New York: NY, 1979, pp. 3-70.

Chapter 8

Radical Formation in Dairy Products: Prediction of Oxidative Stability Based on Electron Spin Resonance Spectroscopy

Dorthe Kristensen, Maiken V. Kröger-Ohlsen, and Leif H. Skibsted

Department of Dairy and Food Science, Food Chemistry, Royal Veterinary and Agricultural University, Frederiksberg C, Denmark

Rate of formation and steady state concentration of radicals in dairy products depend on the composition of the raw milk, including fatty acid profile, balance between pro- and antioxidants, and enzyme activity, which are influenced by the feeding regimen, breed, age, and health of the individual cow. Course of cooling, processing conditions (especially heat treatment parameters), packaging conditions, storage temperature, and exposure to light or chemical contaminants also influence radical formation in dairy products. The oxidative status of dairy products has traditionally been evaluated by peroxide value (PV), thiobarbituric acid reactive substances (TBARS), or sensory analysis. However, recent investigations indicate that different Electron Spin Resonance (ESR) techniques may be further developed into useful methods to predict the oxidative stability of dairy products.

Introduction

Milk and dairy products are vulnerable to oxidation during processing and storage. This can result in a decrease in nutritional value along with formation of strong off-flavors and potentially toxic lipid and cholesterol oxidation products. Unsaturated fatty acids, especially in the phospholipids, have a strong tendency to autoxidize, and redox active metal ions can further catalyze formation of radicals from lipid hydroperoxides *(1)*. Singlet oxygen seems to be involved as a precursor for formation of the first hydroperoxides *(2)*, but enzymatic processes may also be important for initiation of oxidation. The further progression of oxidation depends on many factors such as the presence of metal ions, other pro-oxidants, or chain-breaking antioxidants. Oxidation not only involves lipids, but proteins and smaller molecules, such as ascorbate, are also oxidized. The common feature is formation of radicals as reactive intermediates. Prediction of the oxidative stability of a given product in relation to the quality of the raw milk, heat treatment, and further processing would clearly be advantageous for manufacturers, and in this context it would be of interest to measure the radicals that are formed prior to or by cleavage of hydroperoxides. Electron Spin Resonance (ESR) spectroscopy is a technique that detects paramagnetic compounds such as radicals, and investigation of oxidation in milk and dairy products by ESR spectroscopy has been found useful in a number of studies. Recently a promising predictive method has been developed to characterize resistance to lipid oxidation that uses relatively mild accelerating conditions in combination with ESR spectroscopy *(3)*.

Radical-generating oxidation reactions that have been found to take place in milk and dairy products will be reviewed. In addition, different ESR-based techniques that have been employed in dairy science will be described. Finally, examples will be given of the application of ESR spectroscopy to studies of oxidation in such products; the perspective will be prediction of storage stability based on the tendency of radicals to form during the early stages of oxidation.

Radicals in Milk and Dairy Products

Radicals are continuously produced in milk and dairy products by various mechanisms, such as autoxidation of lipids and other components as a result of processing, light exposure, and enzyme action.

Lipid derived radicals (L·, LO·, and LOO·) are produced during the initiation and propagation stages of the autoxidation of unsaturated lipids. The source of the primary radical in lipid autoxidation is still an open question. However, the reaction may be initiated by cleavage of preformed lipid hydroperoxides (LOOH) by transition metal ions, which form lipid alkoxyl radicals (LO·), or by metal-catalyzed cleavage of hydrogen peroxide (H_2O_2), which forms hydroxyl radicals (·OH, Fenton reaction) that subsequently oxidize lipid molecules by hydrogen atom abstraction *(4)*.

Lipid hydroperoxides are the primary oxidation products of the autoxidation process, but they are also generated by addition of singlet oxygen ($^1O_2$) to double bonds of unsaturated lipids. $^1O_2$ may be produced photochemically in a reaction involving riboflavin as a photosensitizer *(5)*. The reaction can produce either superoxide ($O_2^{\cdot-}$) or $^1O_2$ depending on the reaction conditions as shown by the two mechanisms in Figure 1 *(2,5)*. A further mechanism for production of $^1O_2$ in milk and dairy products is the reaction of H_2O_2 with residual hypochlorite (ClO^-) from cleaning and desinfection procedures *(1)*. On light exposure, milk develops a "burnt feather" flavor characteristic for oxidation of sulphur containing amino acids, followed by a "cardboard-like" flavor characteristic for lipids; this shows that protein oxidation is preceding lipid oxidation *(6)*.

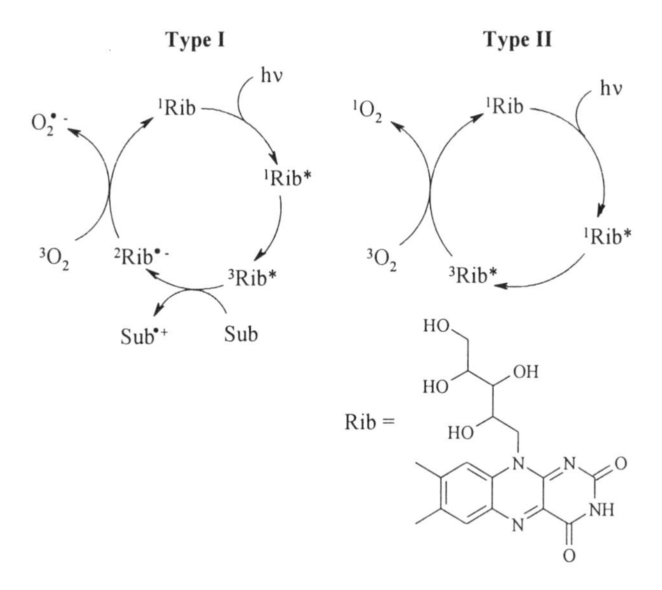

Figure 1. Mechanisms for light induced sensibilisation of riboflavin.

High heat treatment of milk exposes free thiol groups that autoxidize to form thiyl radicals, $O_2^{\cdot-}$ and H_2O_2. Thiols may exert an indirect prooxidant effect through this mechanism *(4)*. The thiols are mainly derived from modification of the milk fat globule membrane and the serum proteins, particularly β-lactoglobulin. In freeze-dried milk proteins, radicals were detected by ESR spectroscopy, and spectral characteristics were consistent with carbon-centered radicals *(7)*.

Ascorbyl radical can be measured in a steady-state concentration in fresh milk. Oxidation of ascorbate by lactoperoxidase has been proposed to be the source of this radical, based on the increase in ESR signal upon an increase in the concentration of H_2O_2 and the decrease in signal upon addition of azide (a lactoperoxidase inhibitor) *(8)*. However, the radical may also stem from autoxidation of ascorbic acid in the presence of transition metals *(9)*.

Enzymatic generation of radicals includes production of $O_2^{\cdot -}$ by xanthine oxidoreductase (XOD), which is found in relatively high amounts in the milk fat globule membrane *(10)* (Figure 2). XOD is synonymous with xanthine dehydrogenase (XD, EC 1.1.1.204) and xanthine oxidase (XO, EC 1.2.3.2). XO catalyzes the oxidation of xanthine (and hypoxanthine) into $O_2^{\cdot -}$ and uric acid under consumption of O_2; whereas, XD requires NAD^+ *(4)*. $O_2^{\cdot -}$ is known to act as a weak oxidant in aqueous media, but it is able to abstract weakly bound hydrogen atoms from compounds such as tocopherols, thiols, and ascorbic acid *(11)*.

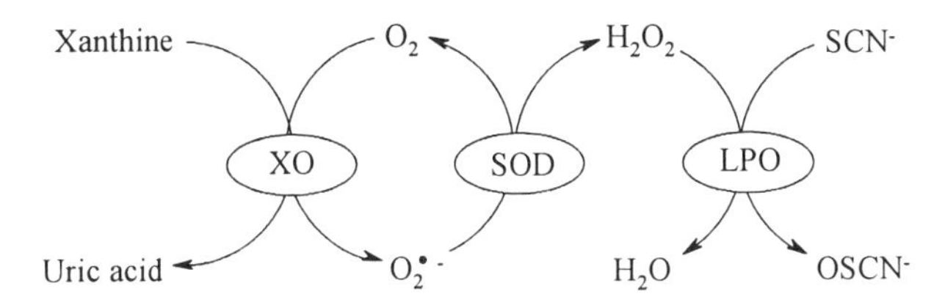

Figure 2. Enzymatic oxygen activation in milk. XO: Xanthine oxidase, SOD: Superoxide dismutase, LPO: Lactoperoxidase.

Radicals may also be produced enzymatically by the lactoperoxidase system, an antimicrobial system naturally present in milk. Lactoperoxidase (EC 1.11.1.7) catalyzes formation of the antimicrobial agent hypothiocyanite ($OSCN^-$) by oxidation of thiocyanate (SCN^-) by H_2O_2 *(12)*. One source of H_2O_2 is the enzyme superoxide dimutase (SOD, EC 1.15.1.1), which catalyzes dismutation of $O_2^{\cdot -}$ to O_2 and H_2O_2 *(10)*. Oxidation of SCN^- and other substrates also produces short-lived radicals, such as $^{\cdot}SCN$ *(13)*, and halide radicals formed in the lactoperoxidase system may add to unsaturated lipids and thereby stimulate lipid oxidation *(14)*.

ESR Techniques in Dairy Science

Traditionally, the oxidative status of dairy products has been evaluated by determination of primary lipid oxidation products, lipid hydroperoxides, expressed as the peroxide value, or by determination of the volatile secondary lipid oxidation products, analyzed by headspace GC or reaction with thiobarbituric acid (TBARS). Oxysterols, especially the 7-ketocholesterol, has also been suggested as an indicator of lipid oxidation *(15)*. Only a few studies have applied ESR spectroscopy, although this technique holds the potential of measuring the precursor for the primary oxidation products. It measures unpaired spin with high sensitivity; therefore, it is a powerful tool in the study of radical reactions. In the following, three different ESR based techniques will be considered: direct measurement, spin trapping, and spin labeling. These techniques for evaluation of oxidation processes in dairy products differ in sensitivity and specificity, and some advantages and limitations of the different ESR techniques are summarized in Table I.

Table I. Advantages and limitations of different ESR-based techniques for evaluation of radical formation and oxidation processes in dairy products.

Technique	*Phase*	*Advantage*	*Limitation*
Direct measurement	Aqueous	Type of radical may be assigned.	Few radicals are stable in aqueous media.
	Solid	No introduction of foreign substances.	Broad signals with little information about the type of radical. Risk for artifact formation in sample preparation including freezing, drying or freeze-drying.
Spin trapping	Aqueous / solid	Short-lived radicals may be studied. Possibility for assignment of radicals.	Only a relative measure of radicals. Introduction of foreign substances with a risk of side reactions (oxidation, reduction). Verification by different spin traps or alternative analytical methods needed for identification.
Spin labeling	Aqueous / solid	Known initial concentration of radical.	Introduction of foreign substances with a risk of side reactions (oxidation, reduction). Mechanism for disappearance of added radicals may not be related to the actual concentration of radicals in the system.

Direct measurement

Direct measurement normally requires no pre-treatment of the sample, and the absolute concentration of the radical can be determined. Under optimal conditions, the identity of the radical may be assigned from spectral characteristics.

The majority of radicals are highly reactive, and the lifetime is too short to allow for direct detection in a liquid phase. A well-known exception to this is the ascorbyl radical that has been investigated in milk and infant formula by direct detection *(8,16)*. Protein-derived and protein-lipid-derived radicals can also be measured directly in products in which the aqueous phase is immobilized e.g., milk powder *(17-19)*, freeze-dried milk, and cheese *(20-22)*. In such products, a broad slightly asymmetric signal is observed (Figure 3), which has no hyperfine structure due to the restricted motion of the molecule. Similar radicals have been observed in dry milk protein products (sodium caseinates) and freeze-dried preparations of β-lactoglobulin, β-casein, α_s-casein and κ-casein *(7)*.

Spin trapping

In spin trapping experiments, relatively stable ESR-active compounds, the *spin adducts,* are formed by reaction of radicals with ESR-silent compounds, *the spin traps,* added to the sample. The most commonly used spin traps are nitroxides and nitrones, which form stabilized radicals by reaction with other radicals *(23)*. Based on the characteristics of the spin adduct (e.g. hyperfine pattern, coupling constants, and g-value), an assignment of the radical in question is often possible. However, due to lack of specificity of the often-used nitroxides, like N-*t*-butyl-α-phenylnitrone (PBN), a valid verification of the radicals trapped depends on identification by techniques such as HPLC-MS. Despite the lack of spectral resolution, spin trapping seems to be a promising technique for prediction of the oxidative susceptibility of dairy products (see later sections).

Spin labeling

A third technique involves addition of a known concentration of a stable radical, a *spin label,* to the sample. The progress of oxidation is followed by the decay of this radical due to reaction with other radicals formed in the sample, producing ESR-silent compounds. However, it should be mentioned that some spin labels may participate in other redox reactions *(21)* or may even act as an antioxidant.

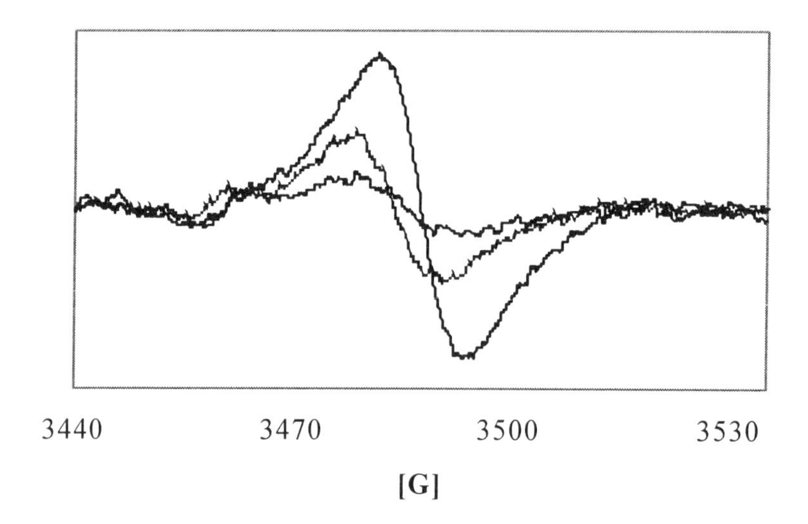

Figure 3. ESR spectrum of processed cheese stored for 15 months at 5, 20, and 37 ℃ (increasing intensity) exposed to fluorescent light (2000 lx). The cheese was freeze-dried prior to analysis. (Adapted with permission from reference 21. Copyright 1999 ACS.)

Oxidative Changes in Milk an Dairy Products: Correlation with Radical Concentrations

Relatively few studies have been carried out on milk and dairy products using ESR spectroscopy, and except for ascorbyl radical, the radicals formed in such products obviously yield broad and poorly resolved spectra, which have provided little structural information. Likewise, the typical spectrum obtained in a spin trapping experiment, using nitroxide spin traps, yields information on radial concentration, not structure. Regardless of the nonspecific nature of such measurements, quantitative determinations of radicals have been found to correlate well with other measurements of oxidation. Particularly, radical concentrations determined at an early stage of oxidation (or in an accelerated test using relatively mild conditions) could be used to determine the susceptibility of certain products to oxidation. ESR spectroscopy is therefore a promising technique for prediction of oxidative stability; this will be illustrated in the following examples.

Milk

The tendency for radicals to form in freeze-dried raw milk, as measured by ESR spectroscopy, was found to be inversely related to the content of the natural antioxidant, α-tocopherol. Milks, with large variations in α-tocopherol concentrations, were obtained through a feeding study where diets included either α-tocopherol-poor roughage or α-tocopherol-sufficient roughage combined with intraperitoneal injection of *rac*-α-tocopherol acetate *(20)*. The potential of ESR for detection of early events in oxidation is evident from this study, as tocopheryl radicals apparently did not interfere with the lipid and/or protein radicals.

Milk Powder

ESR spectroscopy has been used to measure the concentration of radicals in milk powder in several experiments *(17-19)*.

Oxidative carry-over effects were demonstrated in instant whole milk powder. Butter oil is used in instant whole milk powder. Following 10 months storage, milk powder, produced using butter oil with inferior oxidative status, had higher radical concentration *(17)*.

In an accelerated storage experiment, development of radicals in whole milk powder was greater for powder prepared using low heat conditions relative to powders prepared using medium or high heat conditions *(18)*. Further, it was found that increasing temperature and water activities during storage resulted in increased radical concentration. This is in accord with findings from long-term storage experiments that used standard analytical methods. In another accelerated storage experiment, a good (inverse) correlation between the sensory score of reconstituted milk, from instant whole milk powder, and concentration of radicals in the powder was demonstrated *(19)*.

Cheese

Spin trapping, with 5,5-dimethyl-l-pyrroline-*N*-oxide (DMPO), and spin labeling, with 2,2,6,6-tetramethylpiperidine-1-oxyl (TEMPO), were used to investigate the influence of light and temperature on the oxidation of processed cheese during an 11 day accelerated storage experiment *(21)*. Based on results obtained by both methods, it was concluded that light is more important than temperature for formation of radicals during short-term storage. The direct measurement of free radicals in freeze-dried preparations of cheese samples stored for 15 months was evaluated as a third method. It was found that for longer storage time, temperature becomes more important than light for formation of radicals and lipid oxidation, a result confirmed by analysis of TBARS (Figure 4).

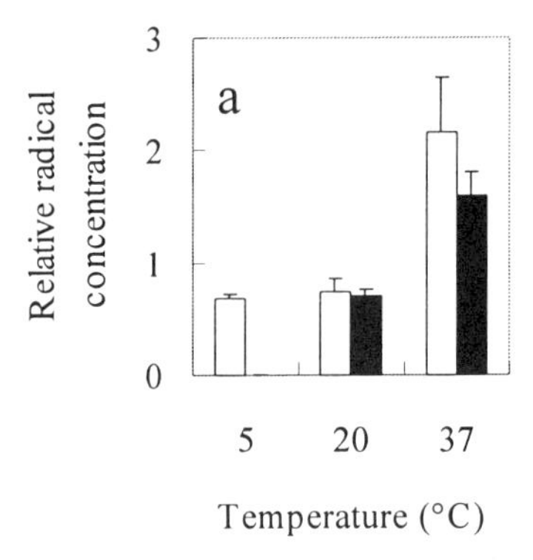

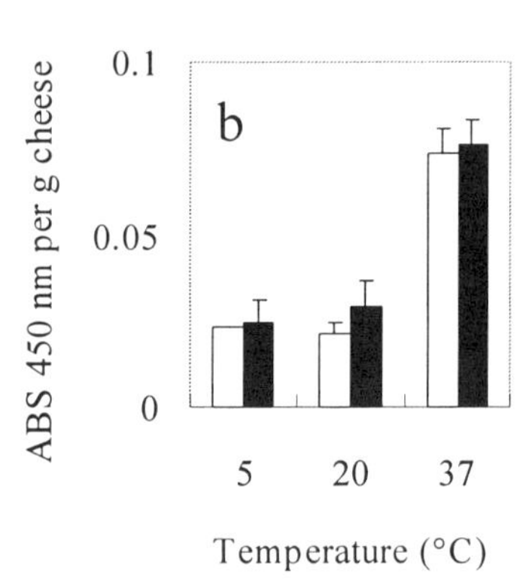

Figure 4. Influence of temperature and light exposure from fluorescent tubes (2000 lx) on (a): signal intensity in the ESR spectrum as a measure of free radicals in processed cheese stored for 15 months. The cheese was freeze-dried prior to recording of the spectra as shown in Figure 3. (b): secondary lipid oxidation products measured as TBARS (450 nm absorbance). □: Light; ■: Dark. (Adapted with permission from reference 21. Copyright 1999 ACS.)

Light induced oxidation in sliced Havarti cheese was monitored for up to 21 days. The cheese was packaged in a modified atmosphere, stored at 5 °C, and exposed to light (1000 lx fluorescent light) or protected against light. Samples were analyzed by standard analytical methods, such as determination of peroxides, color measurement, sensory evaluation, and ESR spectroscopy *(22)*. The tendency of radicals to form was evaluated by ESR spectroscopy of freeze-dried cheese samples. It was found that radical concentration decreased during the initial stages of storage and reached a minimum after four days. Cheeses exposed to light reached the minimum most quickly, suggesting that light accelerates a further transformation of radicals into primary and secondary oxidations products.

Based on spectral characteristics and the similarity of the systems, radicals observed in milk powder and in freeze-dried cheese (Figure 3) may be similar to radicals detected in protein-containing lipid emulsions that were freeze-dried and stored in air for five weeks *(24)*. The ESR signal increased up to about seven days of incubation and was considerably weaker after 21 days. The disappearance of radicals coincided with an increase in the fluorescence intensity of chloroform extracts of the freeze-dried emulsion, indicating formation of tertiary oxidation products (reactions between lipid oxidation products and protein/amino acids).

Butter and Dairy Spread

An accelerated test procedure was developed to evaluate the oxidative stability of food lipids with varying degrees of saturation. The procedure is based on ESR spectroscopy and the spin trap N-t-butyl-α-phenylnitrone *(3)*. Four different food lipids were analyzed: lipid fraction from mayonnaise enriched with fish oil, rapeseed oil, dairy spread made from rapeseed oil and cream, and butter. A good correlation was found between degree of saturation and the delay in radical formation, induction period, under accelerated conditions. The order of stability of lipid was as anticipated mayonnaise < rapeseed oil < dairy spread < butter.

The induction period for formation of PBN spin adducts in milk fat, isolated from butter, is shown in Figure 5. The course of radical formation resemblances that typically observed in other accelerated oxidation tests, such as Rancimat and Oxidograph methods. This ESR procedure has the advantage of milder conditions compared to other methods, combined with a short assay time.

Perspectives

ESR spectroscopy is a new and promising technique in dairy science. In addition to numerous specific applications, it may prove to be useful in predicting the oxidative stability of dairy products. Areas for further investigation include the identification of radicals formed during production and storage of milk and dairy products, and also transfer of radicals between enzymes, lipids, proteins, etc. New spin traps, designed with specific hydrophilic/hydrophobic balance to trap specific radicals, provide good opportunities for evaluation of such radical transfer

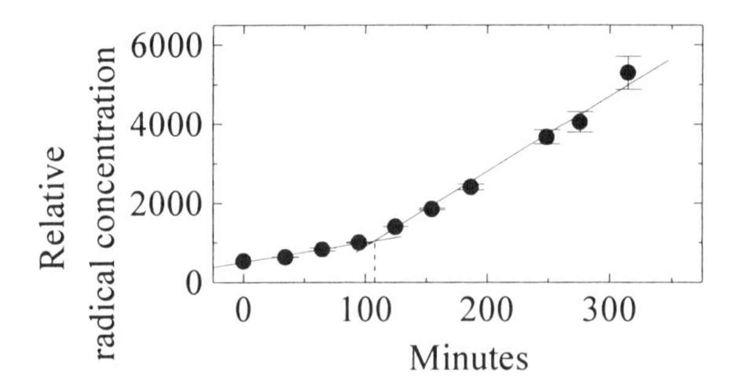

Figure 5. Formation of PBN spin adducts in lipid isolated from butter at 80 °C. The method is described in ref. (3), and the induction time is indicated. Unpublished results.

reactions. Application of new spin traps for accelerated and long-term storage experiments should also be tested.

Finally, the relation between radical concentration, as measured by ESR, and subsequent development of oxidation should be explored further, and the potential of ESR spectroscopy to predict oxidation should be further established in storage experiments and accelerated tests.

Acknowledgement

This research was performed as part of a collaboration project between LMC-Center for Advances Food Studies and the Danish Dairy Research Foundation as part of the FØTEK program.

References

1. O´Connor, T. P.; O´Brien, N. M. In *Advanced Dairy Chemistry: Lipids*; Fox, P. F., Ed.; Chapman & Hall: London, 1995; Vol.2, pp 309-347.
2. Aurand, L. W.; Boone, N. H.; Giddings, G. G. *J. Dairy Sci.* **1977,** *60*, 363-369.
3. Thomsen, M. K.; Kristensen, D.; Skibsted, L. H. *JAOCS* **2000,** *in press*.
4. Halliwell, B; Gutteridge, J. M. C. *Free Radicals in Biology and Medicine;* 2nd ed.; Clarendon Press: Oxford, UK, 1989; 543 pp.
5. Korycka-Dahl, M.; Richardson, T. *J. Dairy Sci.* **1978,** *61,* 400-407.
6. Skibsted, L. H. *Int. Dairy Fed. Bull.* **2000,** *346,* 4-9.
7. Hansen, P. M. T.; Harper, W. J.; Sharma, K. K. *J.Food Sci.* **1970,** *35*, 598-600.
8. Nakamura, M. *J. Biochem.* **1994,** *116*, 621-624.
9. Buettner, G.R. *J. Biochem. Biophys. Meth.* **1988,** *16,* 27-40.

10. Björck, L.; Claesson, O. *J. Dairy Sci.* **1979,** *62,* 1211-1215.
11. Korycka-Dahl, M.; Richardson, T. *J. Dairy Sci.* **1980,** *63,* 1181-1198.
12. Aune, T. M.; Thomas, E. L. *Eur. J. Biochem.* **1977,** *80,* 209-214.
13. Løvaas, E. *Free Radic. Biol. Med.* **1992,** *13,* 187-195.
14. Kanner, J.; Kinsella, J. E. *Lipids,* **1983,** *18,* 204-210.
15. Nielsen, J. H.; Olsen, C. E.; Lyndon, J.; Sørensen, J.; Skibsted, L. H. *J. Dairy Res.* **1996,** *63,* 615-621.
16. Champagne, E. T.; Hinojosa, O.; Clemetson, C. A. B. *J. Food Sci.* **1990,** *55,* 1133-1136.
17. Stapelfeldt, H.; Mortensen, G.; Skibsted, L. H. *Milchwiss.* **1997,** *52* , 266-269.
18. Stapelfeldt, H.; Nielsen, B. R.; Skibsted, L. H. *Int.Dairy* J. **1997,** *7,* 331-339.
19. Stapelfeldt, H.; Nielsen, B. R.; Skibsted, L. H. *Milchwiss.* **1997,** *52,* 682-685.
20. Stapelfeldt, H.; Nielsen, K. N.; Jensen, S. K.; Skibsted, L. H. *J. Dairy Res.* **1999,** *66,* 461-466.
21. Kristensen, D.; Skibsted, L. H. *J. Agric Food Chem.* **1999**, *47,* 3099-3104.
22. Kristensen, D.; Orlien, V.; Mortensen, G.; Brockhoff, P.; Skibsted, L. H. *Int. Dairy J.*, **2000,** *10,* 95-103.
23. Buettner, G. R. *Free Radic. Biol. Med.* **1987,** *3,* 259-303.
24. Howell, N. K.; Saeed, S. In *Advances in Magnetic Resonance in Food Science;* Belton, P. S.; Hills, B. P.; Webb, G. A., Eds., The Royal Society of Chemistry: Bodwin, UK; 1999, pp 135-143.

Chapter 9

The Influence of Ascorbic Acid and Uric Acid on the Oxidative Stability of Raw and Pasteurized Milk

Jacob H. Nielsen, Henrik Østdal, and Henrik J. Andersen

Department of Animal Product Quality, Danish Institute of Agricultural Science, Research Centre Foulum P.O. Box 50, DK-8830 Tjele, Denmark

Spontaneous oxidation is a well-described phenomenon in milk, which is supposed to proceed depending on the anti-oxidative potential of raw milk. Ascorbate and urate are often considered as antioxidants in biological systems due to their reducing abilities. However, both ascorbate and urate are able to reduce transition metal ions and hereby promote cyclic Fenton chemistry in the presence of hydroperoxides. Consequently the role of both ascorbate and urate can be considered as double edge swords in the anti-oxidative balance of biological systems. The reviewed data clearly show that ascorbate and urate contribute to the anti-oxidative balance in raw and pasteurized milk with peroxidases and xanthine oxidase being the most prominent pro-oxidative enzymes in the degradation of ascorbate and urate. The complexity of the anti-oxidative roles of ascorbate and urate is illustrated by the ability of thiols groups to regenerate ascorbate from dehydro-ascorbate, and the ability of ascorbate to regenerate urate from the urate radical.

Introduction

In milk, as in many other foods, oxidative reactions are a prime cause of flavor deterioration and loss in nutritional quality (1). Spontaneous oxidation in milk giving rise to oxidized flavor (off-flavor) is a well-described phenomenon (2; 3), which is supposed to proceed depending on factors inherent in the milk itself. These inherent factors include fatty acid composition, content of low-molecular weight antioxidants, pro- and anti-oxidative enzyme systems, and transition metal ion content. Moreover, external factors such as handling, agitation, temperature, exposure to light, and contamination by metals and microorganisms are known to trigger additional deteriorative oxidative reactions in milk.

Even though bacteriological factors still are important in the ongoing deteriorative processes in milk, the increased focus on hygiene throughout the whole production chain in most industrial countries makes chemical changes, especially the oxidative processes, the important limiting factors with regard to freshness and suitability of the milk for further processing.

The delicate balance between the overall anti- and pro-oxidative processes, which is also known to exist in most other biological systems, determines the oxidative stability of milk. Any disturbance in this balance in advantage of pro-oxidative activity, most often ascribed as oxidative stress, will unavoidably mediate deteriorative processes which *i*) are already perceived in the fresh milk, e.g. oxidized flavor, (3) *ii*) are recognized by poor technological quality of the milk, e.g. altered rheological properties (4) or *iii*) give rise to inferior sensory quality of highly processed dairy products (5).

Figure 1 gives a schematic representation of inherent factors in milk contributing to the anti- and pro-oxidative balance together with external factors which have been suggested to promote oxidative stress in milk.

The composition of milk reflects to a certain extent the composition of the feed given to the cow. Consequently, inherent factors in milk of relevance to the anti-oxidative balance will change depending on the feed composition and thereby influence the oxidative stability of the milk.

It has been extensively shown that specific feeding regimes can increase the concentrations of both polyunsaturated lipids and pro-oxidative metal species, which both make the milk more prone to autoxidation (6; 7; 8; 9; 10). Thus several studies have been performed to limit autoxidation in milk through optimized feeding (11; 12). Moreover, numerous studies have shown that antioxidants in the feed are transferred to the milk and thereby improve the oxidative stability of milk. It has been reported that increasing concentrations of dietary vitamin E can effectively reduce the intensity of oxidized flavor in milk (13; 14). Recent studies confirm that α-tocopherol protects milk fat from oxidation (15). In contrast, earlier studies have not been able to show that

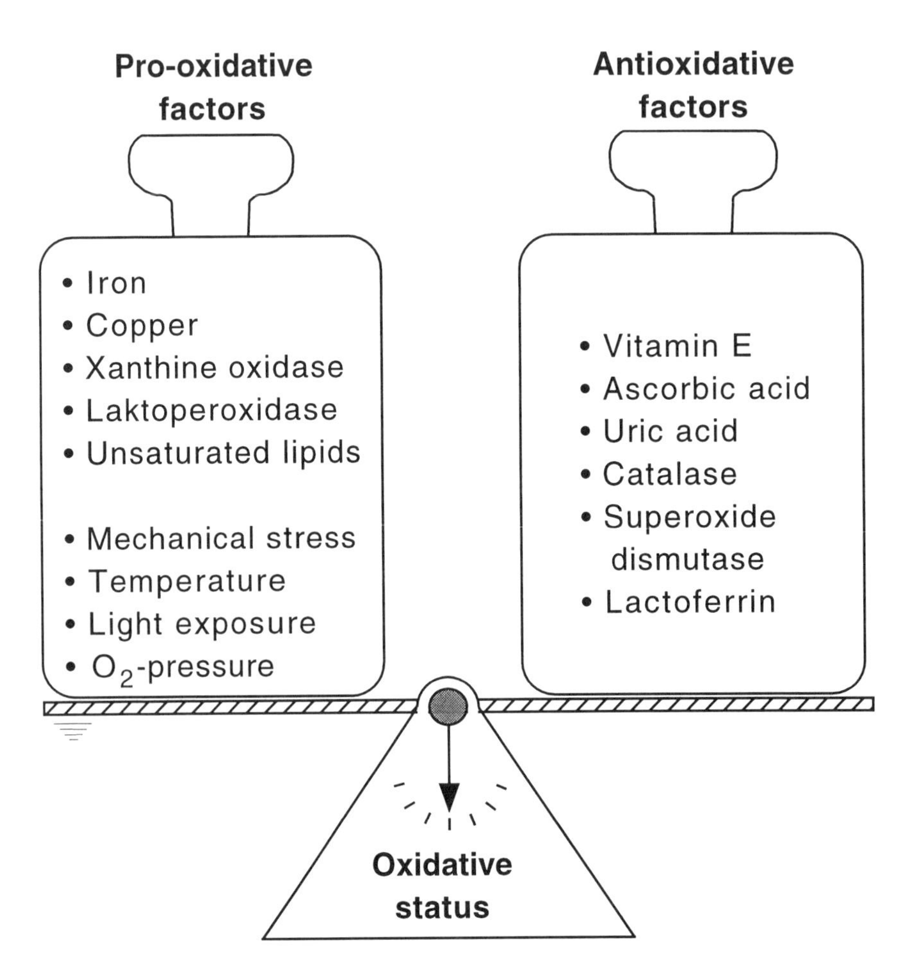

Figure 1. Factors affecting the oxidative balance of milk; high content of antioxidative factors will result in milk with increased oxidative stability compared with milk containing either low content of antioxidative factors or high content of pro-oxidative factors

oxidized flavor in milk is related to the concentration of tocopherols in the raw milk (16; 17). This controversy supports the statement that it is difficult to correlate spontaneous oxidation in milk to a single factor (3; 18)

The present paper discusses factors influencing the concentration and reaction of the water-soluble low-molecular weight antioxidants, ascorbic acid and uric acid, in raw and pasteurized milk together with their interactions in relation to potential deteriorative processes in milk.

Ascorbic Acid

The ascorbic acid concentration of raw milk and pasteurized milk has been reported to be about 12-25 mg/L (19; 20; 21) respectively. The pKa of ascorbic acid is 4.19 which means that the ascorbic acid will be present as ascorbate in milk.

Ascorbate in milk originates either directly from the feed given to the cow or is synthesized from glucose in the liver (22).

In raw milk ascorbic acid is spontaneously oxidized during storage at 4 °C (23). Ascorbic acid is known to oxidize concurrently with consumption of oxygen (24). In pasteurized milk, the oxidation of ascorbic acid is found to proceed with the same rate in closed and open bottles, which suggests that enough oxygen is dissolved in pasteurized milk to allow maximum rate of autoxidation (25). This suggests that ascorbic acid could act as an anti-oxidant in milk by reducing the presence of oxygen which is necessary for oxidative processes. However, for more than five decades ascorbic acid has mainly been reported as a potential pro-oxidant in milk, as its oxidation has been linked with the development of oxidized flavor especially when copper ions are present (26; 27; 28). Recent studies show that ascorbate is an effective antioxidant in milk if oxidation is induced by light exposure or by activation of lactoperoxidase (29). This dual effect has likewise been reported for biological systems in general (28; 30; 31; 32).

The pro-oxidative activity of ascorbic acid in milk has mainly, as mentioned above, been linked to transition metals due to its ability to reduce metal ions and thereby make redox-cycling of transition metal ions possible (9). Nielsen and co-workers (23) have recently shown that iron does not show pro-oxidative activity against ascorbic acid as opposed to copper ion within concentrations found in milk. It is supposed that iron is chelated by lactoferrin, citrate and proteins in the milk.

Ascorbic acid is also known to interact with activated radical species formed by enzyme systems in milk e.g. xanthine oxidase and lactoperoxidase (33; 27).

Xanthine oxidase catalyses the formation of hypoxanthine to xanthine and the further oxidation of xanthine to urate. These reactions proceed with simultaneous formation of super oxide (Figure 2). Xanthine oxidase is found in milk in high concentrations and originates from the milk fat globule membrane (MFGM). Embedded in the membrane, the enzyme is actually a dehydrogenase, but upon release from the MFGM the enzyme is converted to xanthine oxidase. This release is induced by mechanical stress, low temperature etc. Previously, xanthine oxidase had been shown to induce formation of ascorbyl radicals (33), and Nielsen and coworkers (23) have recently reported that addition of hypoxanthine to raw milk results in a rapid oxidation of ascorbic acid. Superoxide formed by the xanthine oxidase (Eq. 1 and Eq. 2) can react directly with ascorbate resulting in the ascorbyl radical (Eq. 3). Moreover dismutation of superoxide forms hydrogen peroxide which can react with transitions metals or peroxidases (mainly lactoperoxidase) forming compound I and compound II species and/or protein radicals on other milk proteins (29) (Eqs. 4-6). These species are known to react with ascorbate resulting in a further formation of ascorbyl radicals (33, 34).

$$\text{Hypoxanthine} \xrightarrow{XO} \text{xanthine} + O_2^{\bullet-} \qquad \text{Eq. 1}$$

$$\text{Xanthine} \xrightarrow{XO} \text{uric acid} + O_2^{\bullet-} \qquad \text{Eq. 2}$$

$$O_2^{\bullet-} + \text{ascorbate} \rightarrow \rightarrow \text{ascorbyl radical} \rightarrow \text{dehydroascorbate} \qquad \text{Eq. 3}$$

$$\text{Compound I} + \text{ascorbate} \rightarrow \rightarrow \text{ascorbyl radical} \rightarrow \text{dehydroascorbate} \qquad \text{Eq. 4}$$

$$\text{Compound II} + \text{ascorbate} \rightarrow \rightarrow \text{ascorbyl radical} \rightarrow \text{dehydroascorbate} \qquad \text{Eq. 5}$$

$$\text{Protein radical} + \text{ascorbate} \rightarrow \rightarrow \text{ascorbyl radical} \rightarrow \text{dehydroascorbate} \qquad \text{Eq.6}$$

Recent studies by Nielsen et al. (23) have shown that simultaneous addition of hypoxanthine and catalase to raw milk preserves ascorbate considerably. This indicates that the formed radical species induced by hydrogen peroxide and milk peroxidases are important for the loss of ascorbate in addition to the direct oxidation of ascorbate through reaction with superoxide. This is supported by the fact that addition of hydrogen peroxide to raw milk initiates a dramatic decrease in the ascorbate content (23). This enzymatically catalyzed pathway is supported by the fact that addition of hydrogen peroxide to UHT-milk only results in a minor oxidation of ascorbate which indicates that Fenton reaction chemistry is only of minor importance. Overall, the present data states that xanthine oxidase- and peroxidase-mediated oxidation are crucial factors in ascorbate oxidation as suggested by (33).

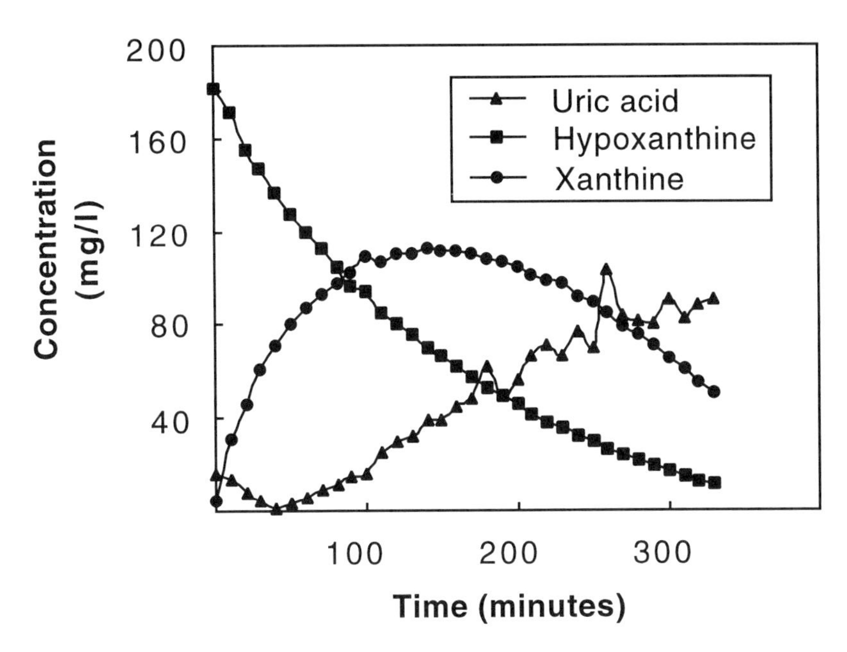

Figure 2. Fate of hypoxanthine in raw milk during storage at 4°C. Hypoxanthine (180 mg/L) was added to raw milk and the concentration of hypoxanthine, xanthine and urate was determined by HPLC.

Uric Acid

The concentration of urate in milk is in the range of 5-20 mg/L (35; 36), which is comparable to that of ascorbic acid (20; 21). The pKa value of uric acid is 5.7, which means that uric acid will be present as urate in milk.

Urate in milk originates from the metabolism of bacterial nucleotides formed by the microbial activity in the rumen. Ruminal microbial activity is influenced by feeding regimes and production conditions, and it has been shown that the uric acid content is influenced by feeding and milk yield (37; 38; 39).

Much work has been performed on the effect of urate in physiologically related (40). Like ascorbate, urate can act as an antioxidant due to its ability to scavenge free radicals (41; 42), but may as well act as a pro-oxidant through reduction of transition metal ions promoting Fenton reaction chemistry (43; 44). Despite this, the activity of urate in relation to oxidative reactions in milk has to our knowledge not received any attention. Østdal et al. (36) have been able to show an antioxidative effect of adding uric acid to milk comparable to that of ascorbic acid. Activation of endogenous peroxidase activity in raw milk by addition of hydrogen peroxide leads to the formation of dityrosine, a typical protein oxidation product (29). Addition of uric acid (10-30 mg/l) prevented lactoperoxidase-induced dityrosine formation by 70-95%. This antioxidative effect of urate was further verified as uric acid addition retarded light-induced lipid oxidation in pasteurized milk (36). Based on these results we propose urate to be a general antioxidant in milk. As mentioned above, the uric acid content in milk is influenced by feeding and production conditions as mentioned above which means that the concentration of urate can be manipulated to achieve an optimal uric acid concentration. The above-mentioned experiments were all performed by adding uric acid to milk, but through feeding we are currently able to manipulate the urate concentration in milk by 25% (data not shown). Ongoing experiments will permit us to investigate the oxidative stability and overall milk quality within the urate concentration range achievable through feeding.

Synergism between Low-Molecular Anti-Oxidants in Milk

Beside having a direct anti-oxidative effect on deteriorative processes, low-molecular weight antioxidants in milk enter complex interactions (synergism) which are important in relation to the overall anti-oxidative balance in milk.

Already three decades ago, it was proposed by Tappel (45) that a synergy between Vitamin E and ascorbic acid existed (46), as the reduction potential of the ascorbate radical/ascorbate system at pH 7.0 is 0.28 V, which is above that of the α-tocopherol radical/α-tocopherol ($E^o = 0.5$) (47); $\alpha\text{-Toc}^{\bullet} + \text{AscH}^{-} \rightarrow \alpha\text{-TocH} + \text{Asc}^{\bullet -}$.

Subsequently, this interaction between the ascorbyl radical and α-tocopherol has been confirmed to occur at a fairly high rate constant (~$1.5 \times 10^6\ M^{-1}s^{-1}$) in different biological systems (48; 49; 50). However, until now it has not been established whether this reaction occurs in milk.

Oxidation of ascorbic acid by reaction with reactive oxygen/nitrogen species in biological fluids - as also shown above for milk - leads to its depletion by the reactions,

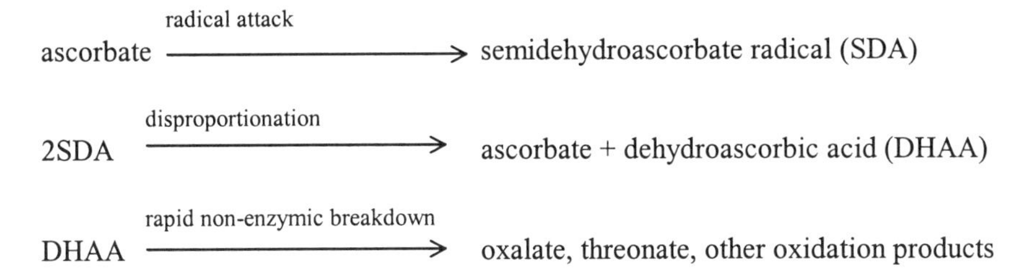

In theory, DHAA can be recycled to ascorbate either through direct chemical reaction or through enzymatic reduction. The latter is known to happen in biological systems either by non-specific enzyme systems (e.g. protein-disulphide isomerase (51)) or by a possible specific GSH/NADPH-dependent semidehydroascorbate reductase(s) (52; 53; 54). Provided similar enzyme systems are active in raw milk, these should be able to recycle ascorbate. However, such enzymatic reduction of DHAA has to the authors' knowledge not yet been reported in raw and low-pasteurized milk. Results given in Figure 3 show that reduction of DHAA to ascorbic acid takes place in pasteurized milk. This DHAA reduction could be mediated by the presence of hydrogen donors of which reduced GSH seems to be the most obvious according to previous reports (55). Results in Figure 3 shows that addition of reduced GSH or cysteine to milk will induce a reduction of DHAA back to ascorbate. Moreover, recent data shows that SH-groups known to be exposed upon heating of β-lactoglobulin can recycle ascorbate in milk (data not shown). This indicates that the reduction of DHAA can be both a enzymatic and non-enzymatic. The reduction potential of the urate/urate radical system ($E^o = 0.59$ V) is considerably higher than that of ascorbate ($E^o = 0.28$ V) at pH 7.0 (48). Hence ascorbate would be expected to reduce the urate radical; $UrH^{\bullet -} + AscH^- \rightarrow UrH^{2-} + Asc^{\bullet -}$.

Recent data show that ascorbic acid in fact seems able to recycle uric acid during light exposure of milk (36). Due to the above-mentioned data regarding the potential anti- and pro-oxidative activity of urate in milk, further studies of the possible consequences of urate radical formation and its interaction with biological anti-oxidants other than ascorbate (such as thiols) are warranted to predict its antioxidative activity in milk and other foods.

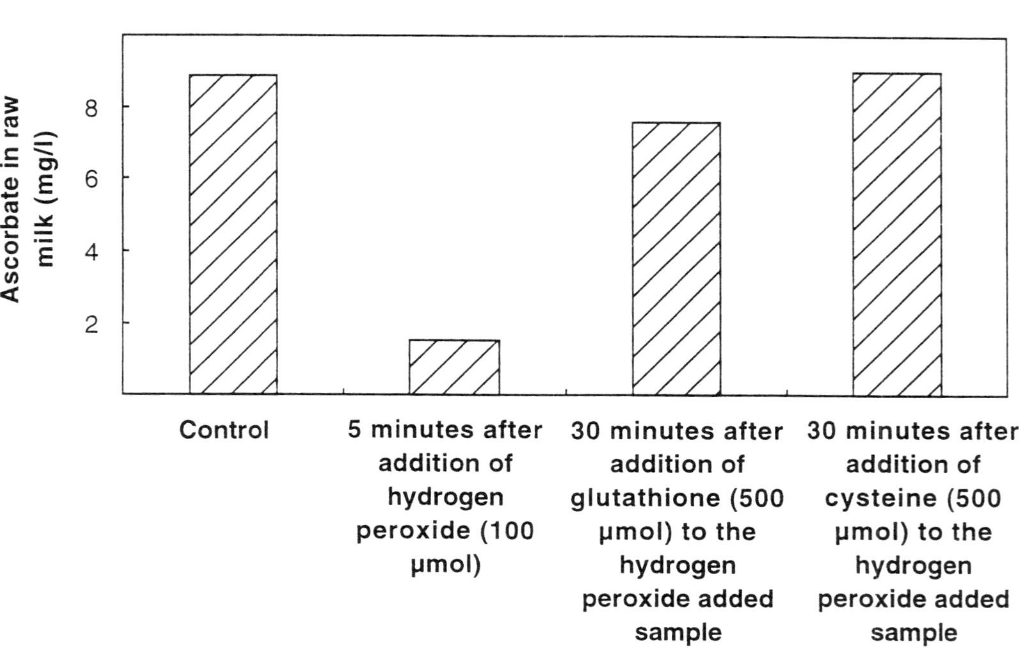

Figure 3. Regeneration of ascorbic acid in raw milk by thiol compounds. Ascorbic acid was oxidized by activation of lactoperoxidase through addition of 100 µM hydrogen peroxide. Regeneration of ascorbic acid was measured 30 minutes after addition of glutathione or cysteine to the hydrogen peroxide added sample. Control equals concentration of ascorbic acid in raw milk prior to the addition of hydrogen peroxide.

Conclusions

Ascorbate and urate are considered to be antioxidants in biological systems due to their reaction with potentially damaging free radicals. However, like ascorbate, urate is able to reduce transition metal ions making these more reactive towards hydroperoxides, e.g. hydrogen peroxide and lipid hydroperoxides. Therefore, the role of ascorbic acid and uric acid is somewhat controversial and the effect of these two antioxidants will especially depend on the type of catalyst inducing the observed oxidative changes.

The data reviewed above clearly show that ascorbate and urate may contribute substantially to the anti-oxidative balance in both pre- and post-pasteurized milk. Moreover, the results clearly show that peroxidases and xanthine oxidase are prominent enzymes in relation to oxidation of ascorbate and urate, and that the oxidative mechanisms responsible for oxidation of ascorbate and urate are very complex.

A further understanding of the anti-oxidative capacity of low-molecular weight anti-oxidants in milk, their fate and the synergism between them should make it possible to set up strategies to improve the anti-oxidative balance in milk through the whole processing chain and thereby guarantee high quality milk for consumption and further processing.

Acknowledgements

The authors wish to thank the Danish Dairy Board and the FØTEK programme for the financial support of the project.

References

1. Allen, J. C.; Joseph, G. Deterioration of pasteurized milk on storage (Review). *J. Dairy Research* 1985, **52**, 469-487.
2. Hill, R. D.; Van Leeuwen, V.; Wilkinson, R. D. *New Zealand J. Dairy Sci. Technol.* 1977, **12**, 69-77.
3. Granelli, K.; Barrefors, P.; Bjoerck, L.; Appelqvist, L-A. *J. Sci. Food Agric.* 1998, **77**, 161-171.
4. Hirano, R.; Hirano, M.; Oooka, M.; Dosako, S.; Nakajima, I; Igoshi, K *J. Food Sci.* 1998, **63**, 35-38. Atamer, M.; Kocak, C.; Cimer, A.; Odabasi, S.; Tamucay, B.; Yamaner, N. *Milchwissenschaft*, 1999, **54**, 553-556.

5. Edmonson, L. F.; Douglas, F. A. Jr.; Rainey, N. H.; Goering, H. K *J. Dairy Sci.* 1972, **55**, 677.
6. Sidhu, G. S.; Brown, M. A.; Johnson, A. R. *J. Dairy Sci.* 1973, **56**, 635.
7. Elvehjem, C. A.; Steenbock, H.; Hart, E. B. *J. Biol. Chem.* 1929, **83**, 27-34.
8. King, R. L.; Dunkley, W. L. *J. Dairy Sci.* 1959, **42**, 420-427.
9. Muhamed, O. E.; Satter, L. D.; Grummer, R. R.; Ehle, F.R. *J. Dairy Sci.* 1988, **71**, 2677-2688.
10. Sidhu, G. S.; Brown, M. A.; Johnson, A. R. *J. Dairy Res.* 1976, **43**, 239-250.
11. Ford, J. E.; Schröder, M. J. A.; Bland, M. A.; Blease, K. S.; Scott, K. J. *J. Dairy Res.* 1986, **53**, 391-406.
12. *St-Laurent, A. M.; Hidiroglou, M.; Snoddon, M.; Nicholson, J. W. G. Can. J. Anim. Sci.* 1990*,* **70**, 561-570.
13. Nicholson, J. W. G.; St-Laurent, A. M.; Mc Queen, R. E.; Charmley, E. *Can. J. Anim. Sci.* 1991, **71**, 135-143.
14. Stapelfeldt, H.; Nielsen, K. N.; Jensen S. K.; Skibsted L. H. *J. Dairy Res.* 1999, **66**, 461-466.
15. Charmley, E. *New Brunswick Milk Marketing Board* 1991, 20.
16. Nicholson, J. W. G. *Bull. of Int. Dairy Fed.* 1993, **281**, 1-12.
17. Barrefors, P.; Granelli, K.; Appelqvist, L-A.; Bjoerck, L. *J. Dairy Sci.* 1995 **78**, 2691-2699.
18. Nagasawam, T.; Kuzuyam, Y.; Shigatam, N. *Vitamin* 1959, **16**, 676-679.
19. Schröder, M. J. A. *J Dairy Research* 1982, **49**, 407-424.
20. Scott, K. J.; Bishop, D. R.; Zechalko, A.; Edwards-Webb, J. D.; Jackson, P. A.; Scuffam, D. *J. Dairy Res.* 1984, **51**, 37-50.
21. Walstra, P.; Jenness, R. *Dairy Chemistry and Physics* 1984**,** John Wiley & Sons, Inc.
22. Nielsen, J. H.; Hald G.; Kjeldesen L.; Andersen H. J. & Østdal H. Oxidation of ascorbate in raw milk *J. Agric. Food Chem.* 2000 (submitted).
23. Khan, N. M. T.; Martell, A. E. *J. Am. Chem. Soc.* 1967, **89**, 4176-4185.
24. Hansson E.; Olsson H. *Fette Seifen,-Anstrichmittel* 1976, **78**, 381-382.
25. Chilson, W.H. *Milk Plant Monthly* 1935, **24** (11), 24-28; (12), 30-32, 34.
26. Kruhovsky, V. N.; Guthrie, E. S. *J. Dairy Sci.* 1945, **28**, 565-579.
27. King, R. L. *J. Dairy Sci.* 1963, **46**, 267-274.
28. Østdal, H.; Bjerrum, M. J.; Pedersen, J. A.; Andersen, H. J *J. Agric. Food Chem.* 2000, **48**, 3939-3944.
29. Halliwell, B.; Gutteridge, J. M. C. *Methods Enzymol.* 1990, **186**, 1-85.
30. Kadiiska, M. B.; Burkitt, M. J.; Xiang, O. H.; Mason, R. P. *J. Clin. Invest.* 1995, **96**, 1653-1657.
31. Porter, W. L. *Tox. Ind. Health* 1993, **9**, 93-122.
32. Nakamura, M. *J. Biochem.* 1994, **116**, 621-624.

33. Østdal, H.; Skibsted, L. H.; Andersen, H. J. *Free Radical Biology & Medicine* 1997, **23** 754-761.
34. Wolfschoon-Pombo, A.; Klostermeyer, H. *Milchwissenschaft*, 1981, **36**, 598-600.
35. Østdal, H.; Andersen, H. J.; Nielsen, J. H. *J. Agric. Food Chem.* 2000, **48** 5588-5592.
36. Giesecke, D.; Ehrentreich, L.; Stangassinger, M. *J. Dairy Sci.* 1994, **77**, 2376-2381.
37. Johnson, L. M.; Harrison, J. H.; Riley, R. E.; *J. Dairy Sci.* 1998, **81**, 2408-2420.
38. Tiemeyer, W.; Stohror, M.; Giesecke, D. *J. Dairy Sci.* 1984, **67**, 723-728.
39. Becker, B. F. *Free Rad. Biol. Chem.* 1993, **14**, 615-631.
40. Schlotte, V.; Sevanian, A.; Hochstein, P.; Weithmann, K. U. *Free Radic. Biol. Med.* 1998, **25**, 839-847.
41. Simic, M. G.; Jovanovic, S. V. *J. Am. Chem. Soc.* 1989, **111**, 5778-5782.
42. Abuja, P. M. *FEBS Lett.* 1999, **446**, 305-308.
43. Bagnati, M.; Perugini, C.; Cau, C.; Bordone, R.; Albano, E.; Bellomo, G. *Biochem. J.* 1999, **340**, 143-152.
44. Tappel A. L. *Am. J. Clin Nutr.* 1970 **23**, 1137-1139.
45. Packer, J. E.; Slater, T. F.; Wilson, R. L. *Nature* 1979, **278** 737-738.
46. Buettner, G. R. *Arch. Biochem. Biophys.* 1993, **300**, 535-543.
47. Burton, G. W.; Wronska, U.; Stone, L.; Foster, D. O.; Ingold, K. U. *Lipids* 1990, **25**, 199-210.
48. Traber, M. G. *Free Rad. Biol. Med.* 1994, **16**, 229-239.
49. Tanaka, K.; Hashimoto, T.; Tokumaru, S.; Iguchi, H.; Kojo, S. *J. Nutr.* 1997, **127**, 2060-2064.
50. Halliwell, B.; Gutteridge, J. M. C. Oxford University Press, 1999.
51. Diliberto, E. J.; Dean, G.; Carter, C.; Allen, P. L. *J. Neurochem.* 1982, **39**, 563-568.
52. Rose, R. C. *Am. J. Physiol.* 1989, **256**, F52-F56.
53. Choi, J-L.; Rose, R. C. *Proc. Soc. Exp. Biol. Med.* 1989, **190**, 369-378.
54. Meister, A. *Biochem. Biophys. Acta* 1995, **1271**, 35-42.

Chapter 10

Pseudoperoxidase Activity of Myoglobin: Pigment Catalyzed Formation of Radicals in Meat Systems

Maiken V. Kröger-Ohlsen, Charlotte U. Carlsen, Mogens L. Andersen, and Leif H. Skibsted

Department of Dairy and Food Science, Food Chemistry, Royal Veterinary and Agricultural University, Frederiksberg C, Denmark

Oxygen transport and storage proteins cause a continous flux of radicals in muscles. In meats these one-electron transfer reactions are mainly related to the heme pigment myoglobin, and they are accelerated by the post-mortem decrease in pH. One radical-generating pathway is the pseudoperoxidase reaction of myoglobin, whereby the pigment is activated by hydrogen peroxide to form a protein-based radical that is deactivated in two one-electron steps, leaving the oxidized substrates also as radicals. Possible substrates are proteins, lipids, and numerous smaller compounds such as ascorbate, reducing cofactors, carotenoids, and plant phenolics. The reactions involved are accelerated by acid, and a ferric protein radical in equilibrium with the Fe(IV)=O group may be the actual reacting species. In radical exchange reactions long-lived protein radicals seem to be formed in processes that could be accelerated by phase transitions in the meat caused by freezing, which could also lead to membranal damage.The nature of the substrate-derived radicals determines whether the reaction has an overall prooxidative or antioxidative effect in the meat system and will thus depend on the depletion of reductants in the meat.

Introduction

Meat contains oxidizable substrates (lipids, proteins, small reducing compounds) and potential prooxidants: heme pigments in high concentrations, from which chelatable iron may be liberated, and also various redox enzymes. *In vivo* these are kept in balance, and oxygen is activated in controlled reactions to act as the final electron acceptor. However, when the animal is slaughtered, a complex process begins where reducing cofactors are depleted, compartmentalization gradually ceases, and metabolites accumulate. Muscle tissue is transformed into meat that eventually deteriorates, both oxidatively and by the action of microorganisms.

An important contribution to the oxidative stress of meat systems comes from a continous flux of radicals generated inside the system, and pigments are likely to play a role in these reactions. Myoglobin is quantitatively the most important pigment in meat, and due to its redox active heme iron and its ability to activate oxygen, it is able to catalyze the formation of radicals by at least three different pathways: *i*) autoxidation of oxymyoglobin, followed by enzymatic reduction of metmyoglobin (*1,2*), *ii*) catalytic breakdown of lipid hydroperoxides (*3*), and *iii*) pseudoperoxidase activity of myoglobin.

The two former pathways have been extensively studied, and although their importance compared to other prooxidative reactions are still matter of discussion, they have long been recognized as radical generating reactions in meat systems (*4*). Far less is known about the third pathway: the pseudoperoxidase reaction of myoglobin. The individual steps involved in this reaction sequence have been studied in model systems for more than 40 years, but only few studies have been made in order to establish their influence on the oxidative status of real meat systems (*5-7*). The reactions involved are characterized by short-lived intermediates of high reactivity, and if formed at all in real meat systems, these myoglobin species will be present in extremely low steady state concentrations. Although Electron Spin Resonance (ESR) spectroscopy is a very sensitive technique for detection of radicals, the steady state concentration of the reactive myoglobin intermediates may be too low to allow for their direct detection in meat. However, using spin trapping and freeze quenching techniques in model systems, a picture is emerging of the nature of the radicals formed, and their role in oxidative processes should be further elucidated using kinetic modelling.

Myoglobin

Myoglobin is a heme protein found in muscle cells and in meat in a concentration of 2-7 mg/g wet weight (*8*). The physiologically active form, myoglobin or deoxymyoglobin (MbFe(II)), binds dioxygen reversibly to form oxymyoglobin (MbFe(II)O_2). MbFe(II)O_2 is autoxidized to the thermodynamically stable form of myoglobin, metmyoglobin (MbFe(III)), with concomittant release of superoxide (*1*). Enzymatic reduction of MbFe(III) takes place *in vivo* and to some degree in meat (*2,9*), but eventually MbFe(III) will accumulate in meat as reducing cofactors are depleted (*10*). Superoxide dismutates to form dioxygen and hydrogen peroxide, and along with hydrogen peroxide from other sources, hydrogen peroxide can activate MbFe(III) into a pseudoperoxidase (*11*). The production of hydrogen peroxide in meat has also been found to increase with time (*4*). The transformations between different myoglobin species have been extensively studied, and an overview of known reactions is given in Figure 1.

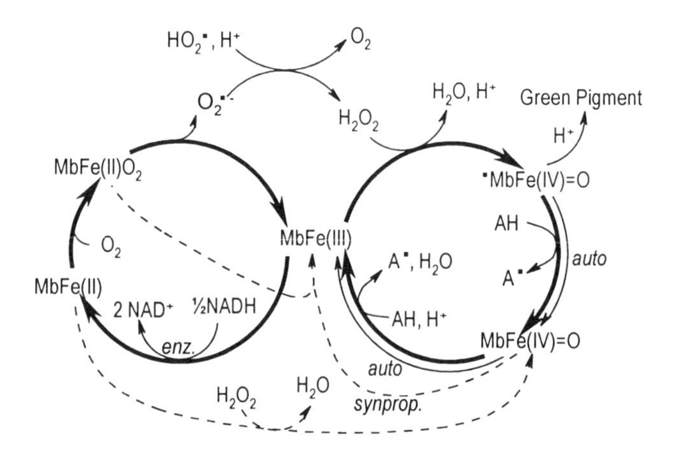

Figure 1. Transformation between myoglobin species. Bold cycle to the left: Color cycle of meat. Bold cycle to the right: Pseudoperoxidase cycle of myoglobin. Dotted pathways: Not treated in this text (93,94).

The Pseudoperoxidase Reaction of Myoglobin

The prosthetic group of myoglobin is identical to the prosthetic group of the heme peroxidases, and generally myoglobin is able to undergo the same reactions as these enzymes; therefore myoglobin is sometimes referred to as a pseudoperoxidase (*12*). Myoglobin differs from the true peroxidases by its rate of reaction with substrates, reacting considerably slower than the true enzymes; furthermore, the myoglobin molecule is altered during the peroxidase reaction cycle (at least under certain conditions), resulting in slower activation and faster deactivation (*11-13*).

The reaction catalyzed by peroxidases is the reduction of hydroperoxides by reducing substrates:

$$H_2O_2 + 2\ AH \rightarrow 2\ A^{\cdot} + 2\ H_2O$$

The enzyme passes through two hypervalent states during this reaction, compound I (or perferrylmyoglobin, ·MbFe(IV)=O, in the case of myoglobin) and compound II (or ferrylmyoglobin, MbFe(IV)=O, in the case of myoglobin) (*12*), as shown in Figure 1. The unpaired electron in ·MbFe(IV)=O is located in the protein part of the molecule (*11,14*) where it is located on different amino acid residues (*15,16*). The two hypervalent myoglobin species contain iron in the oxidation state +4 with an oxo group in the sixth coordination site (*12,17,18*). They are both powerful oxidizing agents: the reduction potential of MbFe(IV)=O is + 0.9 V (pH = 7) (*19,20*), and the reduction potential of ·MbFe(IV)=O is even higher although the exact value is unknown (*21*). As seen in Figure 1, radicals are further formed in the two steps of deactivation yielding MbFe(III) and two molecules of oxidized substrate.

Activation of Myoglobin

The first step in the pseudoperoxidase reaction of myoglobin is the activation of MbFe(III) by hydrogen peroxide, and in the laboratory a mixture of ·MbFe(IV)=O and MbFe(IV)=O is obtained together with residual MbFe(III). The reaction mechanism is poorly understood, but it is generally assumed that formation of ·MbFe(IV)=O precedes formation of MbFe(IV)=O (*11*). A proton is liberated in the process, leaving ·MbFe(IV)=O as a neutral radical (*20,22*). A possible precursor of the protein-based radical of ·MbFe(IV)=O – a porphyrin radical cation – has been generated by laser flash photolysis, but its formation by reaction of MbFe(III) with hydrogen peroxide is so slow (compared to its reduction by the globin moiety to yield the protein-based radical) that it is not observed when myoglobin is activated by this procedure (*23*).

The reaction is sensitive to pH. According to reports in the literature, a stable green pigment is formed when the reaction is carried out at mildly acidic pH (an excess of hydrogen peroxide may also be required) (*24-26*). The green pigment contains a modified porphyrin ring and is resistant to both oxidation and reduction. It is most likely formed in competition with other reaction products, since substrate-derived radicals can be detected when MbFe(III) is activated by hydrogen peroxide in the presence of oxidizable substrates at pH as low as 5.5 (*27*). pH of meat is in the range of 5.5-6.5, and further studies of the activation of myoglobin in this pH-range is clearly of interest in relation to fresh meat products.

Deactivation of Perferrylmyoglobin

In the absence of external reductants, ·MbFe(IV)=O decays spontaneously to MbFe(IV)=O and MbFe(III) by pathways that are poorly understood (*11,14,28-32*). One of the protein-based radicals of ·MbFe(IV)=O, a tyrosyl phenoxyl radical, has been reported to have a half-life of 45 seconds (*33*), and another of these radicals, a tryptophan-centered radical, has been found to react with dioxygen to form a peroxyl radical with a half-life of 7 seconds (*33-35*).

Several products of the spontaneous decay of ·MbFe(IV)=O contain modifications of the porphyrin ring and the globin moiety (*28-32*). Some of these changes are visible by optical spectroscopy, and they have been reported not to occur when reducing compounds are present during the activation of MbFe(III), indicating that fast reduction of ·MbFe(IV)=O prevents modification of the porphyrin and the globin moiety (*36-42*). Indeed, the reduction of ·MbFe(IV)=O by the water soluble carotenoid crocin (*21*) and ascorbate (*43*) has been found to be faster than its formation from MbFe(III) and hydrogen peroxide. In general, studies of reaction of ·MbFe(IV)=O are hampered by its slow formation and fast decay, and accordingly only a lower limit for the rate of reduction of ·MbFe(IV)=O can be estimated from kinetic studies when ·MbFe(IV)=O is generated by activation of MbFe(III) by hydrogen peroxide ($\sim 4 \cdot 10^5$ $M^{-1}s^{-1}$ at pH = 6.8 and 25 °C for reduction of MbFe(IV)=O by crocin (*21*)).

Deactivation of Ferrylmyoglobin

Ferrylmyoglobin exists in a protonated and a deprotonated form, $^{+}$HMbFe(IV)=O and MbFe(IV)=O, with $pK_a \sim 5.0$ (*25,44-49*). Reduction of $^{+}$HMbFe(IV)=O is faster than reduction of MbFe(IV)=O by one or two orders of magnitude (*46,48,49*), and it seems that the Fe(IV)=O group itself is not very reactive but must be converted to an activated form by protonization prior to reduction (*50-52*).

Protonization has been suggested to take place at two different histidyl residues: at the distal histidine, which is situated above the heme and forms hydrogen bonds to heme ligands such as O_2 (*51*), and at the proximal histidine, which is positioned below the heme and is covalently bound to the heme iron (*45*). Yet another possibility is direct protonization of the Fe(IV)=O group to yield a ferric protein-based radical, $^{\cdot+}$MbFe(III)-OH, which could be the actual reactive species of ferrylmyoglobin (*48*). This is similar to findings for cytochrome *c* peroxidase, where compound II is a pH-dependent equilibrium mixture of the Fe(IV)=O species and a ferric tryptophan-based radical with the latter being stabilized at low pH ($pK_a \sim 5.3$) (*53*). This is also similar to microperoxidase-8, where the Fe(IV)=O species is in equilibrium with a ferric porphyrin radical which is stabilized at pH < 6 (*50*).

Autoreduction of ferrylmyoglobin

MbFe(IV)=O is relatively stable in the absence of external reductants, but eventually it decays to modified MbFe(III) in a rather slow reaction known as the autoreduction reaction (*11,54*). Amino acid residues of the globin are oxidized during the reaction, and specifically tyrosine residues are believed to be involved (*54*). The autoreduction is specific acid catalyzed and follows first-order kinetics at pH < 7 (*24,46*), but at higher pH values the reaction shifts to a higher (mixed) reaction order (*55*).

One-electron reduction of the Fe(IV)=O center of ferrylmyoglobin by the globin molecule would be expected to produce a globin-based radical (*54*), and two such radicals have been detected by freeze quenching and low-temperature ESR spectroscopy (*55*). The identities of these radicals are still unknown; however, their spectra are not derived from hemichrome as seen by comparison with the very broad ESR spectrum of this compound (*56,57*).

The importance of the autoreduction of ferrylmyoglobin for generation of radicals in real meat systems is probably negligible, since the reaction is significantly slower than reduction of ferrylmyoglobin by a number of common cellular constituents (e.g. two or more orders of magnitude slower than reduction by ascorbate (*48*)).

Deactivation by small reducing substrates

A large number of compounds have been shown to reduce ferrylmyoglobin: traditional antioxidants such as ascorbate (*38,40,42,43,48,49,58*), α-tocopherol (*37*), NADH (*46*), glutathione (*33,59*), flavonoids (*60*), phenolic acids (*42,49,61,62*), as well as amino acids and oligopeptides (*63,64*), and also inorganic substrates such as iodide, nitrite, and thiocyanate (*65,66*). The general reaction scheme shown in Figure 2 is based on detailed kinetic studies of reduction of ferrylmyoglobin by ascorbate and chlorogenate (*49*).

It accounts for the observed pK_a value of ferrylmyoglobin as well as for the saturation-like kinetics that was also observed in these studies (*48,49*). The latter effect probably stems from complex formation between ferrylmyoglobin and the substrate, with the complex being reduced by a second molecule of the substrate in parallel with direct reduction of the non-complexed pigment. The redox properties of ferrylmyoglobin are modulated by formation of these complexes, as the rate of reduction of the complexes are significantly lower than the rate of reduction of the uncomplexed pigment (more than 50 times in the case of ascorbate). Combinations of ascorbate and chlorogenate reduced ferrylmyoglobin with a less-than-additive efficiency, an effect that is also predicted by the general reaction scheme in Figure 2 (*49*).

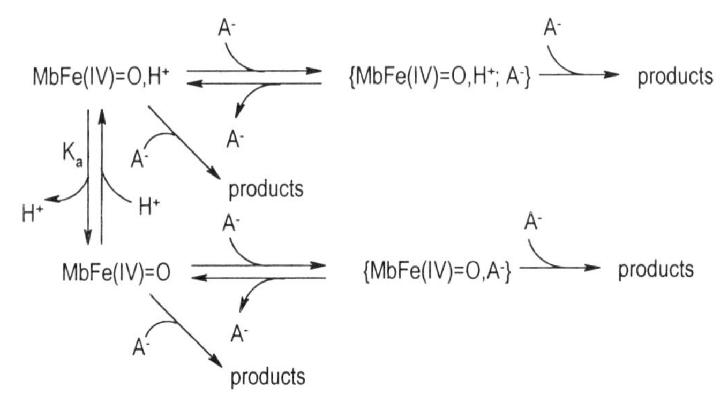

Figure 2: General reaction scheme for reduction of ferrylmyoglobin

Deactivation of hypervalent myoglobin by proteins

Relatively few reports are given in the literature of pseudoperoxidase activity of myoglobin towards proteins. It has been found that myoglobin is a better catalyst for protein oxidation than the classical enzyme horseradish peroxidase, which is usually more effective than myoglobin towards smaller substrates (*67,68*). Disulfide and dityrosine cross-links are formed when LDL apolipoprotein B (*68*), myosin (*67,69*), and other myofibrillar proteins (*70*) are oxidized by $^{\cdot}$MbFe(IV)=O, as summarized for the tyrosine cross-links in myosin (*67*):

$$\mathrm{MbFe(III)} + \mathrm{H_2O_2} \rightarrow {}^{\cdot}\mathrm{MbFe(IV){=}O} + \mathrm{H_2O} + \mathrm{H^+}$$
$$ {}^{\cdot}\mathrm{MbFe(IV){=}O} + \mathrm{myosin(Tyr)} \rightarrow \mathrm{MbFe(IV){=}O} + \mathrm{myosin(Tyr^{\cdot})}$$
$$ n\ \mathrm{myosin(Tyr^{\cdot})} \rightarrow \mathrm{myosin(Tyr)_n}$$

Oxidation of proteins (β-lactoglobulin (*47*), papain and ficin (*71*)) has also been demonstrated for MbFe(IV)=O alone with the same pH-dependence as for reduction of MbFe(IV)=O by smaller compounds.

Several proteins – e.g. β-lactoglobulin and γ-globulin – form long-lived tyrosyl radicals (half-lives of 13-18 minutes) when oxidized by $^{\cdot}$MbFe(IV)=O (*72-74*). The stability of these radicals may depend on a hidden position in the protein (*74*). The significance of such radicals in relation to oxidation in meat systems is unknown, since on one hand they may serve as intermediate radicals scavengers, but on the other hand they may be able to spread oxidative damage by diffusion to other parts of the system (*73,74*).

Deactivation of hypervalent myoglobin by lipids

Initiation of lipid peroxidation in isolated muscle membranes by hypervalent myoglobin was first demonstrated by Kanner and Harel (*5,75*) and has later been observed by other authors (*76,77*). Hypervalent myoglobin has also been found to oxidize low density lipoproteins (*78-80*), palmitoyl-linoleoyl-phosphatidylcholine in large unilamellar vesicles (*81*), and free polyunsaturated fatty acids (*82-85*).

The reaction mechanism is still a matter of debate (*56,83*), but the following reactions have been proposed for peroxidation of lipids by hypervalent myoglobin (*5,75,86*), written here as hydrogen abstraction reactions:

$$^{\cdot}MbFe(IV){=}O + LH \rightarrow HMbFe(IV){=}O + L^{\cdot}$$

and

$$MbFe(IV){=}O + LOOH \rightarrow MbFe(III)\text{-}OH + LOO^{\cdot}$$

Newman et al. (*76*) suggested that the peroxyl radical of $^{\cdot}MbFe(IV){=}O$ that was later assigned to a tryptophan residue (*35*) is responsible for initiation of lipid peroxidation. In contrast, Rao et al. (*83*) found that MbFe(IV)=O binds polyunsaturated fatty acids in the heme crevice and subsequently abstracts a bisallylic hydrogen atom prior to addition of O_2 to form the lipid peroxyl radical, and that MbFe(IV)=O rather than the protein-based radical of $^{\cdot}MbFe(IV){=}O$ is the initiator of lipid peroxidation.

The majority of studies of the peroxidase activity of myoglobin towards lipids have been carried out at pH ~ 7.4, whereas pH of interest to meat is in the range of 5.5-6.5, and high concentrations of hydrogen peroxide have also been used as compared to what may be found in real meat systems. It should be kept in mind, however, that changes of pH in the range 5.5-7.4 has a profound effect on the interaction of MbFe(III) with both hydrogen peroxide (*24-26*) and free fatty acids (*56*), and that the ratio of hydrogen peroxide to MbFe(III) affects the amount of hypervalent species formed as well as the amount of iron liberated from the pigment (*11,13,87-89*). In agreement with the pH-dependent behavior of hypervalent myoglobin, the stimulation of muscle membrane lipid peroxidation by hypervalent myoglobin has been found to depend on pH, with a maximum at pH ~ 6.5 (*5*).

Kanner et al. made an important point (*6*). They showed that although hypervalent myoglobin was able to initiate lipid peroxidation in a muscle membrane model system, the cytosolic extract of the same muscles contained reducing compounds that effectively inhibited peroxidation of membrane lipids by hypervalent myoglobin. Presumably, the hypervalent myoglobin oxidized reducing compounds in the cytosol in competition with oxidation of membrane lipids. Both low-molecular-weight and high-molecular-weight compounds were found to inhibit lipid peroxidation, in line with the findings mentioned earlier that hypervalent myoglobin is able to oxidize a large number of substrates.

Concluding Remarks

In meat, the fast reduction of any hypervalent myoglobin by the reducing compounds present in the cytosol will result in formation of radicals, and the nature of these may determine whether the pseudoperoxidase reaction has an overall prooxidative or antioxidative effect. An antioxidative effect *in vivo* and in meat has been suggested by several authors (*24,36,61,90*), because the hypervalent myoglobin species are formed in a hydrogen peroxide-consuming step and the pseudoperoxidase cycle therefore prevents hydrogen peroxide from entering the Fenton reaction. This implies deactivation of hypervalent myoglobin by compounds which form relatively stable radicals such as phenoxyl radicals or ascorbyl radicals upon one-electron oxidation. On the other hand, some radicals formed by reduction of hypervalent myoglobin may be able to initiate further oxidation, e.g. the radicals formed by oxidation of nitrite or iodide which have been found to stimulate lipid peroxidation (*91*) and oxidation of other cellular components (*92*) when formed by reduction of activated lactoperoxidase.

Although the radicals formed by reaction of reducing compounds with hypervalent myoglobin may be harmless, the pseudoperoxidase activity contributes to the depletion of antioxidants in the system and thereby affects the oxidative status of the meat system.

References

1. Satoh, Y.; Shikama, K. *J. Biol. Chem.* **1981,** *256,* 10272-10275.
2. Hagler, L.; Coppes, R. I.; Herman, R. H. *J. Biol. Chem.* **1979,** *254,* 6505-6514.
3. Tappel, A. L. *J. Biol. Chem.* **1955,** *217,* 721-733.
4. Kanner, J. *Meat Sci.* **1994,** *36,* 169-189.
5. Kanner, J.; Harel, S. *Arch. Biochem. Biophys.* **1985,** *237,* 314-321.
6. Kanner, J.; Salan, M. A.; Harel, S.; Shegalovich, I. *J. Agric. Food Chem.* **1991,** *39,* 242-246.
7. Rhee, K. S.; Ziprin, Y. A.; Ordonez, G. *J. Agric. Food Chem.* **1987,** *35,* 1013-1017.
8. Livingston, D. J.; Brown W. D. *Food Technol.* **1981,** *35,* 244-252.

9. Mikkelsen, A.; Skibsted, L. H. *Z. Lebensm. Unters. Forsch.* **1992,** *194,* 9-16.
10. Skibsted, L. H. Chemical changes in meat and meat products during storage, transportation and retail display – theoretical considerations. In: *Meat quality and meat packaging*; Taylor, S. A.; Raimundo, A.; Severini, M.; Smulders, F. J. M., Eds.; ECCEAMST, Utrecht, Netherlands, 1996; pp. 169-181.
11. King, N. K., Winfield, M. E. *J. Biol. Chem.* **1963,** *238,* 1520-1528.
12. Chance, M.; Powers, L.; Kumar, C.; Chance, B. *Biochemistry* **1986,** *25,* 1259-1265.
13. Yonetani, Y.; Schleyer, H. *J. Biol. Chem.* **1967,** *242,* 1974-1979.
14. Davies, M. J. *Free Rad. Res. Comms.* **1990,** *10,* 361-370.
15. Evans, S. V.; Brayer, G. D. *J. Mol. Biol.* **1990,** *213,* 885-897.
16. Gunther, M. R.; Tschirret-Guth, R. A.; Witkowska, H. E.; Fann, Y. C.; Barr, D. P.; Ortiz de Montellano, P. R.; Mason, R. P. *Biochem. J.* **1998,** *330,* 1293-1299.
17. Sitter, A. J.; Reczek, C. M.; Terner, J. *Biochim. Biophys. Acta* **1985,** *828,* 229-235.
18. La Mar, G. N.; de Ropp, J. S.; Latos-Grazynski, L.; Balch, A. L.; Johnson, R. B.; Smith, K. M.; Parish, D. W.; Cheng, R. J. *J. Am Chem. Soc.* **1983,** *105,* 782-787.
19. Koppenol, W. H.; Liebman, J. F. *J. Phys. Chem.* **1984,** *88,* 99-101.
20. George, P.; Irvine, D. H. *Biochem. J.* **1955,** *60,* 596-604.
21. Jørgensen, L. V.; Andersen, H. J.; Skibsted, L. H. *Free Rad. Res.* **1997,** *27,* 73-87.
22. Everse, J. *Free Radic. Biol. Med.* **1998,** *24,* 1338-1346.
23. Candeias, L. P.; Steenken, S. *Photochem. Photobiol.* **1998,** *68,* 39-43.
24. Tajima, G.; Shikama, K. *Int. J. Biochem.* **1993,** *25,* 101-105.
25. King, N. K.; Winfield, M. E. *Aust. J. Bio. Sci.* **1966,** *19,* 211-217.
26. Fox, J. B.; Nicholas, R. A.; Ackerman, S. A.; Swift, C. E. *Biochemistry* **1974,** *13,* 5178-5186.
27. Kröger-Ohlsen, M. V.; Andersen, M. L.; Skibsted, L. H. *Free Rad. Res.* **2000,** *32,* 313-325.
28. Catalano, C. E.; Choe, Y. S.; Ortiz de Montellano, P. R. *J. Biol. Chem.* **1989,** *264,* 10534-10541.
29. Sugiyama, K.; Highet, R. J.; Woods, A.; Cotter, R. J.; Osawa, Y. *Proc. Natl. Acad. Sci. USA* **1997,** *94,* 796-801.
30. Tew, D.; Ortiz de Montellano, P. R. *J. Biol. Chem.* **1988,** *263,* 17880-17886.
31. Rice, R. H.; Lee, Y. M.; Brown, W. D. *Arch. Biochem. Biophys.* **1983,** *221,* 417-427.
32. Giulivi, C.; Cadenas, E. *Arch. Biochem. Biophys.* **1993,** *303,* 152-158.
33. Kelman, D. J.; DeGray, J. A.; Mason, R. P. *J. Biol. Chem.* **1994,** *269,* 7458-7463.
34. Giulivi, C.; Cadenas, E. *Free Radic. Biol. Med.* **1998,** *24,* 269-279.

35. DeGray, J. A.; Gunther, M. R.; Tschirret-Guth, R.; Ortiz de Montellano, P. R.; Mason, R. P. *J. Biol. Chem.* **1997,** *272,* 2359-2362.
36. Arduini, A.; Eddy, L.; Hochstein, P. *Arch. Biochem. Biophys.* **1990,** *281,* 41-43.
37. Giulivi, C.; Romero, F. J.; Cadenas, E. *Arch. Biochem. Biophys.* **1992,** *299,* 302-313.
38. Arduini, A.; Mancinelli, G.; Radatti, G. L.; Damonti, W.; Hochstein, P.; Cadenas, E. *Free Radic. Biol. Med.* **1992,** *13,* 449-454.
39. Mordente, A.; Santini, S. A.; Miggiano, G. A. D.; Martorana, G. E.; Petitti, T.; Minotti, G.; Giardina, B. *J. Biol. Chem.* **1994,** *269,* 27394-27400.
40. Galaris, D.; Cadenas, E.; Hochstein, P. *Arch. Biochem. Biophys.* **1989,** *273,* 497-504.
41. Xu, F.; Mack, C. P.; Quandt, K. S.; Shlafer, M.; Massey, V.; Hultquist, D. E. *Biochem. Biophys. Res. Commun.* **1993,** *193,* 434-439.
42. Laranjinha, J.; Almeida, L.; Madeira, V. *Free Radic. Biol. Med.* **1995,** *19,* 329-337.
43. Galaris, D.; Korantzopoulos, P. *Free Radic. Biol. Med.* **1997,** *22,* 657-667.
44. Wittenberg, J. B. *J. Biol. Chem.* **1978,** *253,* 5694-5695.
45. Foote, N.; Gadsby, P. M. A.; Greenwood, C.; Thomson, A. J. *Biochem. J.* **1989,** *261,* 515-522.
46. Mikkelsen, A.; Skibsted, L. H. *Z. Lebensm. Unters. Forsch.* **1995,** *200,* 171-177.
47. Østdal, H.; Daneshvar, B.; Skibsted, L. H. *Free Rad. Res.* **1996,** *24,* 429-438.
48. Kröger-Ohlsen, M.; Skibsted, L. H *J. Agric. Food Chem.* **1997,** *45,* 668-676.
49. Carlsen, C. U.; Kröger-Ohlsen, M. V.; Bellio, R.; Skibsted, L. H. *J. Agric. Food Chem.* **2000,** *48,* 204-212.
50. Low, D. W.; Winkler, J. R.; Gray, H. B. *J. Am. Chem. Soc.* **1996,** *118,* 117-120.
51. Fenwick, C. W.; English, A. M.; Wishart, J. F. *J. Am. Chem. Soc.* **1997,** *119,* 4758-4764.
52. Fenwick, C.; Marmor, S.; Govindaraju, K.; English, A. M.; Wishart, J. F.; Sun, J. *J. Am. Chem. Soc.* **1994,** *116,* 3169-3170.
53. Coulson, A. F. W.; Erman, J. E.; Yonetani, T. *J. Biol. Chem.* **1971,** *246,* 917-924.
54. Uyeda, M.; Peisach, J. *Biochemistry* **1981,** *20,* 2028-2035.
55. Kröger-Ohlsen, M. V.; Andersen, M. L.; Skibsted, L. H. *Free Rad Res.* **1999,** *30,* 305-314.
56. Baron, C. P.; Skibsted. L. H.; Andersen, H. J. *Free Radic. Biol. Med.* **2000,** *28,* 549-558.
57. Svistunenko, D. A.; Sharpe, M. A.; Nicholls, P.; Wilson, M. T.; Cooper, C. E. *J. Magn. Reson.* **2000,** *142,* 266-275.
58. Giulivi, C.; Cadenas, E. *FEBS Lett.* **1993,** *332,* 287-290.

59. Galaris, D.; Cadenas, E.; Hochstein, P.*Free Radic. Biol. Med.* **1989,** *6,* 473-478.
60. Jørgensen, L. V.; Skibsted, L. H. *Free Rad. Res.* **1998,** *28,* 335-351.
61. Kanner, J.; Frankel, E.; Granit, R.; German, B.; Kinsella, J. E. *J. Agric. Food Chem.* **1994,** *42,* 64-69.
62. Castellucio, C.; Paganga, G.; Melikian, N.; Bolwell, G. P.; Pridham, J.; Sampson, J.; Rice-Evans, C. *FEBS Lett.* **1995,** *368,* 188-192.
63. Romero, F. J.; Ordoñez, I.; Arduini, A.; Cadenas, E. *J. Biol. Chem.* **1992,** *267,* 1680-1688.
64. Decker, E. A.; Chan, W. K. M.; Livisay, S. A.; Butterfield, D. A.; Faustman, C. *J. Food Sci.* **1995,** *60,* 1201-1204.
65. Kröger-Ohlsen, M. V.; Skibsted, L. H. *Food. Chem.* **2000,** *70,* 209-214.
66. Kröger-Ohlsen, M. V.; Skibsted, L. H. *Food Chem.* **1999,** *65,* 9-13.
67. Hanan, T.; Shaklai, N. *Free Rad. Res.* **1995,** *22,* 215-227.
68. Miller, Y. I.; Felikman, Y.; Shaklai, N. *Arch. Biochem. Biophys.* **1996,** *326,* 252-260,
69. Bhoite-Solomon, V.; Kessler-Icekson,G.; Shaklai, N. *Biochem. Int.* **1992,** *26,* 181-189.
70. Martinaud, A.; Mercier, Y; Marinova, P.; Tassy, C.; Gatellier, P.; Renerre, M. *J. Agric. Food Chem.* **1997,** *45,* 2481-2487.
71. Mikkelsen, A.; Skibsted, L. H. *Z. Lebensm. Unters. Forsch.* **1998,** *206,* 199-202.
72. Irwin, J. A.; Østdal, H.; Davies, M. J. *Arch. Biochem. Biophys.* **1999,** *362,* 94-104.
73. Østdal, H.; Skibsted, L. H.; Andersen, H. J. *Free Radic. Biol. Med.* **1997,** *23,* 754-761.
74. Østdal, H.; Andersen, H. J.; Davies, M. J. *Arch. Biochem. Biophys.* **1999,** *362,* 105-112.
75. Harel, S.; Kanner, J. *J. Agric. Food Chem.* **1985,** *33,* 1188-1192.
76. Newman, E. S. R.; Rice-Evans, C. A.; Davies, M. J. *Biochem. Biophys. Res. Commun.* **1991,** *179,* 1414-1419.
77. Gatellier, P.; Anton, M.; Renerre, M. *J. Agric. Food. Chem.* **1995,** *43,* 651-656.
78. Dee, G.; Rice-Evans, C.; Obeyesekera, S.; Meraji, S.; Jacobs, M.; Bruckdorfer, K. R. *FEBS Lett.* **1991,** *294,* 38-42.
79. Hogg, N.; Rice-Evans, C.; Darley-Usmar, V.; Wilson, M. T.; Paganga, G.; Bourne, L. *Arch. Biochem. Biophys.* **1994,** *314,* 39-44.
80. Laranjinha, J.; Vieira, O.; Almeida, L.; Madeia, V. *Biochem. Pharmacol.* **1996,** *51,* 395-402.
81. Maiorino, M.; Ursini, F.; Cadenas, E. *Free Radic. Biol. Med.* **1994,** *16,* 661-667.
82. Grisham, M. B. *J. Free Rad. Biol. Med.* **1985,** *1,* 227-232.

83. Rao, S. I.; Wilks, A.; Hamberg, M.; Ortiz de Montellano, P. R. *J. Biol. Chem.* **1994,** *269,* 7210-7216.
84. Galaris, D.; Sevanian, A.; Cadenas, E.; Hochstein, P. *Arch. Biochem. Biophys.* **1990,** *281,* 163-169.
85. Hamberg, M. *Arch. Biochem Biophys.* **1997,** *344,* 194-199.
86. Kanner, J.; Harel, S. *Lipids* **1985,** *20,* 625-628.
87. Davies, M. J. *Biochim. Biophys. Acta* **1991,** *1077,* 86-90.
88. Gutteridge, J. M. C. *FEBS Lett.* **1986,** *201,* 291-295.
89. Harel, S.; Kanner, J. *Free Rad. Res. Comms.* **1988,** *15,* 21-33.
90. Yang, W.; de Bono, D. *FEBS Lett.* **1993,** *319,* 145-150.
91. Kanner, J.; Kinsella, J. E. *Lipids* **1983,** *18,* 204-210.
92. Reszka, K. J.; Matuszak, Z.; Chignell, C. F.; Dillon, J. *Free Radic. Biol. Med.* **1999,** *26,* 669-678.
93. Whitburn, K. D. *J. Inorg. Biochem.* **1985,** *24,* 35-46.
94. Yusa, K.; Shikama, K. *Biochemistry* **1987,** *26,* 6684-6688.

Chapter 11

NO• Production during Thermal Processing Of Beef: Evidence for Protein Oxidation

K. M. Schaich

Department of Food Science, Rutgers University, 65 Dudley Road, New Brunswick, NJ 08901-8520

Low temperature EPR studies were conducted on beef cubes thermally-processed in MRE-type flexible pouches to observe release of "free" iron or other metals that could contribute to lipid oxidation and meat quality degradation during storage. Instead of detecting unbound iron, the surprising result was detection of NO• adventitiously trapped by hemes in the meat, yielding the pentacoordinate Fe^{2+}-nitroso-heme EPR signal. NO• is produced via reaction of arginine with H_2O_2 or lipid hydroperoxides, and clearly reflects oxidation of the protein. Changes in the globin component of the heme spectra under different heating conditions indicated that denaturation and degradation of the heme apoprotein also was occurring. In control experiments, NO• were not present in raw meat, but grew in as heating of the meat progressed. When the beef cubes were processed in saline instead of water, the Fe^{2+}-nitroso-heme EPR signal was replaced by an unidentified nitrogen-centered free radical. Chemical analyses for protein carbonyls and differential solvent solubility support the NO• pathway and a potentially significant role for protein oxidation in production of meat textures, flavors and off-flavors.

Introduction

During thermal processing muscle proteins undergo a number of physical changes and chemical reactions that contribute to development of characteristic meat qualities, including softening of texture and formation of a wide range of flavors *(1,2)*. At the same time, heat–induced disruptions in cellular structures lead to decompartmentalization of potential reactants and release of metals and other oxidation catalysts that influence storage stability and production of off-flavors and odors *(3)*. In general, meat textures have been attributed to proteins, while desirable meat flavors are thought to arise from amino acids or amino acid-lipid interactions. Off-flavors formed upon reheating (referred to as warmed-over flavor) have been attributed to lipid oxidation products *(4)*, although causal correlations have been difficult to establish.

Since the early work of Watts *(5,6)*, the catalytic activity of hemes has been recognized as a major contributor to warmed-over flavor in muscle foods. Tappel *(7)* first proposed that hemes act primarily by decomposing lipid hydroperoxides, facilitating chain branching reactions. Hultin's research *(8,9)* has suggested that free iron released from hemes may be the actual active catalyst rather than intact heme iron. Marnett and coworkers *(10,11)* have shown that ferryl and perferryl iron complexes of hemes can directly oxygenate lipids to ketones and other carbonyl products without initiating radical reactions. However, many questions about degradation mechanisms in thermally processed meats still remain unanswered.

In this study electron paramagnetic resonance (EPR) was used to gain direct evidence of heme structures, iron forms, and free radicals that may help elucidate degradative processes in cooked beef.

Materials and Methods

One inch cubes were cut from fresh beef chuck and sealed with water in military MRE laminate packages approximately 4" x 6". The packs were processed at a range of temperatures, then cooled and frozen until analysis (Center for advanced Food Technology, Rutgers University, unpublished data).

For EPR analysis, the cooking liquid was drained from the beef cubes, the cubes were rapidly minced into fine pieces, 2 g were transferred to 10 mm ID quartz EPR tubes, and the tubes were capped and frozen in liquid nitrogen.

EPR analyses were conducted on a Varian E-12 EPR spectrometer interfaced to and controlled by a MassComp 5500 minicomputer and equipped with an X-band (9.5 GHz) microwave bridge, 100 kHz modulation, and variable rate signal averaging. Microwave frequencies were measured by a

Hewlett Packard 505C frequency counter; magnetic field scans were routinely calibrated using a solution of Fremy's salt (peroxylamine disulphonate). The measurement temperature of 77 K was maintained using a Hewlett Packard variable temperature controller and liquid nitrogen/gas transfer assembly connected to a dewar in a wide-stack ENDOR cavity. Low temperature measurements were necessary for detection of hemes and iron complexes and to limit broadening of any peroxyl radicals that may have been present. Spectra were generated by repeatedly collecting and signal averaging 30 second scans over at least 10 minutes, and were presented as first derivatives of the absorption at resonance *(12)*. A DPPH crystal mounted provided an internal field marker for normalizing successive scans. Data was collected, stored, and analyzed by computer.

Results

Heat and oxidation both should induce free radical formation and changes in heme structure and metal availability, all of which may be actively involved in the molecular alterations affecting degradation of food quality. It was expected that free radicals formed during thermal processing would be quenched by the cooking water and thus undetectable by EPR, while changes in the iron as it was released from macromolecular binding sites and changed spin states in its free form should be detected easily. Nevertheless, two free radical species were clearly detected.

In beef cubes thermally processed in water, weak radicals were detectable at room temperature, but were not strong enough to identify or quantitate. At 77K (liquid nitrogen temperature) signals from a nitroso-heme complex with two principal components appeared (Figure 1). The set of three sharp, narrow lines in the center of the spectra is typical of pentacoordinate Fe^{2+}-heme-NO• complexes *(13,14)*, while the broad downfield signal with moderate to no fine structure arises from electron interactions with the heme apoprotein *(14 -16)*.

These signals resulted from adventitious trapping of NO• by the heme pigments naturally present in the beef cubes. The reaction of NO• with hemes is well-known and is used as a method for NO• detection *(16,17)*. Comparable spectra have been reported for nitrosomyoglobin *(18)*, α-NO• hemoglobin under anaerobic conditions *(17)*, and for heme-NO• complexes generated synthetically *(14,18)*, enzymatically in biological tissues producing NO• *(19)*, and chemically in nitrite-cured meats *(19-21)*.

The signals were stable to air but decreased with standing at room temperature. The heme component lost structure and intensity while the relative contribution of the NO• triplet increased with processing time. The former is associated with an increasing degree of heme denaturation and destruction, particularly in the apoprotein. The decreases in hyperfine structure and signal intensity with heating result in part from scission of the iron-imidazole bond *(14,17,19-21)*. The increased NO• triplet indicates increasing oxidation of amino groups on the protein. These attributions can be seen clearly in the NO•

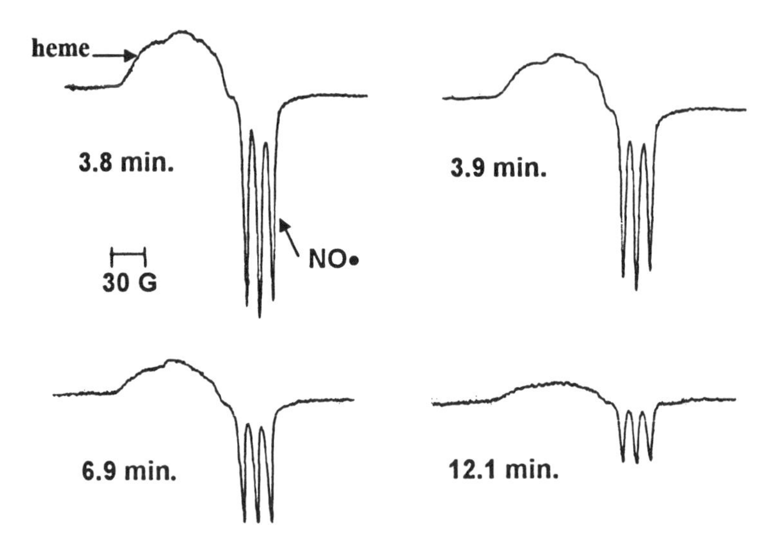

Figure 1. EPR signals of beef cubes in water thermally processed in sealed laminated pouches at 250 °C for varying times. Signals were recorded at 77 K.

signals from actively metabolizing rat keratinocytes, in which the heme is intact and high levels of NO• are being produced enzymatically (Figure 2). The signal marked with the arrow is observed also in fresh cured meats and synthetic complexes and is associated with hexacoordinate iron and a radical localized on the heme imidazole *(16)*.

NO•-heme spectra were not observed in beef cubes processed in saltwater. In these samples, signals were very weak single lines at field positions and g-values characteristic of protein free radicals (Figure 3). Further experimentation will be necessary to distinguish how salt changes reaction mechanisms or alters reaction environments, but several reasonable explanations may be offered. Chloride ions have a high affinity for iron so may compete with NO• for ligand positions on the iron. Thus, it is possible that

NaCl or a contaminant in it inhibit NO• formation, competitively react with it, or increase heme destruction, thereby eliminating the NO• trapping capability. High iron contamination in salt can provide a new catalyst for free radical reactions and increase the importance of oxidizing lipids as radical generators. A NaCl-induced change in the ordered structure of water effectively changes the solvent environment of the system. Especially at 77K, this could shift the conformational equilibrium of the heme and spin state of the iron, affecting EPR detection. It could also alter the electron tunneling in the heme, thereby modifying the final localization of the free electrons *(18)*.

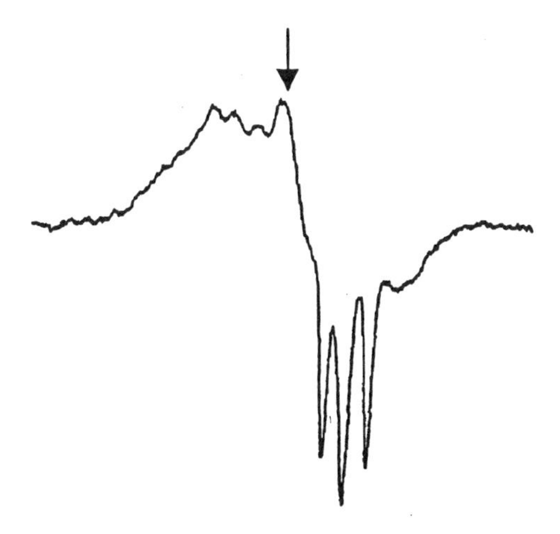

Figure 2. NO•-Hb EPR signals from actively metabolizing rat keratinocytes. The arrow marks the peak especially associated with an intact with hexacoordinate iron and a radical localized on the heme imidazole.

One important point needs to be emphasized here – the detection of NO• is highly temperature sensitive. If the N_2 gas flow dropped and the temperature began to rise, the NO• signal disappeared and another unidentified signal appeared at near room temperature. Thus, careful control of temperature is critical in these measurements.

To determine whether this phenomenon was generalizable, fresh ground beef purchased from a local grocery store was heated at the same temperature as the beef cubes (250 °F) for up to 30 min. Immediately after opening when the meat was still bright red, there was no evidence of radicals (Figure 4). The next sample taken showed the beginnings of a free radical, and this line grew

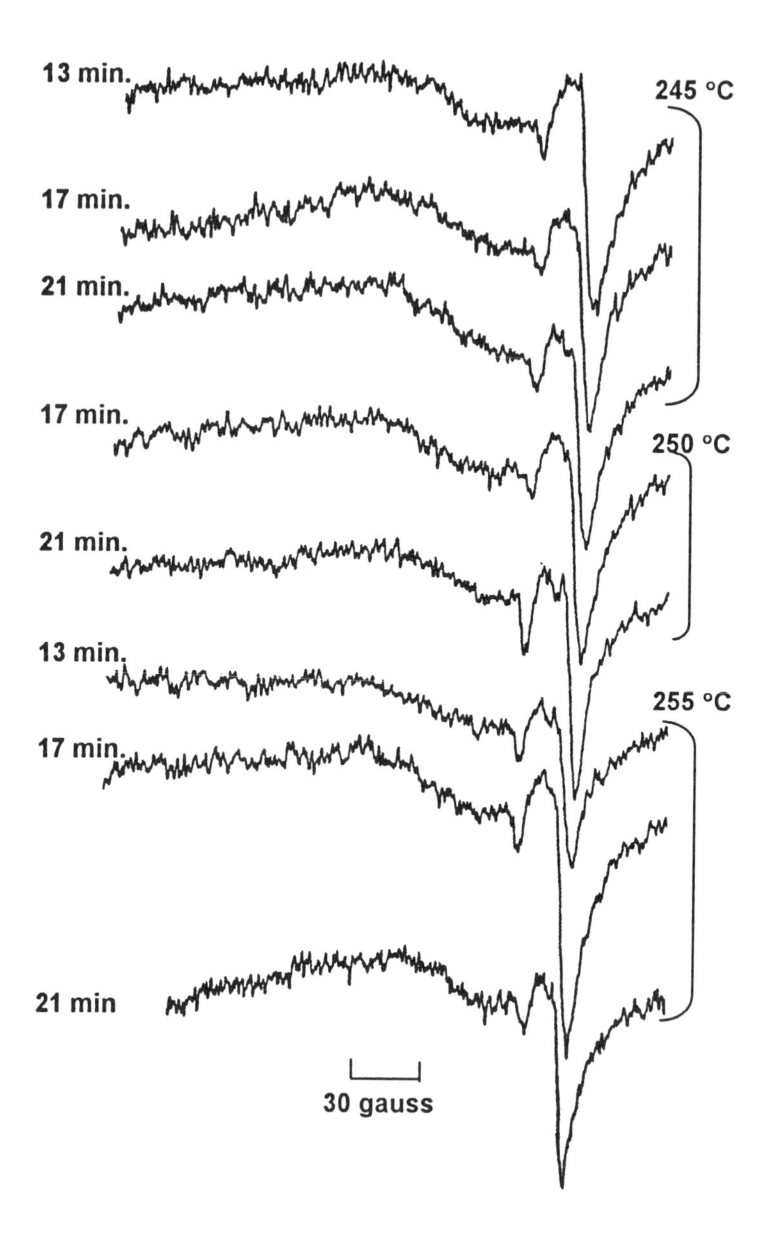

Figure 3. Low temperature EPR signals from beef cubes processed in saline. NO•-heme and coordinated iron signals found in beef cooked in water are replaced by unidentified free radical signals centered at g = 2.005-2.006 with additional unresolved lines or other radical species. Weakly complexed iron is also evident in the broad background.

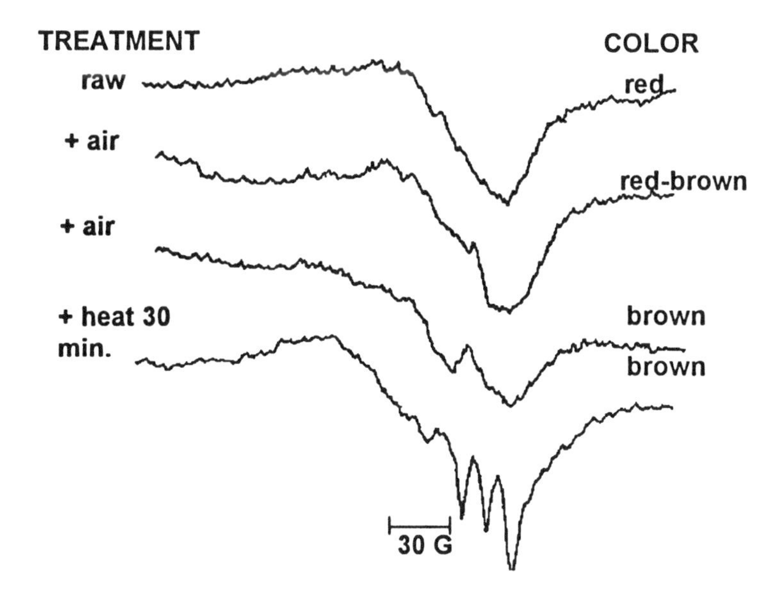

Figure 4. EPR signals from fresh ground beef raw, exposed to air, and heated at 250 °C for 30 minutes. Signals recorded at 77 K (LN_2).

even larger in a sample that has turned brown. The NO• signals appeared only after heating began and increased with heating time. They were weaker than in the beef cubes because the ground beef was heated in air rather than sealed packages. NO• binds rapidly to hemoglobin and myoglobin, but is readily displaced by oxygen.

These are seminal observations that provide clear evidence of protein oxidation in processed beef. NO• have received considerable attention recently for their involvement in physiological and toxicological processes, but have not previously been reported in foods without nitrite. NO• are known to be produced in living tissues by enzymatic oxidation of arginine (Figure 5). They are also produced when arginine is oxidized via Fenton reactions with H_2O_2 (J. Laskin, Toxicology, Rutgers University, and K. Schaich, unpublished) or with lipid hydroperoxides (Schaich, unpublished) in model systems, as shown in the reaction scheme below. H_2O_2 and LOOH form during heating of muscle tissue, and are the likely mediators of NO• generation during processing.

In support of this source of NO•, protein carbonyl production increased in samples with strong NO• -heme EPR signals. As the heme component degraded and NO• component increased, protein carbonyl levels and total protein solubility decreased. However, at the same time there was a marked

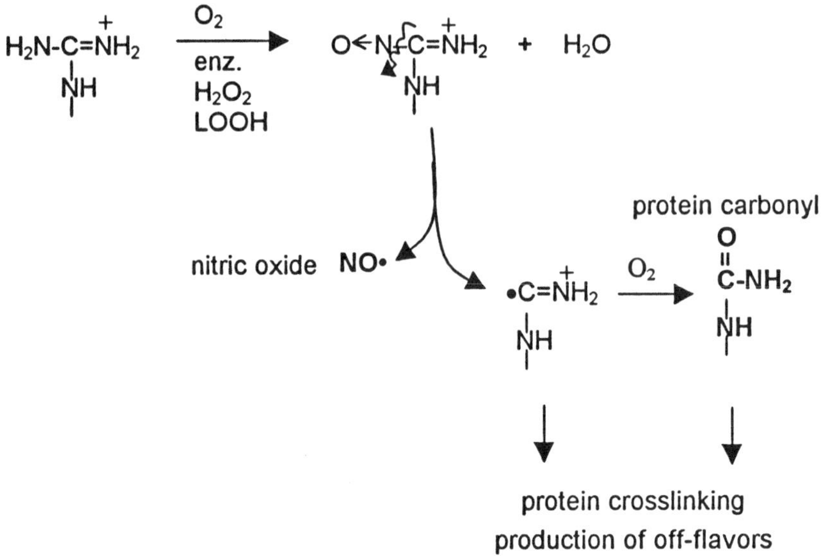

Figure 5. Reaction scheme for formation of nitric oxide radicals via oxidation of arginine side chains on proteins.

increase in the TFA-extractable protein, which is made up primarily of crosslinked proteins (Schaich, unpublished). Thus, we currently speculate that the unidentified EPR signal observed may have originated from side chain radicals left after release of NO•, or new radicals initiated by oxidizing lipids, and that recombination of these radicals as well as condensation of protein carbonyls and amino groups led to protein crosslinking.

The intermediate radicals and protein carbonyls are likely precursors for reactions leading to crosslinking and to production of flavors and off-flavors. A role for protein oxidation in WOF was proposed by Schaich *(22)*, and corroborating evidence was provided by Spanier and coworkers *(23,24)*. They showed a strong correlation between lipid oxidation, protein carbonyls, and loss of a flavor-enhancing low molecular weight peptide (Lys-Gly-Asp-Glu-Glu-Ser-Leu-Ala) named Beefy Meaty Peptide with production of off-flavors.

The recognition that proteins oxidize during thermal processing opens intriguing new directions for more detailed study that, if verified, may reveal new strategies for increasing shelf life and improving quality of thermally-processed beef.

Summary and conclusions

EPR evidence was presented for NO• produced during heating of beef muscle and trapped by hemes in the tissue. This is the first documentation of NO• production in foods without nitrite, and results implicate protein oxidation as a degradation mechanism occurring during thermal processing of protein foods.

References

1. Ledward, D. A. Meat. In *Effects of Heating on Foodstuffs*, Priestley, R.J., Ed., Applied Science Publishers, Ltd, London, 1979, pp. 157-194.
2. Schricker, B. R.; Miller, D. D. *J. Food Sci.* **1983**, *48*, 1340.
3. Buchowski, M. S.; Mahoney, A. W.; Carpenter, C. E.; Cornforth, D.P. *J. Food Sci.* **1988**, *53*, 43.
4. St. Angelo, A. J.; Bailey, M.E., Eds. *Warmed-over Flavor of Meat*, Academic Press, Orlando, FL, 1987.
5. Watts, B. M. *Adv. Food Res.* **1954**, *5*, 1.
6. Younathon, M. T.; Watts, B. M. *Food Res.* **1959**, *24*, 728.
7. Tappel, A. L. *Arch. Biochem. Biophys.* **1953**, *44*, 378.
8. Decker, E. A. ; Hultin, H. O. *J. Food Sci.* **1990**, *55*, 947.
9. Huang, C.-H.; Hultin, H. O. *J. Food Biochem.* **1992**, *16*, 1-14.
10. Marnett, L. J. *Inform*, **1990**, *1*, 338.
11. Dix, T. M..; Marnett, L. J. . *Biol. Chem.* **1985**, *260*, 5351.
12. Schaich, K. M. 2000. EPR methods for studying free radicals in foods, this volume.
13. Wayland, B. B.; Olson, L. W. J. *Chem. Soc,. Chem. Commun.* **1973**, 897-898.
14. Bonnett, R.; Charalambides, A. A.; Martin, R. A. *J. Chem. Soc. Perkins I* **1980**, 974.
15. Neto, L. M.; Nascimento, O. R.; Tabak, M., Caracelli, I. *Biochim. Biophys. Acta* **1988**, *956*, 189.
16. Singel, D.J.; Lancaster, J.R., Jr. In *Methods in Nitric Acid Research,* M. Feelisch, J.S. Stamler, Eds, Wiley, NY, 1996, pp. 343.
17. Kosaka, H.; Shiga, T. In *Methods in Nitric Acid Research,* M. Feelisch, J.S. Stamler, Eds., Wiley, NY, 1996, pp. 373-381.
18. Nascimento, O. R.; Neto, L. M.; Wajnberg, E. *J. Chem. Phys.* **1991**, *95*, 2265-2268.
19. Bonnett, R.; Chandra, S.; Charalambides, A. A.; Sales, K. D.; Scourides, P. A. *J.C.S. Perkin I*, **1980**, 1706-1710.

20. Pegg, R. B.; Shahidi, F.; Gogan, N. J., DeSilva, S. I. *J. Agric. Food Chem.* **1996**, *44*, 416-421.
21. Killday, B.; Tempests, M. S.; Bailey, M. E.; Metral, C. J. *J. Agric. Food Chem.* **1988**, *36*, 909-914.
22. K. M. Schaich. *CRC Crit. Rev. Food Sci. Nutr.* **1980**, *13*, 189.
23. Spanier, A.; Miller, J.A. *J. Muscle Foods* **1996**, *7*, 355-366.
24. Spanier, A. M.; Boylston, T. D. *Food Chem.* **1994**, *50*, 251-260.

Natural Antioxidants

Chapter 12

Antioxidants in Plants and Oleaginous Seeds

Fereidoon Shahidi

Department of Biochemistry, Memorial University of Newfoundland, St. John's, Newfoundland A1B 3X9, Canada

Plants are a rich source of phenolic and polyphenolic compounds that serve as secondary metabolites to protect them from oxidative stress of photosynthesis and wound and also as antifeedant against herbivores. In addition, these phenolics serve as good filters against UV light. The mixture of compounds present in different plant sources provides an excellent opportunity for exploitation to control oxidative deterioration of food lipids. The type, concentration and complexity of phenolics in different plants present a challenge for their isolation, identification and application. Examples will be provided to demonstrate the isolation, testing, and activity determination of a number of natural extracts from selected oilseeds.

Natural antioxidants are present in foods as their endogenous constituents or are added to them to preserve their quality. Antioxidants that occur naturally in plant foods belong to the phenolic group of compounds and include phenolic acids, phenylpropanoids, flavonoids and isoflavones, including catechins and anthocyanins/anthocyanidins, as well as phytates, sterols and carotenoids, among others. In addition, vitamins C and E as well as phospholipids may act as antioxidants in foods. In oilseeds, phenolic antioxidants are present in the free, esterified, glycosidic or bound form and may be simple or complex in their chemical structures.

It has clearly been demonstrated that there is an association between the consumption of plant foods and reduced risk in certain degenerative diseases as well as the ageing process (1,2). Plant-based foods contain a variety of bioactive substances, including phenolic compounds which may be responsible for health promotion ascribed to them. In live plants, these phenolic compounds protect against oxidative stress and attack by herbivores. They also act as UV filters and wound healing agents. Thus, plant foods containing these bioactive components may prevent damage to DNA, proteins, sugars and lipids upon consumption and hence protect the body from a variety of degenerative diseases such as cancer, cardiovascular disease, cataract, immune disorders and brain dysfunction (*1-4*).

Natural antioxidants also protect foods from oxidation by quenching of free radicals via donation of an electron or a hydrogen atom, or by deactivation of prooxidant metal ions and singlet oxygen. However, the efficacy of antioxidants in controlling oxidation of food lipids depends on the system to which they are applied (ie. bulk oil versus emulsion or low water activity foods such as cereals) as well as fatty acid constituents of lipids involved, their minor constituents and storage condition of the products. In addition, application of mixed antioxidants or natural extracts might prove beneficial because of possible synergism amongst components involved. This contribution reviews the latest progress in the isolation, and identification as well as uses of naturally occurring phenolic antioxidants from oilseeds.

Biosynthesis of Plant Phenolics. Phenolic compounds in plant foods are secondary metabolites which are derived from phenylalanine, and in some plants from tyrosine via enzymatic deamination assisted by ammonia lyase (Figure 1). Phenylpropanoids, the first products of deamination of phenylalanine and/or tyrosine consist of a phenyl ring (C_6) and a 3 carbon side chain (C_3). These C_6-C_3 compounds may subsequently undergo hydroxylation in the phenyl ring and possibly subsequent methylation. This would lead to the formation of a large number of products which include cinnamic acid, *p*-coumaric acid, caffeic acid, ferulic acid and sinapic acid (*5*).

The C_6-C_3 compounds so produced may undergo further changes which would lead to the elimination of a 2-carbon moiety (acetate) group from cinnamic acid to eventually afford benzoic acid derivatives (Figure 2). Similar to the phenylpropanoid series, hydroxylation and possible subsequent methylation may lead to the formation of hydroxybenzoic acid, vanilic acid, syringic acid and gallic acid (*5*). Meanwhile, dimerization of phenyl propanoids at the 8,8-positions leads to the formation of lignans and polymerization of C_6-C_3 units results in the production of lignin which is the cell wall constituent of plants.

Condensation of *p*-coumaric acid and 3 or 4 molecules of mevalonic acid in conjugation with coenzyme A leads to the formation of resveratrol or tetrahydroxychalcones (THC) under the influence of stilbene synthase and chalcone synthase, respectively (Figure 3). Subsequent chalcone isomerase-assisted cyclization of THC would lead to the formation of a tricyclic molecule, the simplest of which is a flavonone. Further reactions of flavonones lead to the formation of flavonols, flavones, catechins, anthocyanidins and isoflavones. These compounds are composed of C_6 - C_3 - C_6 units and are also the building block of condensed tannins, referred to as proanthocyanidins (*6*).

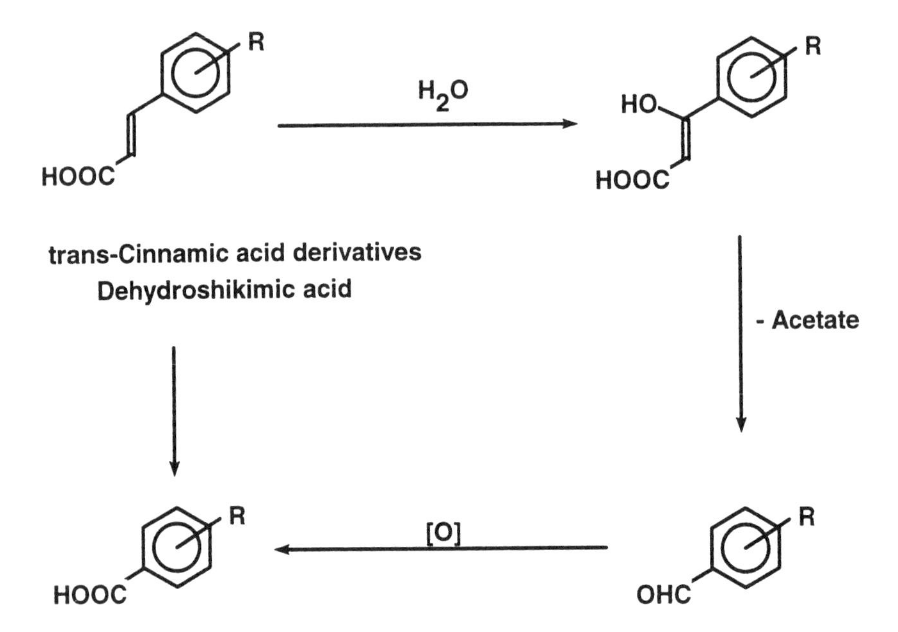

Figure 1. Biosynthesis of phenylpropanoids (C_6-C_3) from phenylalanine and tyrosine. PAL, phenylalanine ammonia lyase; TAL, tyrosine ammonia lyase.

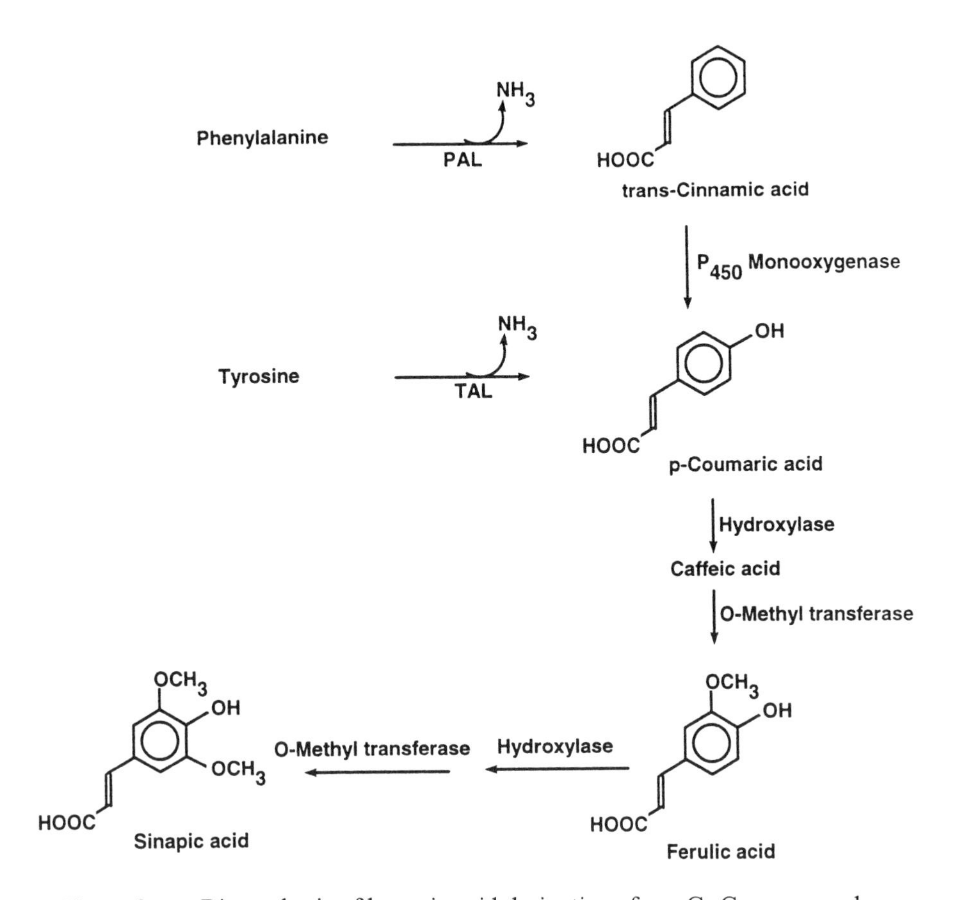

Figure 2. Biosynthesis of benzoic acid derivatives from C_6-C_3 compounds or dehydroshikimic acid.

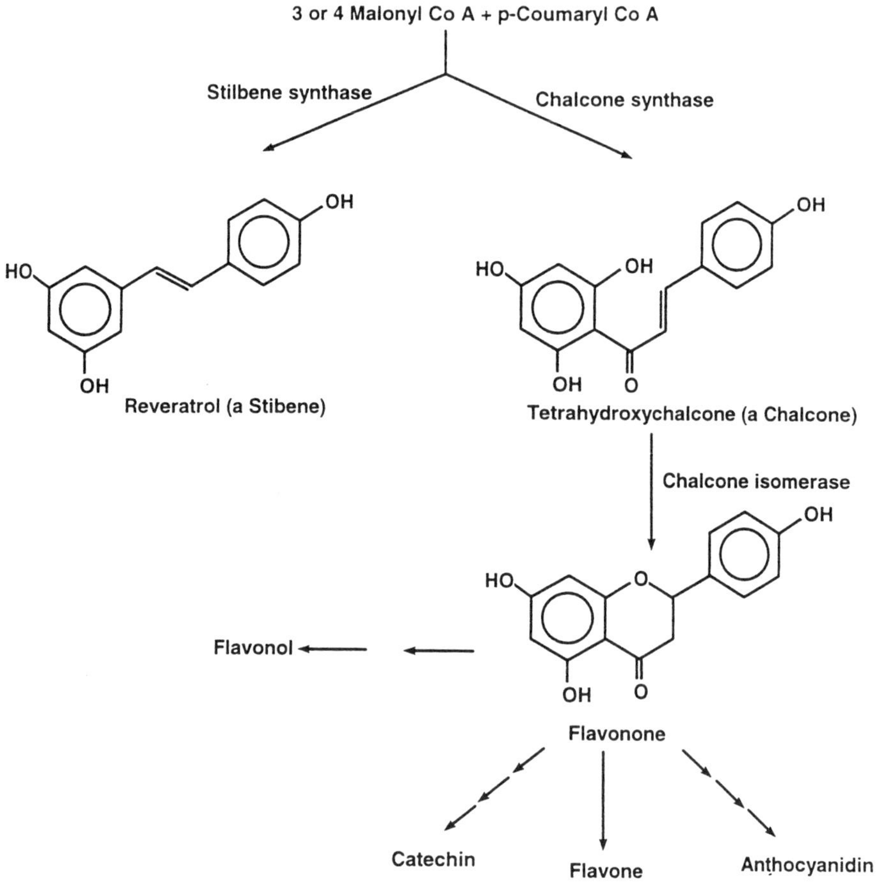

3 or 4 Malonyl Co A + p-Coumaryl Co A

Stilbene synthase

Chalcone synthase

HO OH OH

Reveratrol (a Stibene)

HO OH OH OH O

Tetrahydroxychalcone (a Chalcone)

Chalcone isomerase

HO O OH OH O

Flavonol

Flavonone

Catechin

Flavone

Anthocyanidin

Figure 3. Biosynthesis of flavonoids and resveratrol from malonayl Co A and *p*-coumaryl CoA (C_6-C_3).

The type of phenolics occurring in different plant foods depend on their source. Oilseeds, similar to other plant foods contain a variety of phenolic and other compounds which may be important as antioxidants. These phenolics generally exist in the free, esterified and etherified with sugars or bound form. They may belong to both simple and complex phenolic forms. The latter group consists of both hydrolyzable and condensed tannins. Condensed tannins are formed via polymerization of flavonoids while hydrolyzable tannins are produced by the reaction of gallic acid with hexose molecules. Each gallic acid may further be condensed with other gallic acids through ester linkages. Thus, oilseeds contain a variety of antioxidative phenolics that may act co-operatively and synergistically to control their in vivo oxidation. When used as seed or as extracts, they may also control oxidation of food lipids.

Approach for Evaluation, Isolation and Identification of Antioxidative Compounds in Oilseeds

Antioxidative compounds from oilseeds such as canola/rapeseed, mustard, flax, borage and evening primrose, soybean, cottonseed, peanut and sesame have been investigated. The antioxidative components present are diverse and may end up in the extracted oil or in their resultant meal after oil extraction. The antioxidative compounds present include tocopherols, sterols, phospholipids, phenolic acids and phenylpropanoids, flavonoids and isoflavonoids, hydrolyzable and condensed tannins, lignans, coumarins, amino acids, peptides and proteins as well as carotenoids (*7*).

To evaluate the efficacy of the source materials in inhibiting lipid oxidation, the ground seeds, as such or defatted, may be tested in a model system (*8-13*). We have found that a meat or fish system would lend itself best for such an evaluation. This approach is particularly of interest to the meat industry as plant materials are often used in fabricated products as protein extenders or as coating material. Application of a powdered seed material at 0.5-2% may be most desirable as at higher levels of application uncharacteristic changes in color and physical characteristics of products may be experienced. Table I shows the inhibition of oxidation of meat lipids by selected oilseeds as reflected in the thiobarbituric acid reactive substances (TBARS) of cooked meat following cooking and during prolonged storage in a refrigerator. The efficacy of canola and mustard meals was most noticeable. The lower activity of certain products, such as flax, may be due to a) low concentration of active phenolic antioxidants in the source material, b) presence of active ingredients in the esterified, glycerylated or bound form, and c) low activity of components involved. Thus, if extracts prepared and purified, one might be able to isolate very effective antioxidants from source materials which may have low activity (*8*).

Table I. Inhibition of Oxidation of Cooked Ground Meat, as Reflected in the Content of Thiobarbituric Acid Reactive Substances (TBARS), by Ground Defatted Oilseeds.[1]

Ground seed	*Addition level, %*	*Inhibition, %*
Canola	1	91.8
	2	95.5
Mustard	1	74.6
	2	96.4
Flax	1	14.6
	2	38.2
Borage	1	71.9
	2	78.6
Evening primrose	1	83.8
	2	84.5

[1]Values are means of seven determinations over a 3-week storage of cooked meat at 4°C. Values for butylated hydroxytoluene at 30 and 200 ppm were 56.4 and 91.0%, respectively.

Extracts may be prepared from ground, defatted oilseeds. In general, methanol, ethanol, acetone, alone or in combination with water, are most frequently used for preparation of extracts. Once prepared, the extracts may be evaluated for their antioxidant activity in a meat, fish, oil or a β-carotene-linoleate model system. Free radical, hydroxyl radical, hydrogen peroxide and superoxide radical scavenging activity of the preparations may also be evaluated (*12-16*). The extracts, in addition to their phenolic compounds, also contain other soluble matters such as sugars and other low-molecular weight compounds. Thus, phenolics content in extracts varies considerably, but on the average is approximately 10% of the total weight of the extract. The antioxidant activity of the extracts may often be less than that of the crude material, perhaps due to the elimination of synergism with other components.

The crude extract may subsequently be fractionated. This is usually done first by employing Sephadex LH-20 column chromatography. Fractions are isolated based on their UV absorption intensity at 280nm (e.g. 15) and after color development at 490 and 725 nm for sugars and phenolics, respectively. However, direct use of high performance liquid chromatography-mass spectrometry (HPLC-MS) might be considered. The fractions from column chromatography may then be tested by thin layer chromatography (TLC), HPLC and possibly by gas chromatography-mass spectrometry (GC-MS) following derivatization.

The antioxidant activity of fractions of crude extracts might be evaluated in a similar manner to that of the crude extracts. Although crude extracts might be of low activity, it is possible that certain fractions exhibit excellent antioxidant potency because of the presence of very active components, albeit in low concentrations. The antioxidant activity of individual components, as separated on TLC, might be evaluated using spraying techniques such as those with β-carotene-linoleate directly on the TLC plates.

Following separation by TLC and/or HPLC or possibly GC, the identity of the isolated components may be evaluated by mass spectrometry. However, in certain cases caution must be exercised for separating isomeric compounds. As an example, we have isolated both (+) catechin and (-) epicatechin from extracts of evening primrose (unpublished results). This was possible only after HPLC separation of a single spot on the TLC plate. In addition, it is necessary to consider possible deglycosylation of compounds during the extraction and subsequent chromatographic procedures. Phenolic compounds often exist in the glycosylated from, but under certain conditions their hydrolysis may occur thus leading to the detection of hydrolyzed products.

The following sections provide specific examples about the free radical scavenging properties of preparations and the antioxidative compounds isolated and identified in selected oilseeds.

Free radical - and hydrogen peroxide - scavenging effects of oilseed meals and their fractions

The extracts of several oilseeds, especially borage and evening primrose, were evaluated for their free radicals, hydroxyl radical, superoxide and hydrogen peroxide scavenging activities. In this, 2,2-diphenyl-1-piperylhydrazyl radical ($DPPH^{\cdot}$) was used to monitor the efficacy of the extracts in terms of their scavenging capacity. Generalized reduction reactions between $DPPH^{\cdot}$ and additives (AH) may be written as:

$$DPPH^{\cdot} + AH \longrightarrow DPPH - H + A^{\cdot}$$
$$DPPH^{\cdot} + R^{\cdot} \longrightarrow DPPH - R$$

We found that all extracts and their fractions thereof possessed the ability to quench free radicals, albeit to different extents (*14*). The effects may be attributed to the hydrogen and electron donating abilities of their phenolics.

The crude extracts of several oilseeds were also found to completely scavenge hydroxyl radicals. The hydroxyl radicals were generated via iron-catalyzed Haber-Weiss reaction and spin-trapped with 5,5-dimethyl-1-pyrroline N-oxide (DMPO).

The resultant DMPO-OH adduct was detected using an electron paramagnetic resonance (EPR) spectrometer. Fig. 4 shows the effect of evening primrose crude extract (EPCE) on the scavenging of the hydroxyl radical. While the characteristic quartet signal for DMPO-OH adduct was observed in the absence of extracts, crude extracts at 200 ppm as well as most of their fractions completely eliminated the signals.

Superoxide radical ($O_2^{\cdot -}$), generated in a hypoxanthine/xanthine oxidase system was also scavenged readily and completely in assay media containing 200 ppm of extracts of borage and evening primrose. Almost all fractions of extracts also effectively scavenged $O_2^{\cdot -}$ as evidenced by the absorption of the ink-bue color of the system at 560 ppm.

All extracts and their fractions showed H_2O_2-scavenging capacity in a concentration-dependent manner. The values were lower for 100 ppm as compared to those for 200 ppm level of extracts. The scavenging of H_2O_2 by extracts and their fractions is attributed to their phenolics which could donate electrons to H_2O_2, thus neutralizing it to water. The efficacy of the fractions of each extract was dependent on the chemical nature of phenolic compounds involved. Thus, crude extracts, their fractions or purified compounds thereof may be incorporated into foods to minimize free radical-mediated lipid peroxidation. They may also be used as alternative drugs/supplements to treat human diseases associated with free radical-mediated tissue damage.

Antioxidants from Oilseeds

Phenolic compounds are ubiquitons in oilseeds and many compounds have so far been identified in different source materials. Generally, mass spectrometry (MS) as such or in connection with gas chromatography (GC-MS) and high performance liquid chromatography (HPLC-MS) and nuclear magnetic resonance (NMR) are used for structure elucidation. Recently, Wanasundara *et al.* (*18*) reported that the most active phenolic antioxidant in canola meal was 1-O-β-D-glucopyranosyl sinapate (Figure 5). Meanwhile, rapeseed meal contained salicylic, cinnamic, p-hydroxybenzoic, verastric, vanillic, gentisic, protocatechuic, syringic, caffeic, sinapic and ferulic acids (*19*). The major condensed tannins of rapeseed meal were composed of cyanidin, pelargonidin, kaempferol and their derivatives. The main antioxidative compounds in soybean meal were genestein, daidzein, glycitein, prunetin, formononetin and 4',6',7'-trihydroxyisoflavone (*20*). Meanwhile dihydroquercetin has been identified in extracts of peanut (*21*). Rutin and quercetin glycosides were shown to be present in cottonseed extracts (*22*).

In recent studies we have identified several low-molecular-weight phenolic acids, namely syringic acid, sinapic acid and rosmarinic acid in borage meal (Figure 5; unpublished). In evening primrose, we positively identified gallic acid, (+) catechin and (-) epicatechin in the extracts (Figure 5) which also included a

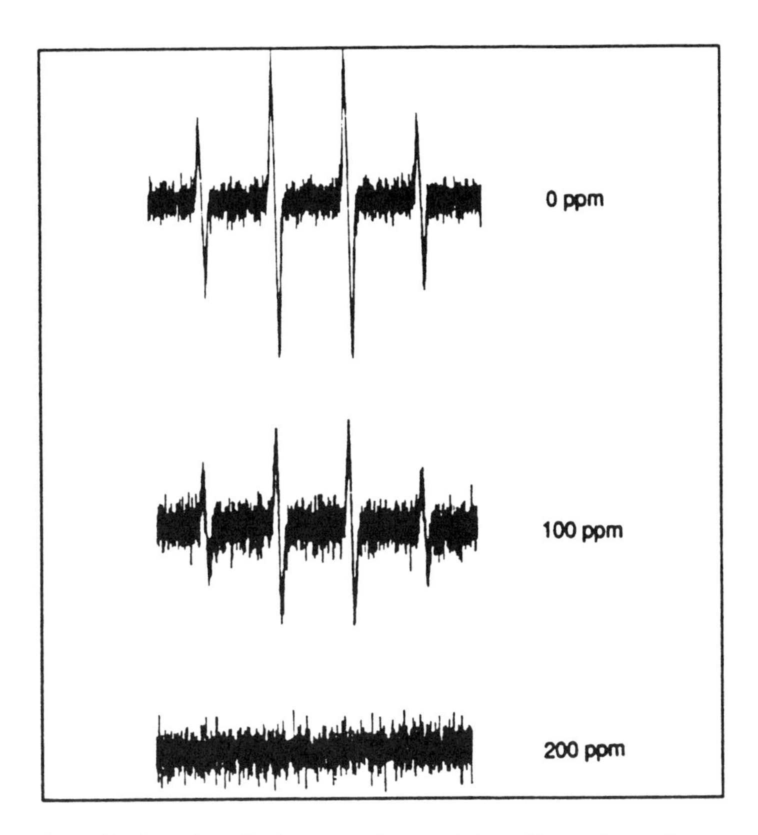

Figure 4. Hydroxyl radical-scavenging activity of evening primrose crude extracts at 100 and 200 ppm (catechin equivalents).

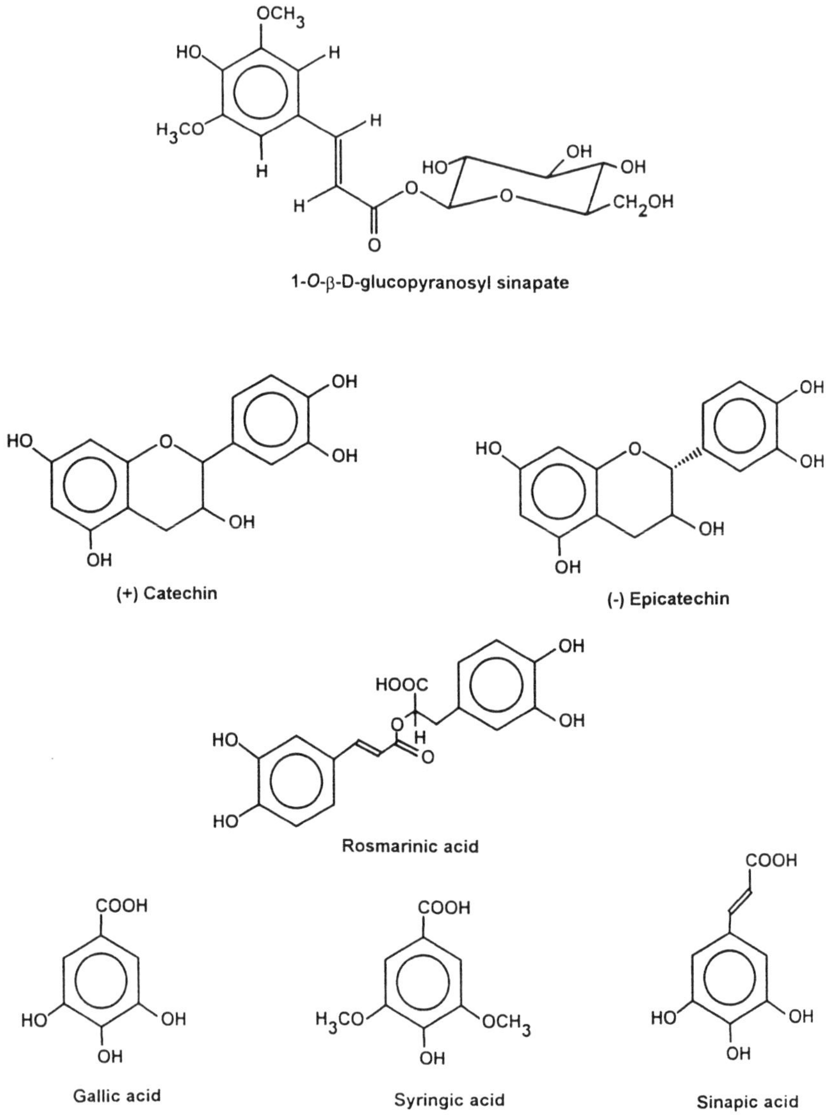

1-*O*-β-D-glucopyranosyl sinapate

(+) Catechin

(-) Epicatechin

Rosmarinic acid

Gallic acid

Syringic acid

Sinapic acid

Figure 5. Chemical structures of selected phenolic antioxidants of canola, borage and evening primrose.

substantial amount of high-molecular-weight compounds including proanthoyanidins (unpublished results). The positive identification and quantitation of a novel high-molecular-weight compound, namely oenothein B (Figure 6) in the extracts of evening primrose meal is in progress. This latter compound, an oligomeric hydrolyzable tannin, was reported by Okuda et al. (*23*) as being present in green parts of evening primrose plant. Presence of polymeric phenolics in evening primrose extracts has also been reported (*24*).

Efficacy of Natural Antioxidants in Different Foods

Plant phenolics provide for a cocktail or soup which contains a variety of compounds with antioxidant activity. These act synergistically and cooperatively with one another, thus protecting the plant and then the food they are added to from oxidative processes. The advantage of the complex mixture of phenolics with varying hydrophilicity / lipophibicity balance (HLB) is their effectiveness in both aqueous emulsion, bulk oil and low water activity products. Porter (*25*) has demonstrated that in food systems with a low surface-to-volume ratio, such as in bulk oils, polar antioxidants with a high HLB, such as propyl gallate and Trolox C, are more effective than non-polar antioxidants such as butylated hydroxytoluene and tocopherols. On the contrary, in foods with a high surface-to-volume ratio, such as oil-in-water emulsions, nonpolar antioxidants with a low HLB are highly effective. Frankel (*26*) has further confirmed this "interfacial phenomenon". Our work on borage and evening primrose crude extracts has demonstrated their excellent effectiveness in both bulk oil and oil-in-water emulsions (*12, 13*).

Conclusions

Natural phenolics in plants and oilseeds provide excellent sources of antioxidant for food application. Presence of a wide variety of phenolics with varying degrees of polarity avails these preparations to exert their effects via different mechanisms. Thus efficacy of natural oilseed extracts in bulk oil, oil-in-water emulsions, low water activity foods and formulated products is excellent. Furthermore, in addition to their safety advantages, food phenolics may augment body's source of natural antioxidants in different ways, including regeneration of α-tocopherol.

References

1. Shahidi, F. *Natural Antioxidants: Chemistry, Health Effects, and Applications,* AOCS Press, Champaign, IL, 1997.
2. Shahidi, F.; Ho, C-T. *Phytochemicals and Phytopharmaceuticals.* AOCS Press, Champaign, IL, 2000.

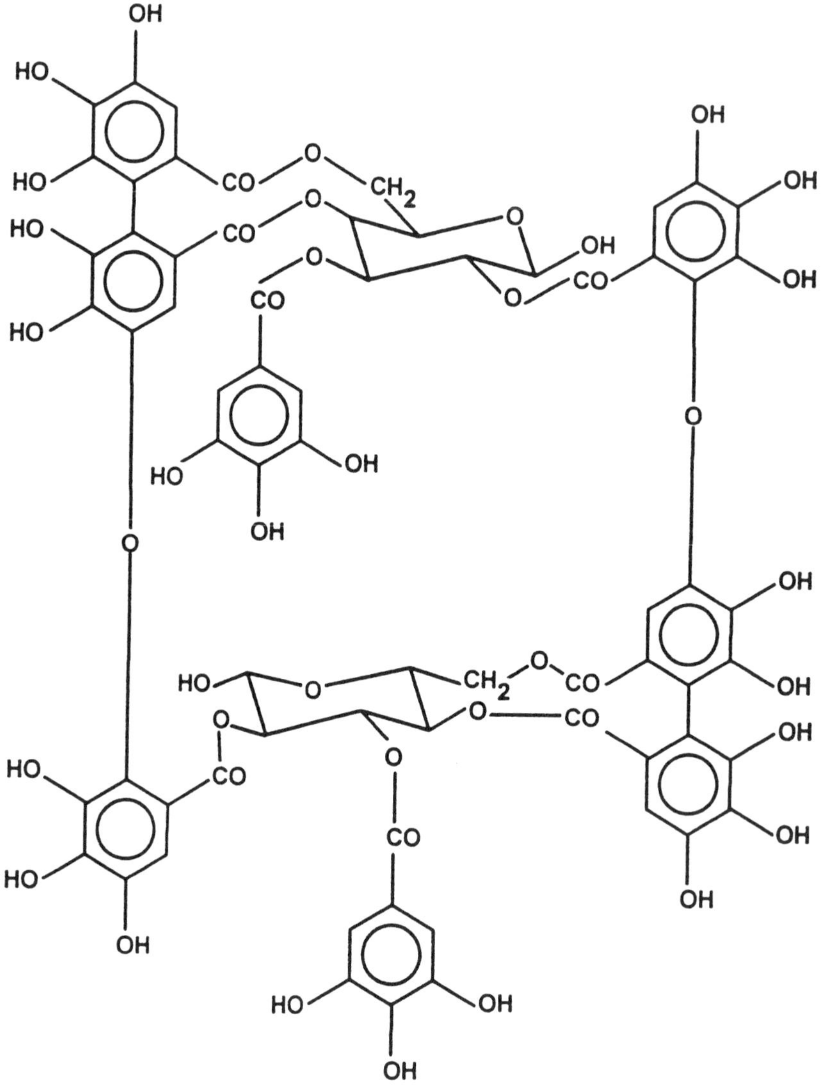

Oenothein B

Figure 6. Chemical structure of oenothein B.

3. Halliwell, B.; Aeschbach, R.; Löliger, J.; Aruoma, O.I. *Food Chem. Toxicol.* **1995**, *33*, 601-617.
4. Ramarathnam, N.; Osawa, T.; Oclin, H.; Kawakishi, S. *Trends Food Sci. Technol.* **1995**, *6*, 75-82.
5. Shahidi, F. *Nahrung* **2000**, *44*, 158-163.
6. Shahidi, F. and Naczk, M. *Food Phenolics: Sources, Chemistry, Effects and Applications*. Technomic Publishing Company, Lancaster, PA, 1995.
7. Shahidi, F. and Shukla, V.K.S. *INFORM* **1996**, *7*, 1227-1232.
8. Shahidi, F. Natural Antioxidants from oilseeds. In *Food Factors and Cancer Prevention*. Ohigashi, H., Osawa, T., Terao, J., Watanabe, S. and Yoshikawa, T.Q. Springer, Tokyo, 1997, pp. 299-303.
9. Shahidi, F., Pegg, R.B., and Saleemi, Z.O. *J. Food Lipids* **1995**, *2*, 145-153.
10. Shahidi, F. and Alexander, D.M. *J. Food Lipids* **1998**, *5*, 125-133.
11. He, Y. and Shahidi, F. *J. Agric. Food Chem.* **1997**, *45*, 4262-4266.
12. Wettasinghe, M.; Shahidi, F. *Food Chem.* **1999**, *67*, 399-414.
13. Wettasinghe, M.; Shahidi, F. *J. Agric. Food Chem.* **1999**, *47*, 1801-1882.
14. Wettasinghe, M.; Shahidi, F. *Food Chem.* **2000**, *70*, 17-26.
15. Shahidi, F.; Wanasundara, U.N.; Amarowicz, R. *Food Res. Int.* **1994**, *27*, 489-493.
16. Shukla, V.K.S.; Wanasundara, P.K.J.P.D.; Shahidi, F. In *Natural Antioxidants: Chemistry, Health Effects and Application*. Shahidi, F. (Ed.), AOCS Press, Champaign, IL. 1997, pp. 97-132.
17. Shahidi, F.; Chavan, U.D.; Naczk, M.; Amarowicz, R. *J. Agric. Food Chem.* **2000**. Submitted.
18. Wanasundara, U.; Amarowicz, R.; Shahidi, F. *J. Agric. Food Chem.* **1994**, *42*, 1285-1290.
19. Kozlowska, H.; Rotkiewicz, D.A.; Zadernowski, R. *J. Am. Oil Chem. Soc.* **1983**, *60*, 1191-1193.
20. Rackis, J.J. In *Soybean Chemistry and Technology*. Smith, A.K. and Gircler, S.J., Eds. AVI Publishing Company, Westport, CT, 1972, pp. 158-202.
21. Pratt, D.E.; Miller, E.E. *J. Am. Oil Chem. Soc.* **1984**, *61*, 1064-1067.
22. Wittern, C.C.; Miller, E.E.; Pratt, D.E. *J. Am. Oil Chem. Soc.* **1984**, *61*, 928-931.
23. Okuda, T.; Yoshida, T.; Hatano, T. *Phytochemistry* **1993**, *32*, 507-521.
24. Lu, F.; Foo, L.Y. In *Nutrition, Lipids, Health and Disease*. Ong, A.S.H.; Niki, E. and Packer, L. Eds. AOCS Press, Champaign, IL, 1995, pp. 86-95.
25. Porter, W.L. In *Autoxidation in Food and Biological Systems*. Simic, M.G. and Karel, M., Eds. Plenum Press, New York, NY., 1980; pp. 295-365.
26. Frankel, E.N., Huang, S-W., Kanner, J. and German, B. *J. Agric. Food Chem.* **1994**, *42*, 1054-1059.

Chapter 13

Inhibition of Citral Deterioration

Hideki Masuda[1], Toshio Ueno[1], Shuichi Muranishi[1], Susumu Irisawa[1], and Chi-Tang Ho[2]

[1]Material R&D Laboratories, Ogawa & Company, Ltd., 15-7, Chidori, Urayasushi, Chiba 279-0032, Japan
[2]Department of Food Science, Rutgers University, 65 Dudley Road, New Brunswick, NJ 08901-8520

Citral is a very important component of lemon flavor. However, citral is liable to be deteriorated by contact with various agents including light, heat, oxygen, and acid. Therefore, the inhibition of deterioration of citral is considered to be an important problem for the stabilization of citrus drinks. For the purpose of inhibiting the deterioration of citral during UV light irradiation under acidic conditions, the ability of UV light absorption and the radical scavenging activity of a variety of additive compounds are studied.

Citral is well-known as the most characteristic component of lemon flavor. The value of citrus drinks containing lemon flavor is dependent on the stability of citral (*1*). However, citral is deteriorated through acid-catalyzed or oxidation mechanisms (*2-7*). The potent off-flavor components formed from citral under acidic and heated conditions have been clarified by aroma extract dilution analysis (AEDA) (*8*). In recent years, longer storage times and displays using transparent containers under fluorescent lamps in stores have significantly increased and citral is easily deteriorated by UV light from fluorescent lamps.

The off-flavor components of citral due to UV light or sunlight have also been studied (*9-14*).

Taking into account the instability of citral, it is important for many flavor and food companies to use food additives which inhibit the deterioration of citral. However, the relationship between the stability of citral and the kinds of additive compounds inhibiting its deterioration has scarcely been reported (*15-18*). In addition, in spite of the wide use of ascorbic acid, the degraded components derived from it can lower the value of a citrus drink (*19-21*). Therefore, it was necessary to find effective additive compounds that inhibit the deterioration of citral.

This study focuses on the effects of the additive compounds on both the stabilization of citral and the reduction of reaction products derived from citral due to UV light.

Experimental

0.1 g of 1% (w/w) solution of each additive compound prepared by dissolving in 50% (w/w) ethanol, was mixed with a 0.1 M citric acid-sodium citrate buffer (99.9 g, pH 3.5 or 7.0) containing 10000 ppb citral. Therefore, the concentration of additive compound in each solution was 10000 ppb. The glass bottle was irradiated at 10°C in a Eyela LST-300 light box with an fluorescent lamp consisting of 3 wavelength-type 40 W×15 (15000 lux) light sources. Instead of light irradiation, the radical initiator, 1000 ppm 2,2'-azobis(2-amidinopropane) dihydrochloride (AAPH), was added to each solution under dark conditions. After the addition of an internal standard, 10 μL of *n*-pentadecane/chloroform solution (1%, w/v), the extraction was carried out with dichloromethane (30 mL × 2), followed by drying. The extract was evaporated (40°C/450 mmHg) and concentrated to about 250 μL in a stream of nitrogen. All tests were run in triplicate and averaged.

The radical scavenging activity was assayed according to the following method (*22-23*). A 0.1 M citric acid-sodium dihydrogenphospate buffer (pH 3.5, 2 mL), 0.04 mM 1,1-diphenyl-2-picrylhydrazyl (DPPH) in 99% ethanol (2 mL), and the additive compound (1000 ppb) were mixed. After allowing the mixture to stand at room temperature for 30 min, the absorbance at 517 nm was measured. The radical scavenging activity of each additive compound (1000 ppb) was defined as the scavenging ratio of the DPPH radical. Hence, the higher the value of the ratio, the higher the radical scavenging activity. All tests were run in triplicate and averaged.

The absorption wavelength (nm) of citral and each additive compound was measured in a quartz glass-cell using a Shimadzu UV-2100 PC UV-VIS scanning spectrometer

A GC analysis was performed using a Hewlett-Packard 5890 SERIES II fitted with FID. A DB-WAX fused-silica capillary column (60 m × 0.25 mm i.d., film thickness: 0.25 μm) was employed. The operating conditions were as follows: initial oven temperature, 60°C, ramped to 210°C at 4°C/min, then held for 42.5 min; injection temperature, 250°C; carrier gas, 1 mL/min He; split ratio, 1/32.

A GC-MS analysis was performed using a Hewlett-Packard 5890 SERIES II fitted with a HP-5972. A DB-WAX fused-silica capillary column (60 m × 0.25 mm i.d., film thickness: 0.25 μm) was employed. The operating conditions were as follows: initial oven temperature, 60°C, ramped to 210°C at 4°C/min, then held for 42.5 min; injection temperature, 250°C; carrier gas, 1 mL/min He; split ratio, 1/50; ionization voltage, 70 eV; ion source temperature, 140°C.

Caffeic acid, ferulic acid, sinapic acid, *p*-coumaric acid, quercitrin, *p*-hydroxy benzoic acid, L(+)-ascorbic acid, gallic acid, protocatechuic acid and vanillic acid were purchased from Nakaraitesc. (-)-Epicatechin was purchased from Aldrich. Rosmarinic acid, (-)-epigallocatechin gallate, and (-)-epigallocatechin were purchased from Funakoshi. Chlorogenic acid was purchased from Wako Pure Chemical. α-Glucosyl rutin was purchased from Toyo Seito. 3,5-Dicaffeoyl quinic acid was isolated by extraction with 50% aqueous ethanol from coffee beans and purified by using column chromatography (silica gel, chloroform-methanol). Citral (a brand name: Natural Citral EOA, neral : geranial = 1 : 1.8) was purchased from Polarome Manufacturing Co., Inc. and purified by vacuum distillation.

Results and Discussion

Three types of volatile products were found to be obtained from citral (neral (**10**) and geranial (**12**)) after UV light irradiation at pH 3.5. Aldehydes **3** and **4**, photocitral B (**5**), aldehyde **6**, and photocitral A (**7**) were formed through photochemical cyclization (*12-13*) (Table I and Figure 1). Both tetrahydrofurans **16** and **17** were obtained by oxidation (*4*) (Table I). The acid-catalyzed cyclization products were furan **2** and alcohols **13-15**, **19** and **21** (*2*, *24-27*) (Table I and Figure 2). As shown in Table II, the relative amount of the acid-catalyzed cyclization products **2**, **13-15**, **19**, and **21** was extremely high (83%) under dark conditions. However, the UV light irradiation increased the photochemical cyclization products **3-7** and oxidation products **16** and **17**. The amount of products of the former and that of the latter were 38% and 27%, respectively.

Figure 3 shows the correlation factor (R) between the residual amount of citral (**10** and **12**) and UV absorbance of each additive compound *vs.* the wavelength at pH 3.5. The correlation factor was shown to have the higher value in the vicinity of 340 nm. Therefore, the effect of the additive compounds on the

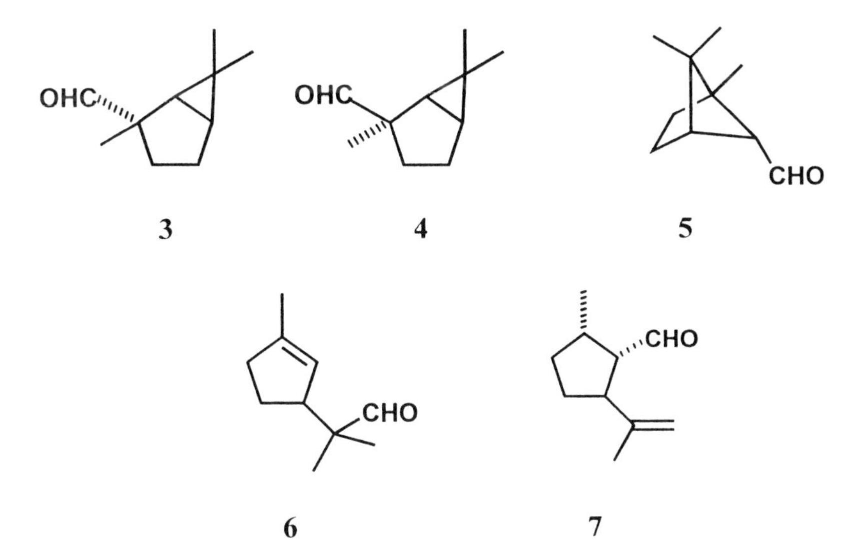

Figure 1. *Structures of photochemical cyclization products formed from citral (10 and 12).*

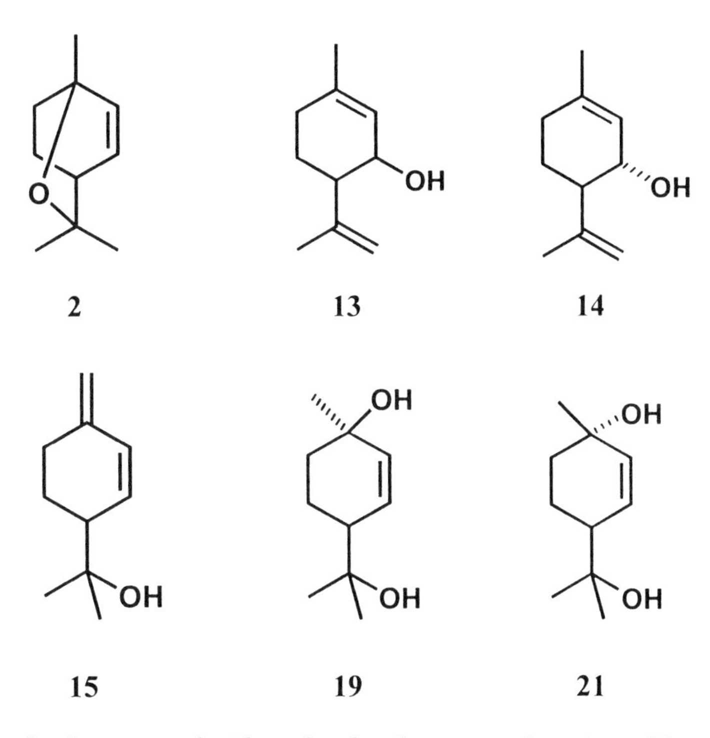

Figure 2. *Structures of acid-catalyzed cyclization products formed from citral (10 and 12).*

Table I. Yield[a] of Volatile Products Formed from Citral (10 and 12) in the Dark[b] or during UV Light Irradiation[b]

no.	*r.t.*[c]	*Products*	*dark*	*UV light*
1	9.76	unknown	9	3
2	11.03	2,3-dehydro-1,8-cineol	8	8
3	16.50	(1*R**,2*S**,5*S**)-2,6,6-trimethylbicyclo[3.1.0]hexane-2-carboxaldehyde	0	49
4	16.77	(1*R**,2*R**,5*S**)-2,6,6-trimethylbicyclo[3.1.0]hexane-2-carboxaldehyde	0	29
5	17.75	(1*R**,4*S**,5*R**)-1,6,6-trimethylbicyclo[2.1.1]hexane-5-carboxaldehyde (photocitral B)	0	77
6	17.98	2-(3-methyl-2-cyclopenten-1-yl)-2 –methylpropionaldehyde	0	208
7	20.27	(1*S**,2*S**,5*S**)-2-isopropenyl-5-methylcyclopentan-1-carboxaldehyde (photocitral A)	0	205
8	23.67	unknown	16	16
9	24.91	unknown	15	17
10	25.23	neral	1510	1017
11	26.42	unknown	22	0
12	26.60	geranial	2449	1021
13	27.16	*cis*-isopiperitenol	10	9
14	27.29	*trans*-isopiperitenol	29	26
15	27.97	*p*-mentha-1(7),2-dien-8-ol	27	14
16	31.96	(2*R**,5*R**)-2-formylmethyl-2-methyl-5-(1-hydroxy-1-methylethyl)-tetrahydrofuran	12	161
17	32.39	(2*S**,5*R**)-2-formylmethyl-2-methyl-5-(1-hydroxy-1-methylethyl)-tetrahydrofuran	21	248
18	35.56	unknown	0	13
19	36.55	*trans-p*-menth-2-ene-1,8-diol	238	223
20	37.88	unknown	0	27
21	38.21	*cis-p*-menth-2-ene-1,8-diol	154	146
22	40.36	unknown	0	32

a) ppb. *b)* A period of 14 days at pH 3.5. *c)* Retention time on DB-WAX.

deterioration of citral (**10** and **12**) was evaluated by comparing the absorbance at 340nm.

A highly positive correlation (R = 0.936) between the residual amount of citral (**10** and **12**) and the absorbance of a variety of the additive compounds at 340 nm at pH 3.5 are shown in Table III. The effects of each additive compound, caffeic acid (**23**), ferulic acid (**24**), sinapic acid (**25**), rosmarinic acid (**26**), chlorogenic acid (**27**), 3,5-dicaffeoyl quinic acid (**28**), *p*-coumaric acid (**29**), α-glucosyl rutin (**30**), quercitrin (**31**), *p*-hydroxy benzoic acid (**32**), (-)-epigallocatechin gallate (**33**), L(+)-ascorbic acid (**34**), (-)-epigallocatechin (**35**), gallic acid (**36**), (-)-epicatechin (**37**), protocatechuic acid (**38**) and vanillic acid (**39**) on the deterioration of citral (**10** and **12**) are shown in Figure 4. Taking into account that the residual amount of citral (**10** and **12**) in the dark and after UV light irradiation was 3959 ppb (1510 plus 2449 ppb) and 2038 ppb (1017 plus 1021 ppb), respectively (Table I), the phenylpropanoids **23-29** and the flavonoids **30** and **31** were found to be effective inhibitors of citral deterioration.

The yield of photochemical cyclization products **3-7** *vs.* the absorbance of a variety of additive compounds at 340 nm at pH 3.5 was found to have a highly negative correlation (R = –0.923) (Table III). Figure 5 shows the effect of each additive compound against the yield of photochemical cyclization products **3-7** at pH 3.5. However, the correlation of oxidation products **16** and **17** (R = –0.257) with the absorbance for a variety of additive compounds at 340 nm at pH 3.5 was lower than that of photochemical cyclization products **3-7** (R = –0.923) (Table III). Interestingly, the relationship between the yield of oxidation products **16** and **17** and DPPH radical scavenging activities of a variety of additive compounds at pH 3.5 was found to have a negative correlation (R = –0.590) (Table III and Figure 6). There appears to be the possibility of participation by the radical species during the oxidation step.

In order to clarify the formation pathway of oxidation products **16** and **17** from citral (**10** and **12**), the effect of UV light, and AAPH, that is, radical initiator, under acidic or neutral conditions was studied (Table IV). The oxidation products **16** and **17** also increased using AAPH instead of UV light under acidic conditions. However, under neutral conditions, epoxides **40** and **41**, the intermediates, were obtained instead of **16** and **17** (*28-29*). The epoxides **40** and **41** were converted into **16** and **17** by the furan-ring formation under acidic conditions. Therefore, as shown in Figure 7, there seemed to be the alternative pathway via epoxidation, hydration, and cyclization compared with Grein's report (*4*).

As for the acid-catalyzed cyclization products **2**, **13-15**, **19** and **21**, there was no correlation with the absorbance of UV light and the DPPH radical scavenging activities of a variety of additive compounds (Table III).

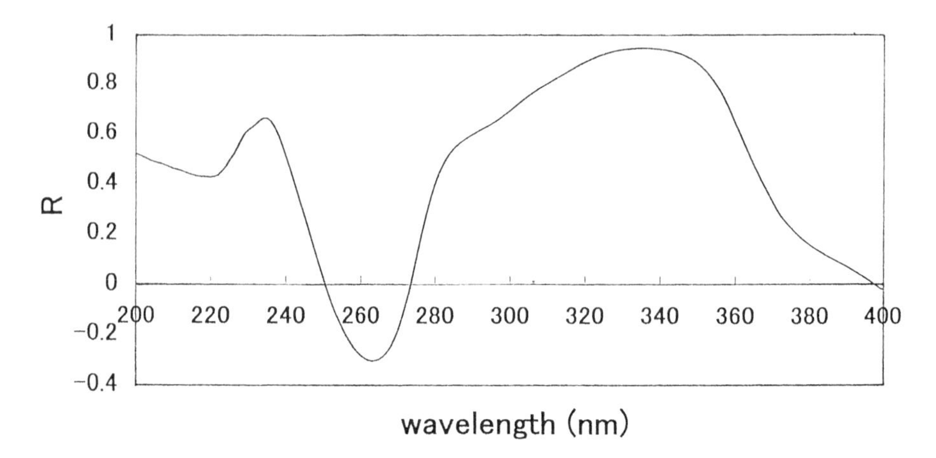

Figure 3. *The correlation factor (R) between the residual amount of citral (**10** and **12**) and UV absorbance of each additive compound* vs. *the wavelength at pH 3.5.*

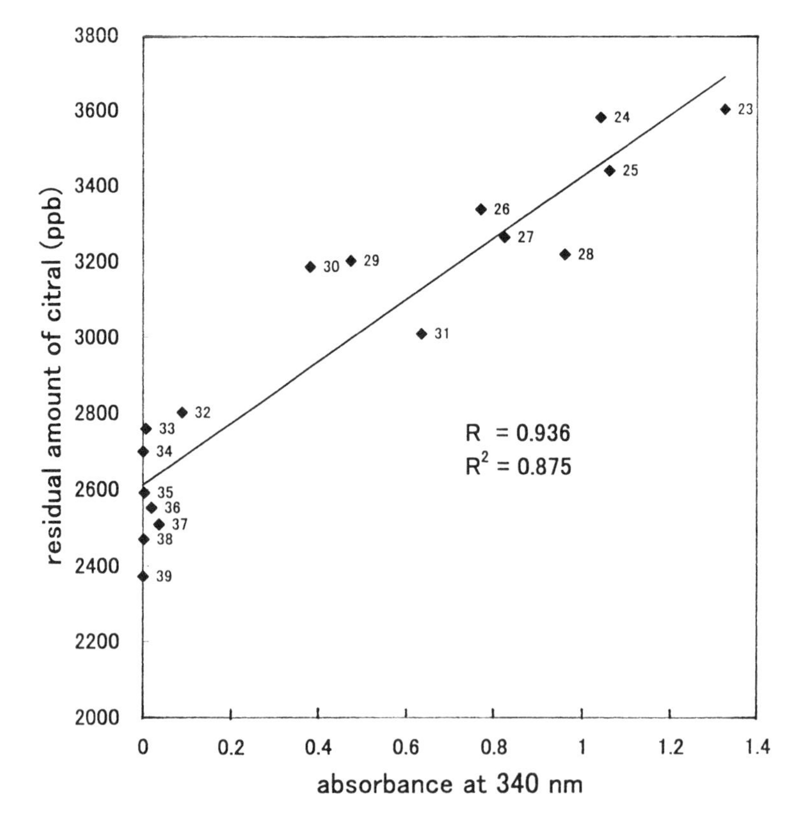

Figure 4. *The residual amount of citral (**10** and **12**)* vs. *the absorbance of each additive compound at 340 nm at pH 3.5.*

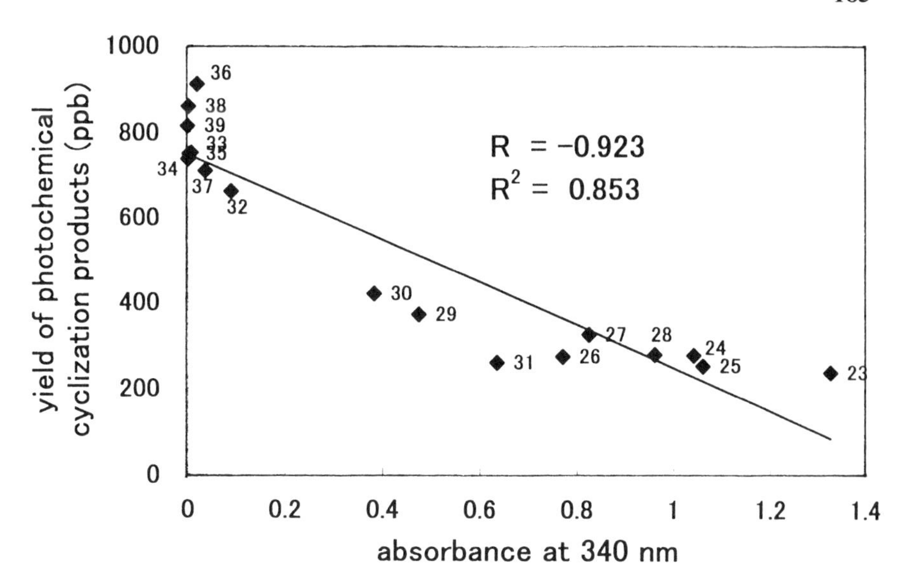

Figure 5. *The yield of photochemical cyclization products from citral (**10** and **12**)* vs. *the absorbance of each additive compound at 340 nm at pH 3.5.*

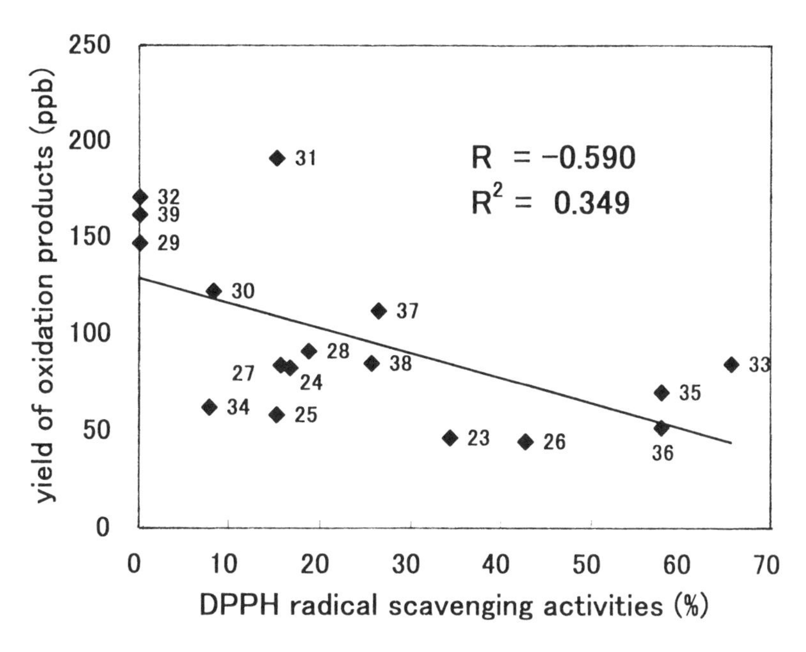

Figure 6. *The yield of oxidation products from citral (**10** and **12**)* vs. *DPPH radical scavenging activity of each additive compound at pH 3.5.*

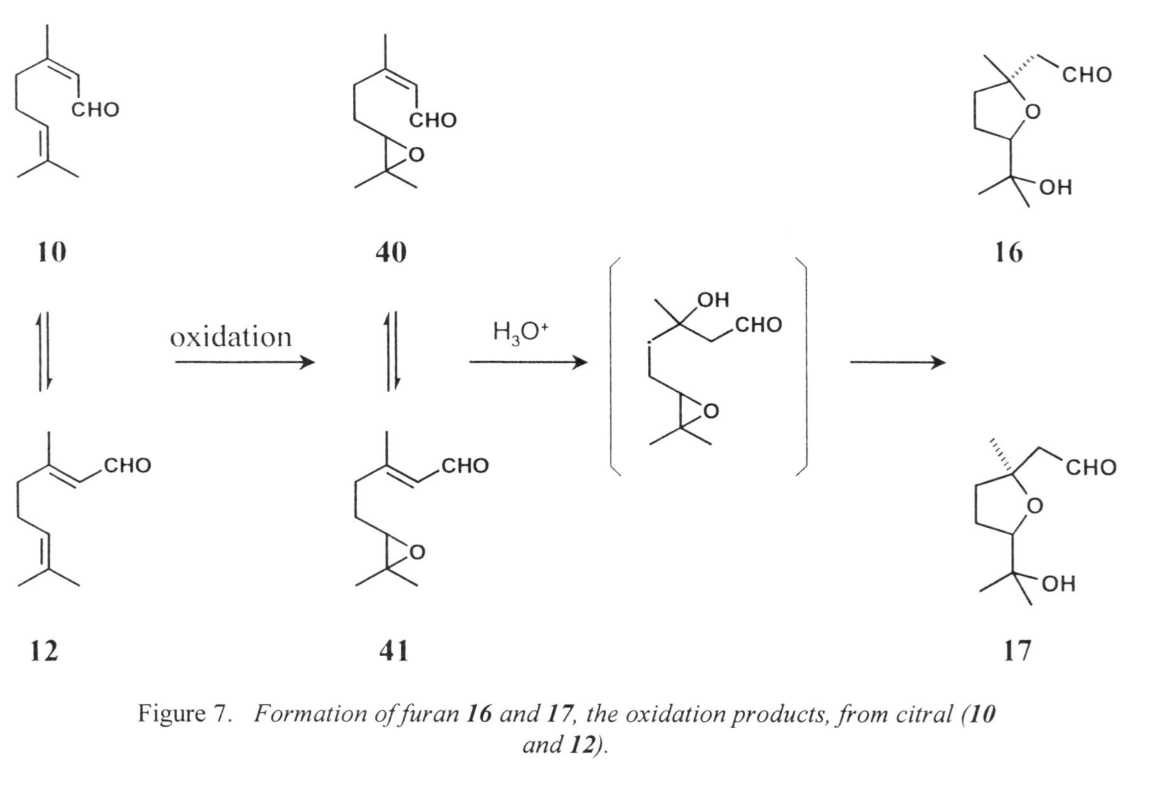

Figure 7. *Formation of furan **16** and **17**, the oxidation products, from citral (**10** and **12**).*

Table II. Concentration[a)] and Relative Amount[b)] of Products Formed from Citral (10 and 12) in the Dark or during UV Light Irradiation[c)] at pH 3.5

Products	*dark*	*UV light*
photochemical cyclization products	0 (0)	568 (38)
oxidation products	33 (6)	409 (27)
acid-catalyzed cyclization products	466 (83)	426 (28)
others	60 (11)	108 (7)
total	559 (100)	1511 (100)

[a)] ppb. [b)] % (in parenthesis). [c)] A period of 14 days.

Table III. Correlation Factor (R) Related to the UV Absorbance and the DPPH Scavenging Activities of Additive Compounds

Products	*RUV*[a)]	*RDP*[b)]
citral (**10** and **12**)	0.936	-0.169
photochemical cyclization products	-0.923	0.288
oxidation products	-0.257	-0.590
acid-catalyzed cyclization products	0.436	0.133

[a)] Correlation factor (R) between the residual amount of citral (**10** and **12**) or the yield of reaction products, and the absorbance of a variety of additive compounds at 340 nm of UV light. [b)] Correlation factor (R) between the residual amount of citral (**10** and **12**) or the yield of reaction products, and the DPPH radical scavenging activities of a variety of additive compounds.

Table IV. Effect of UV Light[a)] and AAPH[a)] under Acidic or Neutral Conditions on the Formation of Oxidation Products from Citral (10 and 12)

pH	*UV light*[b)]	*AAPH*[c)]	*citral (**10** and **12**)*[d)]	***40** and **41***[e)]	***16** and **17***[e)]
3.5	-	-	5277	-	29
3.5	15000	-	4258	trace[f)]	236
3.5	-	1000	5027	14	339
7.0	-	-	6203	35	-
7.0	15000	-	5078	71	trace[f)]
7.0	-	1000	5415	313	8

[a)] A period of 7 days. [b)] lux. [c)] ppm. [d)] ppb of residual citral (**10** and **12**). [e)] ppb. [f)] less than 1 ppb.

Conclusions

The photochemical cyclization products, the oxidation products, and the acid-catalyzed cyclization products comprised about 90% of the volatile products derived from citral during UV light irradiation under acidic conditions. A highly positive correlation was found between the amount of residual citral and the UV light absorption ability of each additive compound. Therefore, there was a highly negative correlation between the yield of the photochemical cyclization products and the UV light absorption ability of each additive compound. As for the oxidation products, the yield had a relatively negative correlation with the radical scavenging activity of each additive compound. The acid-catalyzed cyclization products seemed to have no relation to the kinds of additive compounds.

Acknowledgements

We are grateful to Mr. Osamu Nishimura and Mr. Kenji Adachi for their helpful advice.

References

1. Freeburg, E. J.; Mistry, B. S.; Reineccius, G. A. *Perfumer & Flavorist* **1994**, *19*, 23-32.
2. Kimura, K.; Nishimura, H.; Iwata, I.; Mizutani, J. *J. Agric. Food Chem.* **1983**, *31*, 801-804.
3. Kimura, K.; Iwata, I.; Nishimura, H. *Agric. Biol. Chem.* **1982**, *46*, 1387-1389.
4. Grein, B.; Schmidt, G.; Full, G.; Winterhalter, P.; Schreier, P. *Flavor and Fragrance J.* **1994**, *9*, 93-98.
5. Clark, Jr. B. C.; Chamblee, T. S., In *Off-flavors in Foods and Beverages*; Charalambous, G., Ed.; Elsevier: Amsterdam, Netherlands, 1992; pp. 229-285.
6. Saleeb, R. K.; Ikenberry, D. A., In *Food Flavors, Ingredients and Composition*; Charalambous, G., Ed.; Elsevier: Amsterdam, Netherlands, 1993; pp. 339-354.
7. Ikenberry, D. A.; Saleeb, R. K., In *Food Flavors, Ingredients and Composition*; Charalambous, G., Ed.; Elsevier: Amsterdam, Netherlands, 1993; pp. 355-369.
8. Schieberle, P.; Ehrmeier, H.; Grosch, W.; *Z. Lebensm Unters Forsch.* **1988**, *187*, 35-39.

9. Schieberle, P.; Grosch, W.; *Z. Lebensm. Unters Forsch.* **1989**, *189*, 26-31.
10. Grosch, W.; Schieberle, P.; In *Flavor Science and Technology*; Martens, M.; Dalen, G. A.; Russwurm Jr. H., Eds.; John Wiley & Sons, Ltd.: NY, USA, **1987**, pp. 119-125.
11. Grosch, W. *Flavor and Fragrance J.*, **1994**, *9*, 147-158.
12. Iwanami, Y.; Tateba, H.; Kodama, N.; Kishino, K. *J. Agric. Food Chem.*, **1997**, *45*, 463-466.
13. Tateba, H.; Iwanami, Y.; Kodama, N.; Kishino, In *Flavor Science -Recent Developments*; Taylor, A. J.; Mottram, D. S., Eds.; The Royal Society of Chemistry: Cambridge, UK, 1996; pp. 82-85.
14. Sawada, M.; Teranishi, K.; Yamada, T. *Nippon Shokuhin Kogyo Gakkaishi* **1999**, *46*, 181-186.
15. Kimura, K.; Nishimura, H.; Iwata, I.; Mizutani, J. *Agric. Biol. Chem.* **1983**, *47*, 1661-1663.
16. Inami, O. *Gekkan Food Chemical* **1997**, 23-27.
17. Washino, T. *Shokuhin to Kagaku* **1995**, 96-102.
18. Hiramoto, T.; Tokoro, K.; Kanisawa, T. In *Flavor Chemistry -30 Years of Progress-*; Taylor, A. J.; Teranishi, R.; Wick, E. L.; Hornstein, I. Eds.; Kluwer Academic/Plenum Publishers, NY, USA, 1999; pp. 107-115.
19. Marshall, M.; Nagy, S., Rouseff, R. In *The Shelf Life of Foods and Beverages*; Charalambous, G., Ed.; Elsevier: Amsterdam, Netherlands, 1986; pp. 237-254.
20. Rouseff, R.; Nagy, S.; Naim, M.; Zahavi, U. In *Off-flavors in Foods and Beverages*; Charalambous, G., Ed.; Elsevier: Amsterdam, Netherlands, 1992; pp. 211-227.
21. Koenig, T.; Gutsche, B.; Hartl, M.; Huebscher, R.; Schreier, P.; Schwab, W. *J. Agric. Food Chem.* **1999**, *47*, 3288-3291.
22. Uchiyama, M.; Suzuki, Y.; Fukuzawa, K. *Yakugakuzasshi* **1968**, *88*, 678-683.
23. Chen, J. H.; Ho, C-T. *J. Agric. Food Chem*, **1997**, *45*, 2374-2378.
24. Peacock, V. E.; Kuneman, D. W. *J. Agric. Food Chem.* **1985**, *33*, 330-335.
25. Clark, B. C.; Powell, C. C.; Radford, T. *Tetrahedron*, **1985**, *33*, 2187-2191.
26. Werkhoff, P.; Guentert, M.; Krammer, G.; Sommer, H.; Kaulen, J. *J. Agric. Food Chem.* **1998**, *46*, 1076-1093.
27. McHale, D.; Laurie, W. A.; Baxter, R. L. In *VII International Congress of Essential Oils*; Japan Flavor and Fragrance Manufactures' Association: Kyoto, Japan, 1977; pp. 250-253.
28. Pfander, H.; Kamber, M.; Battegay-Nussbaumer, Y. *Helv. Chim. Acta* **1980**, *63*, 1367-1376.
29. Nali, M.; Rindone, B.; Tollari, S.; Valletta, L. *Gazzetta Chimica Italiana* **1987**, *117*, 207-212.

Chapter 14

Antioxidant Activity of Tannins and Tannin–Protein Complexes: Assessment In Vitro and In Vivo

Ken M. Riedl[1], Stephane Carando[1], Helaine M. Alessio[2], Mark McCarthy[1], and Ann E. Hagerman[1,*]

[1]Departments of Chemistry & Biochemistry and [2]Physical Education, Health and Sport Studies, Miami University, Oxford, OH 45056

The common food constitutents known as tannins are high molecular weight phenolic compounds comprised of gallic acid esters or flavan-3-ol polymers. Although tannins exhibit antioxidant activity in simple chemical assays, previous studies have not established the nutritional relevance of this activity for tannins found in foods such as fruits, wine or tea. We have demonstrated that both hydrolysable and condensed tannins scavenge free radicals in a kinetically complex fashion involving both a fast and a slow scavenging step. Structure/activity studies for monomeric and polymeric phenolic compounds showed that 4 moles of radical were scavenged per *ortho*-substituted diphenol group. The fast scavenging reaction was inhibited by complexation of the tannin to protein, but the overall capacity of the tannin-protein complex for scavenging was similar to that of the free tannin. Protein-tannin complexes were converted to a refractory form during reaction with free radicals, presumably because of oxidation and covalent bond formation. This study is the first to demonstrate antioxidant activity of tannins bound to proteins, a condition relevant to dietary ingestion. Tannin-protein complexes in the gastrointestinal tract may provide persistent antioxidant activity. We evaluated the ability of the tannin epigallocatechin gallate, found in green tea, to serve as an effective antioxidant in a 6.5 week feeding trial. At the end of the experiment, the rats were sacrificed at rest or after a bout of acute exercise, and the levels of antioxidants and oxidized products were determined in various tissues. In the control animals, acute exercise increased the oxidative stress in kidney and liver. In the animals that consumed tea, neither organ exhibited oxidative stress. Ingestion of tea may provide effective antioxidant protection, which over a long term may decrease incidence of oxidative damage to various tissues.

Consumption of diets rich in fruits, vegetables and grains is correlated with a reduced risk of diseases such as cancer and heart disease in humans *(1),(2)*. The benefit of plant foods is not solely due to the levels of vitamins or other nutritive factors they contribute to the diet, but may be due to activities of non-nutritive factors found in many plants. For example, many plant secondary components are antioxidants, and it has been hypothesized that antioxidant activity underlies the stated dietary correlation *(3)*.

The ability of an antioxidant to effectively scavenge free radicals depends on the chemistry of both the initiating radical (superoxide, hydroxyl, alkoxyl, or peroxyl radical) and the antioxidant (e.g. GSH, ascorbate, tocopherol). In addition to having a high scavenging rate for the target radical, an effective antioxidant forms a radical species that is more stable than the initiating radical. If the antioxidant is further oxidized or rearranged to a non-radical species, that species should itself be harmless. The phase-solubility of an antioxidant is an important determinant of the scavenger's ability to scavenge lipid, cytosolic, or protein-bound radical species *(4)*. The unique set of attributes of an antioxidant can potentially lead to a unique role for the antioxidant in the plant that produces it, in the animal that consumes it, or in the processed foods that contain it.

Phenolic Antioxidants

Phenolic compounds are ubiquitous components of fruits, vegetables, and grains. The term phenolic is used in reference to many distinct classes of compounds including low molecular weight flavonoids, phenolic esters and acids, as well as high molecular weight polyphenols known as tannins *(5)*. There are two main classes of tannins in higher plants, hydrolysable (polygalloyl-esterified glucose) and condensed (polymers of flavan-3-ols). Tannins are historically noted for their ability to bind and precipitate proteins as well as to chelate transition metals *(6)*. *Ortho*-substituted diphenolics can donate up to four electrons to reduce oxidizing radicals. Phenolic radicals are relatively stable because of the delocalizing effect of the aromatic ring *(7)*.

The antioxidant activities of phenolic natural products have been recognized for some time. However, an understanding of the contribution that dietary polyphenolic compounds might make to human health has not been achieved. Tannins comprise a significant fraction of the natural products consumed in the human diet, with an estimated intake of more than one g per day *(8)*. Thus the potential contribution of these compounds to physiological antioxidant capabilities is substantial.

Evidence for the antioxidant activity of polyphenolic compounds has largely focused on structure-activity trends for low molecular weight flavonoids. For instance, monomeric flavonoids containing a catechol or galloyl group and a double bond between carbons 2 and 3 of the heterocycle exhibit the most potent antioxidant activity *(9)*. In contrast to monomeric species, polymeric flavonoids (condensed tannins) have remained largely unexplored although their polymeric nature gives rise to unique properties. For example, the numerous phenolic groups per molecule allow tannins to scavenge radicals at extremely high rates when compared to monomeric flavonoids *(10,11)*. Likewise, the multiple phenolic groups result in enhanced scavenging capacity (Riedl and Hagerman, unpublished results). Size and degree of galloylation affect the phase-solubility of tannins *(12)* and may thus determine their efficacy in lipid or aqueous environments *(13)*. Finally, the most obvious difference between simple flavonoids and tannins is that tannins bind to proteins, with especially high affinity for proline-rich proteins such as gelatin or salivary proline-rich proteins (PRPs) *(6)*.

Dietary Tannins

The protein-bound state of tannin is particularly relevant to the alimentary canal and food matrix in which there is an abundance of protein. Some tannins, such as condensed tannins (proanthocyanidins), are not absorbed from the gastrointestinal tract *(14)*, and may be complexed to salivary PRPs during their passage through the alimentary system *(15)*. Other tannins such as epigallocatechin gallate (EGCG) are absorbed after ingestion (*16*) but are presumably bound to proteins such as serum albumin (Carando and Hagerman, unpublished results) during lifetime in the blood plasma. Chemically labile tannins such as gallotannins may release small, easily absorbed phenolics that may remain protein-bound during transport to various organs (*17*). Realistically the antioxidant activity of tannin should be studied in a protein-bound state. Thus one primary goal of our work is to elucidate the effect of protein binding on tannin antioxidant activity. We hypothesized that protein binding would affect the reactivity of tannin and would alter the products of scavenging.

Assessment of Tannins as Antioxidants in vitro

Finding an appropriate assay for tannin antioxidant activity has been challenging, primarily due to the propensity of tannin to bind or react with common reagents used in assays. Many antioxidant assays employ an enzyme-catalyzed reaction to generate radicals or use an oxidation-sensitive protein to detect stress. The inherent limitation of these methods is that tannin may directly interact with the radical generator or sensor in addition to scavenging the radicals themselves. Both modes of inhibition can be viewed as antioxidant behavior but the contribution due to each cannot be determined in many assays. Since tannins bind strongly to many proteins, protein-based systems (e.g. ORAC (*18*), metmyoglobin assay (*19*)) are particularly inappropriate for studying tannin antioxidant activity. Similarly, tannin can chelate and/or reduce the ferric ion of the radical generator in the deoxyribose assay leading to ambiguous results (*10*). Fortunately, Rice-Evans et al. devised an assay based strictly on chemical generation of an azobis tetra ammonium sulfate (ABTS) radical cation (*20*). The key advantages include separate radical generation and scavenging steps and avoidance of protein or metal reagents. These benefits apply to analysis of all antioxidants, but are particularly useful with respect to tannins because they allow examination of the effect of protein binding on the antioxidant activity of tannins.

Assessment of Tannins as Antioxidants in vivo

In parallel with the in vitro system described above we are also determining the effect of tannin-containing diets on the antioxidant status of animals in vivo. It has been suggested that consumption of green tea is beneficial (*21*), because it contains substantial levels of the flavonoid-based antioxidant EGCG, a relatively low molecular weight polyphenolic compound that is absorbed from the gastrointestinal tract into the blood stream (*16*). We used green tea, which is readily consumed by rats without apparent detrimental effects (*22*), as a convenient method for administering EGCG as part of the normal diet without need for gavage or injection. Thus in our in vivo study, the experimental tannin presumably formed the usual tannin-protein complexes during ingestion and passage through the gastrointestinal tract; and underwent the normal physiological processes of uptake into and clearing from the plasma.

Examining oxidative stress in animal systems depends on identifying sensitive biomarkers, and often includes an experimentally imposed oxidative stress to enhance the response. We oxidatively stressed the experimental groups of animals with acute aerobic exercise immediately before sacrifice (*23*). We measured a broad range of biomarkers of oxidative damage and of antioxidants in various tissues from the sacrificed animals. We hypothesized that ingestion of EGCG via green tea would protect the animals from exercise-induced oxidative stress, and that the protection would be exhibited as a change in the ratio of antioxidant to oxidized molecules (*24*).

$ABTS^{+\cdot}$ Scavenging by Tannin

Antioxidant activity was determined by following the conversion of the ABTS radical cation to the colorless ABTS species (*20*). Both the kinetics of the scavenging reaction and the capacity of the antioxidant were determined with this assay. Phenolics examined included several condensed tannins: the procyanidin (epicatechin$_{16}$(4–>8)catechin) from *Sorghum* grain (*12*); a hexameric procyanidin from willow (*25*); pentamers (*26*) from blackbrush and bitterbush); several hydrolysable tannins (pentagalloyl glucose (*27*); polygalloyl glucose (*28*); and oenothein B (*27*)). The corresponding monomers catechin and methyl gallate were also tested. Effects of the proteins bovine serum albumin or gelatin on reaction of the *Sorghum* procyanidin with ABTS radical cation were evaluated.

Kinetics and Capacity of Quenching

All of the tannins examined exhibited both fast-acting antioxidant activity and slow acting antioxidant activity (Riedl and Hagerman, unpublished results). This behavior contrasts with previously described antioxidants. In general, low molecular weight compounds (tocopherol, ascorbate, Trolox) including small phenolics like methyl gallate are fast-acting antioxidants, quenching pre-formed radicals quickly and exhibiting no further activity. Some proteins have been reported to have slow acting antioxidant activity, reaching their full capacity to scavenge radicals only after a relatively long reaction time. Tannins are unique in having both fast and slow acting antioxidant activity. Furthermore, tannins have an extraordinarily high capacity for quenching free radicals. When the reaction was carried out to completion (about 24 hours), the *Sorghum* procyanidin was able to scavenge 108 moles radical cation per mole of procyanidin at pH 7.4.

Structure-Activity Relationships

Figure 1 illustrates ABTS radical cation scavenging by a series of condensed and hydrolysable tannins and the corresponding phenolic monomers. Scavenging was determined 10 min after mixing the phenolic with the radical cation, so these values represent the fast and part of the slow radical scavenging but do not represent the total capacity of the phenolics to scavenge radicals. The ability of the polyphenolic compounds to scavenge ABTS radical during the initial 10 min of reaction ranged from 4 moles radical cation per mole of polyphenolic compound (catechin, methyl gallate) to 73 moles radical cation per mole of polyphenolic compound (*Sorghum* procyanidin).

There is a strong relationship between the number of *ortho*-substituted phenolic rings on the polyphenolic compound and its ability to scavenge the radical (Figure 1). The *Sorghum* procyanidin, with 17 *ortho*-substituted phenolic rings, scavenges far more radical than the other procyanidins (sandbar willow, 5.8 *ortho*-rings; blackbush, 4.8 *ortho*-rings; bitterbrush, 4.5 *ortho*-rings; catechin, 1 *ortho*-ring). The hydrolysable tannins follow the same relationship, with oenothein B (8 *ortho*-substituted rings) and polygalloyl glucose (7.6 *ortho*-rings) scavenging similar amounts of radical. Pentagalloyl glucose, with 5 *ortho*-substituted rings scavenges five times more radical than the monomer methyl gallate.

There is a strong linear correlation (r^2 = 0.96) between the ability of the polyphenolic compounds to scavenge ABTS radical cation and the number of *ortho*-substituted phenolic rings in the polyphenolic compound. The slope of the linear fit indicates that the polyphenolic compounds scavenge 4.3 moles radical cation per mole of *ortho*-substituted phenolic ring. This corresponds well to the expected 4 moles of electrons donated by an *ortho*-substituted phenolic.

Protein-Tannin Complexes

Binding to protein significantly inhibited the fast step of the antioxidant activity for the *Sorghum* procyanidin (Riedl and Hagerman, unpublished results.) For example, interaction between bovine serum albumin and procyanidin has a sharp pH optimum, with maximal interaction at pH 4.9 and very little interaction at higher (pH 7.4) or lower (pH 3) pH (*29*). The fast phase of ABTS radical cation quenching by procyanidin was inhibited significantly by bovine serum albumin at pH 4.9 but not at pH 7.4 or pH 3. Similarly, inhibition of radical scavenging was maximum at procyanidin:protein ratios that optimized the interaction between the tannin and bovine serum albumin (*12*). Gelatin, a PRP

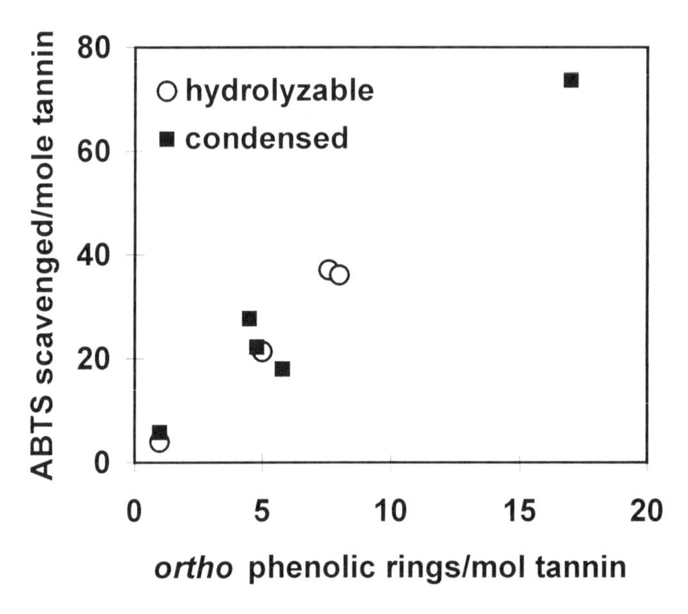

Figure 1. ABTS radical cation scavenging by condensed and hydrolysable tannins as a function of the number of ortho-substituted rings in each tannin. Tannins (0.5-7 nmol) were added to 55 nmol of colored radical cation and the loss of color was monitored at 734 nm. The amount of radical remaining at 10 min was calculated based on the extinction coefficient of the radical cation (12867 M^{-1} cm^{-1}) in order to determine moles of radical cation scavenged per mole of polyphenolic compound.

with high affinity for procyanidin (*30*), inhibited the rapid scavenging reaction more effectively than bovine serum albumin. The inhibition by gelatin paralleled the pH and ratio trends dictating complexation, with maximal inhibition at slightly acidic to neutral pH values where precipitation is maximal (Figure 2).

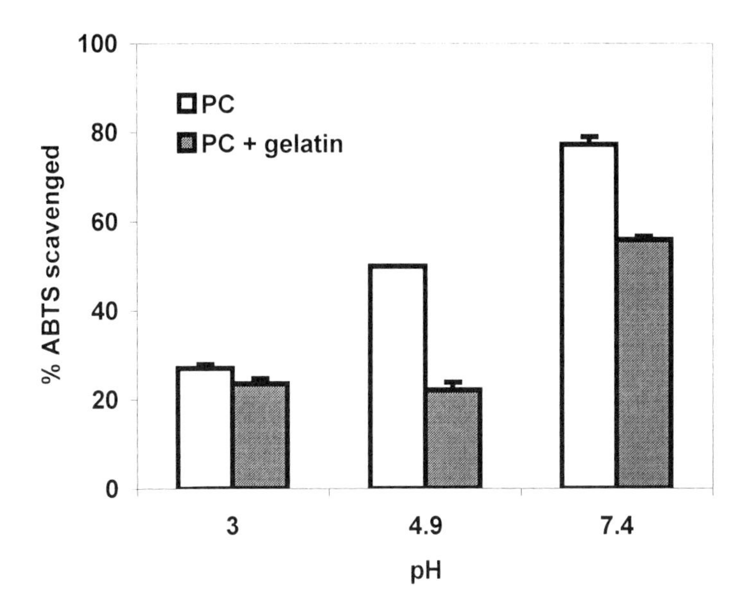

Figure 2. ABTS radical cation scavenging as a function of pH by either procyanidin or by procyanidin-gelatin complexes. Procyanidin (0.8 nmol) with or without 0.09 nmol gelatin was mixed with 55 nmol of the colored radical cation and the loss of color was immediately determined. Bars show the average of 3 determinations; standard deviations are indicated.

Both bovine serum albumin and gelatin inhibited the fast phase of scavenging by the tannin but neither protein affected the long-term capacity of the polyphenolic compound as a radical scavenger. Inhibition by the protein was transient and completely overcome during the hours-long slow reaction. Participation of the protein in the radical scavenging process was clearly demonstrated by the conversion of the tannin-protein complex to a recalcitrant, unreactive form during the scavenging reaction (Figure 3).

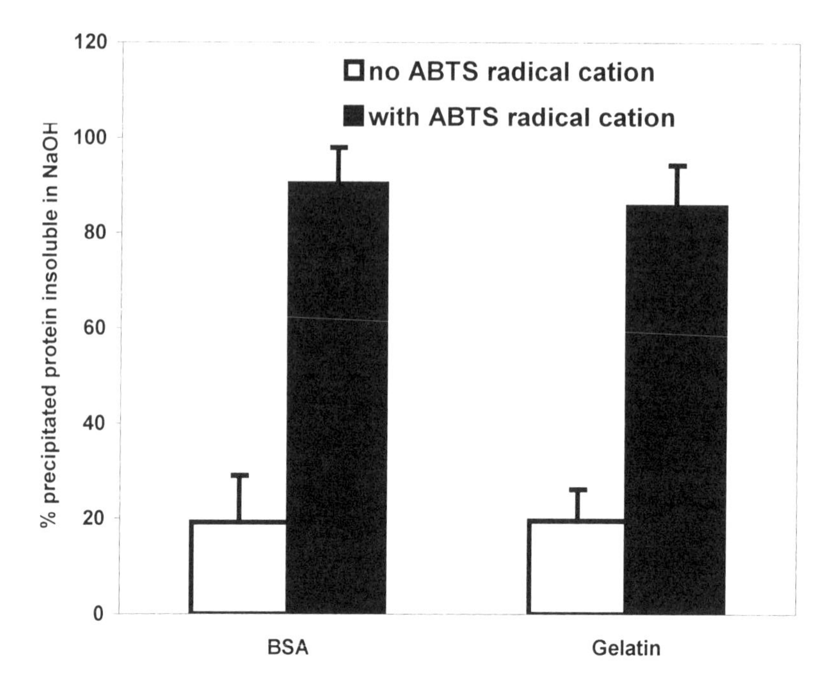

Figure 3. Procyanidin-protein complexes formed in the absence of radicals are readily reversed by treatment with protein denaturants such as detergent, base, or urea. Complexes exposed to ABTS radical cation reacted with the radical, converting it to the colorless ABTS molecule, and converting the procyanidin-protein complex to a form that could not be reversed by any treatment. Protein in the absence of procyanidin did not react with the radical cation. Values are the averages of three determinations; standard deviations are iindicated.

Covalent bond formation between the oxidized polyphenolic compound and the protein, or extensive oxidation of the protein-procyanidin complex during reaction with the radical, are the most likely mechanisms for this reaction. We are pursuing characterization of the oxidized complex at the molecular level.

Tannin as an Antioxidant in vivo

In the in vivo study, young rats were maintained for 6.5 weeks on diets of rat chow and either water or green tea ad lib. The animals were subdivided into groups (n = 6) and were sacrificed either at rest or immediately after a bout of aerobic exercise on an animal treadmill. Plasma and tissues (liver, heart, kidney, red and white muscle) were collected from the animals and were immediately frozen in liquid nitrogen. Detailed results of the analyses of these samples have been reported elsewhere (*J. Agric. Food Chem.* submitted).

Oxidative Stress and Exercise

Our experiment was designed to compare the oxidative stress imposed by acute exercise on animals consuming the polyphenolic compound, EGCG, with that on animals consuming water. For each tissue, antioxidant status was defined as the ratio of oxidized biomolecules to total antioxidants. Exercise-induced oxidative stress for each tissue was then calculated by comparing the antioxidant status of the exercised animals to the antioxidant status of the sedentary animals. Oxidative stress might result from either depletion of antioxidants or from accumulation of oxidized products; our method of comparison is not dependent on the absolute levels of either component of the oxidative status, but instead on the ratio of components (*24*). We determined oxidative status for each tissue collected from the resting animals, and for each tissue collected from the exercise-stressed animals, by measuring total antioxidants with the ORAC assay standardized with Trolox; although this method did not work well for assessment of purified polyphenolic compounds, it is appropriate for the complex mixture of biomolecules found in animal tissues(*18*). We measured the products of lipid oxidation with the thiobarbituric acid-reactive substances (TBARS) assay standardized with malondialdehyde (*31*).

EGCG Suppresses Oxidative Stress

The animals that consumed water were clearly oxidatively stressed by the exercise treatment (Figure 4), in agreement with many previous studies of exercise (*23*). In particular, the kidney and liver were susceptible to oxidative stress in the animals consuming water, perhaps because restricted blood flow to these organs during exercise impeded the normal supply of antioxidants or limited the normal clearing process (*32, 33*).

The animals consuming tea were well protected from these effects of acute exercise, with no indication of oxidative stress in any tissue (Figure 4). This confirms the hypothesis that consumption of dietary polyphenolic compounds may provide substantial antioxidant protection to animals. The potent antioxidant activity of EGCG appears to be manifested when the compound is ingested by a normal dietary route, and at levels typical of human consumption.

Additional studies on the mechanism of protection and on the types and levels of polyphenolic compounds required for protection are underway in our laboratories.

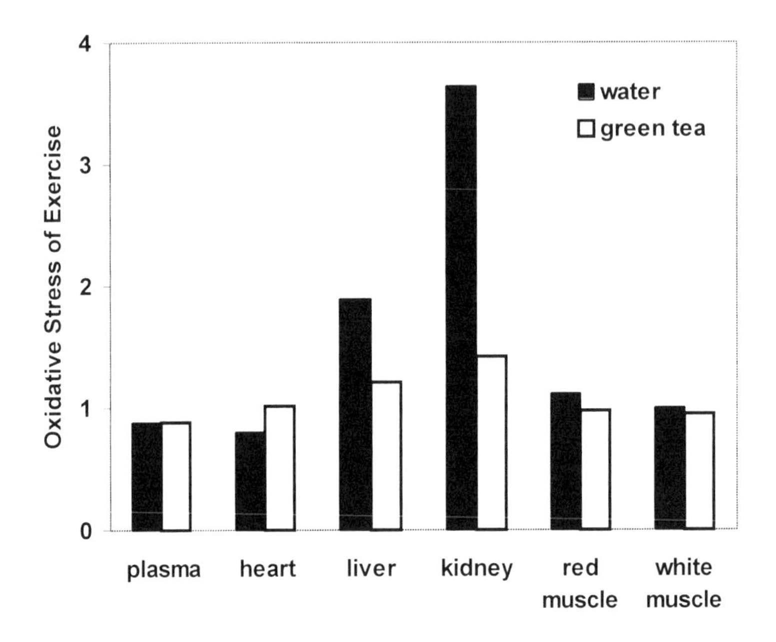

Figure 4. Oxidative stress for various tissues collected from rats in control or green tea diet groups. For each group, 6 rats were maintained on the indicated diet for 6.5 weeks. The animals were sacrificed at rest or after acute aerobic exercise. Antioxidants and products of lipid oxidation were determined for the indicated tissues, and antioxidant status was calculated as the ratio of oxidized products to antioxidants. For each tissue, oxidative stress was calculated by comparing antioxidant status for the exercised animals to antioxidant status for resting animals.

References

1. Hansson, L. E.; Nyren, O.; Bergstrom, R.; Wolk, A.; Lindgren, A.; Baron, J.; Adami, H. O. *Int. J. Cancer* **1994,** *57*, 638-644.
2. Appel L.J.; Moore T.J.; Obarzanek E.; Vollmet W.M.; Svetkey L.P.;

Sacks F.M.; Bray G.A.; Vogt T.M.; Cutler J.A..; Windhauser M.M.; Lin P.H.; Karanja N. *New Eng. J. Med.* **1997,** *336*, 1117-1124.
3. Huang, M. T.; Ferraro, T.; Ho, C. T. In *Food Phytochemicals for Cancer Prevention I. Fruits and Vegetables;* Huang, M. T.; Osawa, T.; Ho, C. T.Rosen, R. T., Eds.; American Chemical Society: Washington, DC, 1994; pp 2-16.
4. Jacob, R. A.; Burri, B. J. *Am. J. Clin. Nutr.* **1996,** *63*, 985S-990S.
5. Haslam, E. *Plant polyphenols. Vegetable tannins revisited*; Cambridge University Press: Cambridge, U.K., 1989.
6. Hagerman, A. E.; Carlson, D. M. *Recent Research Developments in Agricultural and Food Chemistry* **1998,** *2*, 689-704.
7. Simic, M. G.; Jovanovic, S. V. In *Food Phytochemicals for Cancer Prevention II*, Ho, C. T.; Osawa, T.; Huang, M. T.Rosen, R. T., Eds.; American Chemical Society: Washington, DC, 1994; pp 20-33.
8. Pierpoint, W. S. In *Flavonoids in biology and medicine III. Current issues in flavonoids research*, Das, N. P., Ed.; National University of Singapore: Singapore, 1990; pp 497-514.
9. Rice-Evans, C. A.; Miller, N. J.; Paganga, G. *Free Rad. Biol.Med.* **1996,** *20*, 933-956.
10. Hagerman, A. E.; Riedl, K. M.; Jones, G. A.; Sovik, K. N.; Ritchard, N. T.; Hartzfeld, P. W.; Riechel, T. L. *J. Agric. Food Chem.* **1998,** *46*, 1887-1892.
11. Bors, W.; Michel, C. *Free Rad. Biol.Med.* **1999,** *27*, 1413-1426.
12. Hagerman, A. E.; Rice, M. E.; Ritchard, N. T. *J. Agric. Food Chem.* **1998,** *46*, 2590-2595.
13. Plumb, G. W.; De Pascual-Teresa, S.; Santos-Buelga, C.; Cheynier, V.; Williamson, G. *Free Rad. Res.* **1998,** *29*, 351-358.
14. Jimenez-Ramsey, L. M.; Rogler, J. C.; Housley, T. L.; Butler, L. G.; Elkin, R. G. *J. Agric. Food Chem.* **1994,** *42*, 963-967.
15. Lu, Y.; Bennick, A. *Arch. Oral Biol.* **1998,** *43*, 717-728.
16. Maiani, G.; Salucci, M.; Azzini, l. E.; Ferro-Luzzi, A. *J. Chromatog.B* **1997,** *692*, 311-317.
17. Scheline, R. R. *Mammalian Metabolism of Plant Xenobiotics*; Academic Press: New York, NY, 1978.
18. Cao G.; Alessio Helaine M.; Cutler R.G. *Free Rad. Biol. Med.* **1993,** *14*, 303-311.
19. Miller, N. J.; Rice-Evans, C. A.; Davies, M. J.; Gopinathan, V.; Milner, A. *Clin. Sci.* **1993,** *84*, 407-412.
20. Re, R.; Pellegrini, N.; Proteggente, A.; Pannala, A.; Yang, M.; Rice-Evans, C. *Free Rad. Biol. Med.* **1999,** *26*, 1231-1237.
21. Trevisanato S.I.; Kim Y-I. *Nut. Rev.* **2000,** *58*, 1-10.
22. Deng, Z. Y.; Tao, B. Y.; Li, X. L.; Chen, Y. F. *J.Agric. Food Chem.*

1998, *46*, 3875-3878.
23. Sen, C. K. *J.Appl. Physiol.*. **1996,** *79*, 675-686.
24. Alessio, H. M.; Goldfarb, A. H.; Cao, G. *International Journal of Sport Nutrition* **1997,** *7*, 1-9.
25. Schofield, J. A.; Hagerman, A. E.; Harold, A. *J. Chem. Ecol.* **1998,** *24*, 1409-1421.
26. Clausen, T. P.; Provenza, F. D.; Burritt, E. A.; Reichardt, P. B.; Bryant, J. P. *J. Chem. Ecol.* **1990,** *16*, 2381-92.
27. Hagerman, A. E.; Zhao, Y.; Johnson, S. In *Antinutrients and Phytochemicals in Foods*, Shahadi, F., Ed.; American Chemical Society: Washington, DC, 1997; pp 209-222 .
28. Hagerman, A. E.; Klucher, K. M. In *Plant Flavonoids in Biology and Medicine: Biochemical, Pharmacological and Structure Activity Relationships*, Cody, V.; Middleton, E.; Harborne, J., Eds. Alan R. Liss, Inc.: New York, NY 1986; pp 67-76.
29. Hagerman, A. E.; Butler, L. G. *J. Agric. Food Chem.* **1978,** *26*, 809-812.
30. Hagerman, A. E.; Butler, L. G. *J. Biol. Chem.* **1981,** *256*, 4494-4497.
31. Ohkawa H.; Ohishi N.; Yagi K. *Anal. Biochem.* **1979,** *95*, 351-358.
32. Radak , Z.; Asano, K.; Inoue, M.; Kizaki, T.; Oh-Ishi, S.; Suzuki, K.; Tahniguchi, N.; Ohno, H. *Eur. J. Appl. Physiol.* **1996,** *72*, 189-194.
33. Draper, H. H.; Polensek, L.; Hadley, M.; McGirr, L. G. *Lipids 19*, 836-843.

Chapter 15

Effect of Roasting Process on the Antioxidant Properties of *Cassia tora* L.

Gow-Chin Yen, Da-Yon Chuang, and Chi-Hao Wu

Department of Food Science, National Chung Hsing University, 250 Kuokuang Road, Taichung 40227, Taiwan

The antioxidant properties of water extracts from *Cassia tora* L. prepared under different degrees of roasting were investigated. The water extracts of *Cassia tora* showed inhibition of peroxidation of linoleic acid, which was higher than that of α-tocopherol. It exhibited good antioxidant activity in liposome oxidation induced by Fenton reaction as well as in enzymatic microsome oxidative systems. It also inhibited the oxidative DNA damage in human lymphocytes, induced by hydrogen peroxide as evaluated by the comet assay. However, the antioxidant activities of water extracts of *Cassia tora* decreased with higher roasting temperature or longer roasting time. The water extracts of roasted *Cassia tora* increased in the degree of browning and produced chemiluminescence when compared with the unroasted sample. The total polyphenolic compounds and anthraquinones of extracts of *Cassia tora* decreased after the roasting process finished. In conclusion, the decrease in the antioxidant activity of water extracts from roasted *Cassia tora* might have been due to the decrease of polyphenolic compounds and the degradation of Maillard reaction products.

Several epidemiological studies have shown an association between individuals who have a diet rich in fresh fruit and vegetables and a decreased risk of cardiovascular diseases and certain forms of cancer (*1,2*). It is generally assumed that the dietary elements responsible for these protective effects are antioxidant nutrients. However, more recent works have highlighted the additional roles of the polyphenolic components of the higher plants (*3*), which may act as antioxidants or agents of other mechanisms that contribute to their anticarcinogenic or cardioprotective actions. Hertog and Hollman (*4*) have suggested that diets rich in phenolic compounds are associated with longer life expectancy. These compounds have also been found to exhibit many health-related properties because of their antioxidant activities. These properties include anticancer, antiviral, anti-inflammatory activities, effects on capillary fragility, and an ability to inhibit human Platte aggregation (*5*).

The chinese herb "Jue-ming-zi", which is the seed of the plant *Cassia tora* L. (Leguminosae), has been used as a laxative and a tonic for several centuries. Traditionally, Jue ming zi has been used to improve visual acuity and used to remove "heat" from the liver. Modern physicians use this herb to treat hypercholesterolemia and hypertension. This herb has been reported to contain many active substances, including chrysophenol, emodin, rhein, etc. (*6*). Choi et al. (*7*) reported that *Cassia tora* exhibited antimutagenic activity. Wu and Yen (*8*) also indicated that the antimutagenicity of *extracts of Cassia tora* was due to a desmutagenic action, but not a biomutagenic action. Su (*9*) and Kim (*10*) reported that methanolic extracts from *Cassia tora* L. exhibited a strong antioxidant activity on lipid peroxidation. Yen et al. (*11*) also indicated that the antioxidant activity of methanolic extracts from *Cassia tora* L. was stronger than that of *Cassia occidentalis* L., and they also identified an antioxidative compound as 1,3,8-trihydroxy-6-methyl-9,10-anthracenedione (emodin) from *Cassia tora*.

The commercial products of *Cassia tora* include both unroasted and roasted samples, and the laxative effect was found to be higher in unroasted *Cassia tora* than in the roasted product. Roasted *Cassia tora* has a special flavor and color, and it is popularly used to make a health drink. Zhang et al. (*12*) reported that some components, for example, chrysophanol, in *Cassia tora* decreased after the roasting process. Nicoli et al. (*13*) found that medium-dark roasted coffee brews had the highest antioxidant properties due to the development of Maillard reaction products. In view of this, the antioxidant activity of *Cassia tora* might also be influenced by the roasting treatment. However, there are no standard conditions for roasting *Cassia tora* and the data concerning the effect of different degrees of roasting on the antioxidant properties of *Cassia tora* are not available. In this paper, the effects of roasting conditions on the antioxidant properties, the change in active compounds of water extracts from *Cassia tora*, and the relationship between antioxidant activity and degree of roasting of *Cassia tora* are presented.

Preparation of *Cassia tora* with Different Degrees of Roasting

To obtain *Cassia tora* L. seeds under different degrees of roasting, washed and sun-dried samples (600 g) were roasted at 150 ℃, 175 ℃, 200 ℃, and 250 ℃ (internal temperature) for 5 min using a roasting machine. The degree of roasting was classified as minimum roasting (150 ℃), medium roasting (175 ℃), commercial roasting (200 ℃), and over roasting (250 ℃). Since the product would become dark when roasted at 200 ℃ for a long time, the effect of roasting time on the antioxidant activity of *Cassia tora* was studied at 175 ℃ for different periods. Water extracts from *Cassia tora* were prepared using a modified method of Yen and Chen (*14*). Each unroasted and roasted sample (20 g) was extracted with boiling water (200 mL) for 10 min. The extracts were filtered through Whatman no. 1 filter paper, and the filtrates were freeze-dried.

Evaluation of Antioxidant Activity of *Cassia tora* With Roasting

Linoleic Acid Peroxidation System

The antioxidant activity of extracts of *Cassia tora* in the linoleic acid peroxidation system was evaluated by the thiocyanate method. The extracts of unroasted *Cassia tora* exhibited good antioxidant activity in the linoleic acid peroxidation system. At a concentration of 0.2 mg/mL, the extracts of unroasted *Cassia tora* inhibited 94% peroxidation of linoleic acid after incubation for 72 h. This percentage was higher than that of α-tocopherol (82%) and ascorbic acid (41%) but lower than that of BHA (98%) and Trolox (99%). Traditionally, the commercial product of *Cassia tora* has been roasted in order to enhance the flavor during infusion. Therefore, to evaluate the effect of different degrees of roasting on the antioxidant activity, the *Cassia tora* was roasted at different roasting temperatures. The inhibition of peroxidation of linoleic acid were in the order of unroasted (94%) = roasted at 150 ℃ (92%) > commercial product (88%) > roasted at 175 ℃ (85%) > roasted at 200 ℃ (82%) > roasted at 250 ℃ (79%). The antioxidant activity of extracts of *Cassia tora* was significantly ($P < 0.05$) decreased with an increase in roasting temperature, except for the unroasted and those roasted at 150 ℃.

There were no great differences ($P > 0.05$) in antioxidant activity when the *Cassia tora* was roasted at 175 ℃ for 1 to 10 min, but there was a significant decrease ($P < 0.05$) when it was roasted for 80 min. Nicoli et al. (*13*) evaluated the effect of the degree of roasting on the antioxidant activity of coffee brews. They found that the antioxidant activity of coffee brews increased with roasting time from 0 to 10 min and then gradually decreased after 10 min of roasting. They also indicated that the increase in antioxidant activity of coffee brews under minimum roasting was due to the formation of Maillard reaction products and their content of phenolic compounds. However, the phenolic compounds in coffee were degraded after over-roasting and caused the decrease in antioxidant activity. These results of Nicoli et al. (*13*) are different from those in our study in antioxidant activity under minimum roasting. The reason might be due to the difference in composition and degree of browning after roasting between coffee and *Cassia tora*.

Liposomes Peroxidation System

The ability of *Cassia tora* extracts to inhibit peroxidation was evaluated in liposomes; oxidation was induced by Fe^{3+} /H_2O_2/ascorbic acid. The samples of unroasted *Cassia tora* and those roasted at 150 ℃ had better antioxidant activity. They exhibited about 40% inhibition effect at a concentration of 0.05 mg/mL, then increased to 56% at a concentration of 1 mg/mL that was higher than that of α-tocopherol (47%). Although the antioxidant activity of extracts of *Cassia tora* roasted at 200 and 250 ℃ was weaker than that of the unroasted, those roasted at 150 ℃ and α-tocopherol, it was higher than that of EDTA and mannitol. The antioxidant activity of all the test samples was in the order of unroasted = roasted at 150 ℃ > α-tocopherol > roasted at 200 ℃ > roasted at 250 ℃ > EDTA > mannitol at a concentration of 0.5 mg/mL. This result is similar to the trend of inhibition of peroxidation of linoleic acid.

Halliwell and Gutteridge (*15*) indicated that free radical scavengers do not inhibit the peroxidation of some membrane lipids (liposome or microsome) induced by Fenton reaction. Damage to the membrane might result from the "site-specific" effect caused by direct binding of iron ion with the membrane. In our study, mannitol showed no inhibition effect while EDTA showed a weak scavenging effect because it cheleted the iron ion on the membrane. Saija et al. (*16*) demonstrated that quercetin has better antioxidant activity than rutin in a biomembrane system. The reason may be that quercetin rather than rutin interacts with that bilayer membrane of phospholipid. Therefore, the antioxidant activity of antioxidants in a membrane system depends on their ability not only to donate a hydrogen atom but also to incorporate into the membrane. α-Tocopherol scavenges the peroxyl radicals formed from the lipid peroxidation in the inner membrane (*17*). Therefore, it can be predicted that there are some water-insoluble components in the unroasted samples and those roasted at 150 ℃, that can be incorporated into the membrane to afford antioxidant activity.

Enzymatic Lipid Peroxidation of Microsomes

The effects of extracts from *Cassia tora* prepared under different degrees of roasting on enzymatic lipid peroxidation of rat liver microsomes induced by NADH/ADP/Fe^{3+} were also evaluated. The antioxidant activity of all the samples of extracts from *Cassia tora* increased in a concentration dependent manner. The inhibitory effects of all the samples were in the order of unroasted (97.5%) = roasted at 150 ℃ (97.2%) > roasted at 200 ℃ (87.6%) > roasted at 250 ℃ (54.4%) at a concentration of 0.2 mg/mL. Trolox exhibited an inhibitory effect of over 90% at a concentration lower than 0.05 mg/mL, but it showed a similar inhibitory effect at a concentration greater than 0.05 mg/mL. The antioxidant activity of extracts from unroasted *Cassia tora* and those roasted at 150 ℃ was equal to that of Trolox at a concentration greater than 0.2 mg/mL. Malterud et al. (*18*) indicated that anthraquinones and anthrone inhibit the formation of thiobarbituric reactive substances in rat liver cell induced by *t*-butyl hydroperoxide. Since anthraquinones are major pharmacological compounds of *Cassia tora* (*19*), they may have an antioxidant effect in this microsomes peroxidation system.

Hydrogen Peroxide-Induced Oxidative DNA Damage in Human Lymphocytes

Single-cell gel electrophoresis (The comet assay) is a rapid and sensitive method for the detection of DNA damage in individual cells and specially for detecting oxidative DNA strand breaks (*20*). In this assay, under alkaline conditions, DNA loops containing breaks lose supercoiling, unwind, and are released from the nucleus forming a "comet tail" after gel electrophoresis. DNA strand breaks are thus visualized by the Comet assay and computer analysis or by visual grading. Recently, it also has been used to detect the effect of dietary components on DNA damage of Mammalian cells (*21*). The effect of extracts from *Cassia tora* on DNA oxidative damage of human lymphocytes induced by hydrogen peroxide was evaluated by single cell electrophoresis. Table **I** shows that hydrogen peroxide (50 μM) caused the DNA damage and had 35.5 of Tail moment. The extracts from *Cassia tora* prepared with different degrees of roasting exhibited the protective effect on DNA damage at 0.1-1.0 mg/mL of concentrations tested. At a concentration of 1 mg/mL, the unroasted sample showed 65.2% inhibitory effect; however, the samples roasted at 150 and 250 ℃ had 56.3 and 32.1% inhibitory effect, respectively. This result means that the antioxidant activity of extracts from *Cassia tora* decreased with an increased roasting.

Table I. Effects of Water Extracts from *Cassia tora* Prepared under Different Degrees of Roasting on Hydrogen Peroxide-Induced Oxidative DNA Damage in Human Blood Lymphocytes*

concentration (mg/ml)	*DNA damage (Tail moment)***		
	unroasted	*150 ℃*	*250 ℃*
control	35.5 ± 2.1^{a}	35.5 ± 2.1^{a}	35.5 ± 2.1^{a}
0.1	26.8 ± 2.7^{b}	31.8 ± 7.0^{a}	30.4 ± 5.9^{ab}
0.25	19.1 ± 3.2^{bc}	19.2 ± 5.2^{b}	25.9 ± 5.2^{ab}
0.5	14.6 ± 6.1^{c}	19.1 ± 0.2^{b}	23.5 ± 5.5^{b}
1	12.3 ± 1.8^{c}	15.5 ± 3.0^{b}	23.9 ± 3.7^{ab}
2	15.6 ± 1.7^{c}	11.5 ± 2.8^{b}	28.5 ± 3.9^{ab}

* Image of 100 randomly selected cells (50 cells from each of two replicate slides) were analysed from each sample. In control group, H_2O_2 (50 μM) was incubated with human blood lymphocytes at 37 ℃ for 0.5 h before COMET assay.

** Results are mean ± SD for n=3. Values with different superscripts in a vertical column are significantly different ($P < 0.05$).

Scavenging Activity on Hydroxyl Radical

The hydroxyl radical rapidly reacted with the nitrone spin trap 5,5-dimethylpyrrolidine *N*-oxide (DMPO). The effect of extracts from *Cassia tora* on DMPO-OH adduct formation was determined using an EPR spectrometer. Signal intensity of the DMPO-OH adduct decreased when the concentration of unroasted *Cassia tora* extracts was increased. At a same concentration (10 mg/mL), the scavenging effect of extracts from *Cassia tora* on hydroxyl radicals was in the order of unroasted > 150 ℃ roasted > 200 ℃ roasted > 250 ℃ roasted (*22*). This trend is also in agreement with the result that the antioxidant activity of the extracts of unroasted samples was greater than that of roasted samples. The scavenging activity of extracts from *Cassia tora* on hydroxyl radicals also increased with an increase in the concentration.

Characteristica of *Cassia tora* with Roasting

Browning Intensity

Maillard reaction products have been reported to possess scavenging activity on reactive oxygen species (*23,24*). The browning of extracts from *Cassia tora* with different degrees of roasting was determined in order to reveal its relation with antioxidant activity. The sample was hydrolyzed with 1 N NaOH-10% Na_2CO_3 at 100 ℃ for 10 min, and intensity of browning was measured at 420 nm. The browning of extracts significantly increased ($P < 0.05$) from unroasted (A_{420}, 0.445) up to roasted at 200℃ (A_{420}, 0.750), but it decreased slightly at 250℃ (A_{420}, 0.695) of roasting. In general, browning increased with increasing reaction time; however, it might be decreased at a longer reaction time owing to the formation of non-soluble compounds of high molecular weight (*25*). Thus, the absorbance may have decreased at a roasting temperature of 250 ℃ because of over-roasting, which led to polymerization of Maillard reaction products.

Yen and Hsieh (*24*) indicated that the antioxidant activity of Maillard reaction products increased with an increase in reaction time. In the present study, the antioxidant activity of extracts from *Cassia tora* decreased with an increase in degree of roasting. Nicoli et al. (*13*) reported that the antioxidant activity of water extracts of coffee increased with an increase in roasting time of up to 10 min. However, the antioxidant activity of coffee extracts decreased with longer roasting time. They suggested that this might be due to the degradation of antioxidative Maillard products under over-roasting. Such can also explain the decrease in the antioxidant activity of extracts from over-roasted *Cassia tora*.

Chemiluminescence Intensity

Many chemical reactions produce chemiluminescence due to energy release, such as in the case of degradation of hydroperoxide, activated macrophage, Fenton reaction and a singlet oxygen changing from excited state to ground state (*26*). Namiki et al. (*27*) indicated that the Maillard reaction also produces chemiluminescence. Since the browning of water extracts of *Cassia tora* increased after roasting, the chemiluminescence of extracts from *Cassia tora* were determined. The chemiluminescene was measured according to a modified method of Chen et al. (*28*) by a chemiluminescence analyzer. The data were expressed as CL counts/10 s. The unroasted *Cassia tora* had very weak chemiluminescence; however, the intensity of chemiluminescence of *Cassia tora* increased with an increasing degree of roasting. The intensity of the chemiluminescence of *Cassia tora* was in the order of roasted at 250 ℃(396.0 counts/10 s) > roasted at 200 ℃(196.2 counts/10 s) > roasted at 150 ℃ (43.3 counts/10 s) > unroasted (22.8 counts/10 s) at a concentration of 16.7 mg/ml. The chemiluminescence produced in the Maillard reaction might come from the free radical products or low molecular weight hydroperoxides, such as dicarbonyl and pyrazinium compounds (*27*). In our study, the change in chemiluminescence intensity of extracts from *Cassia tora* correlated with its degree of browning. Therefore, it can be said that roasting *Cassia tora* increased browning and produced some free radical products or hydroperoxides, which resulted in the decrease in antioxidant activity.

Changes in Active Compounds of *Cassia tora* with Roasting

Total Phenolic Compounds

Tsushida et al. (*29*) reported that the antioxidant activity of vegetable extracts is related to their content of phenolic compounds. *Cassia tora* contains many phenolic compounds, such as emodin, rhein, chrysophanol, and obtusin, some of which belong to anthraquinones (*19*). Some anthraquinones have been reported to have antioxidant activity (*18*). The concentration of phenolic compounds was measured according to the method of Tang et al. (*30*) and calculated using gallic acid as standard. The total polyphenols in extracts of *Cassia tora* decreased significantly ($P > 0.05$) during roasting. The amount of polyphenols in extracts of *Cassia tora* decreased from 180.64 mg/g (unroasted) to 103.55 mg of Gallic acid/g extracts of *Cassia tora* after roasting at 250 ℃. This means that polyphenols were degraded during the roasting process. The decrease of polyphenols contents in *Cassia tora* after roasting is correlated to the decrease in antioxidant activity. Therefore, it can be suggested that the decrease in antioxidant activity of extracts of roasted *Cassia tora* was related to the decrease in polyphenols.

Anthraquinones

Anthraquinones have been reported to be the main active components in *Cassia tora*, including aloe-emodin, anthrone, aurantiobtusin, chrysophanic acid, emodin, obtusifolin, physcion, rhein, etc. (*19*). Thus, the content of anthraquinones in extracts of *Cassia tora* was determined to understand their role on antioxidant actions. Total anthraquinones, including free anthraquinones (anthraquinones aglycon) and bound anthraquinones (anthraquinones *O*-glycosides), were determined according to the method of Koshioka and Takino (*31*). As results in Table **II** show, the total content of anthraquinones in extracts of *Cassia tora* were in the order of unroasted (88.2 mg/g) > 150 ℃ roasted (70.7 mg/g) > 200 ℃ roasted (26.9 mg/g) > 250 ℃ roasted (14.9 mg/g) ($P < 0.05$) (*22*). This result indicates that anthraquinones were degraded by thermal treatment. The data also indicate that most of the anthraquinones in extracts of *Cassia tora* are in a bound form and contain glycosides. This is in agreement with the studies of Fairbirn and Moss (*32*) who reported that the anthraquinones in *Cassia* plants are partly free but mostl are present as glycosides (with aglycon occurring usually as a reduced form, e.g., anthrones). Malterud et al. (*18*) reported that anthrones (reduced form) have better antioxidant activity than do anthraquinones on lipid peroxidation and free radical scavenging. However, anthrones have reducing power and are easily oxidized to anthraquinones, which causes anthrones to have prooxidant activity.

Table II. Contents of Anthraquinones in Water Extract from *Cassia tora* under Different Degrees of Roasting*

sample	*free (mg/g)*	*bound (mg/g)*	*total (mg/g)*
unroasted	0.89 ± 0.05^{a}	87.35 ± 0.45^{a}	88.24^{a}
150 ℃, 5 min	0.78 ± 0.08^{ab}	69.90 ± 0.02^{b}	70.68^{b}
200 ℃, 5 min	0.65 ± 0.06^{bc}	26.25 ± 1.27^{c}	26.90^{c}
250 ℃, 5 min	0.54 ± 0.10^{c}	14.37 ± 0.98^{d}	14.91^{d}

* Values in a column with different superscripts are significantly different at $P < 0.05$.

(adapted and reproduced with permission from reference *22*)

The individual anthrqauinones content in extracts of *Cassia tora* were also measured according to the method of van den Berg and Labadie (*33*) that was through acid hydrolysis, chloroform extraction, and determination by HPLC. Three anthraquinones, chrysophenol, emodin, and rhein, were detected in extracts of *Cassia tora* under the experimental conditions used in our study (*34*). The unroasted sample contains the highest anthraquinones content, the content of rhein, chrysophanol, and emodin was 10.42, 0.61, and 0.28 mg/g extracts, respectively. The anthraquinones content decreased with increased roasting. The content of those three anthraquinones for the sample roasted at 150 ℃ was

4.8, 0.14 and 0.10 mg/g extracts, respectively. However, the extracts of *Cassia tora* prepared by roasting at 250 ℃ did not show any detectable anthraquinones. Zhang et al. (*12*) indicated that anthraquinones in *Cassia tora* were degraded to a free form (aglycon) by roasting treatment. The content of these three anthraquinones has only one-eighth of the total content of anthraquinones compared with the results reported by the Yen and Chung (*22*). Most individual anthraquinones or anthrones have shown antioxidant activity in linoleic acid peroxidation systems (*35*). Thus, the decrease in antioxidant activity of roasted *Cassia tora* was related to the decrease in anthraquinoids.

Conclusions

The extracts of unroasted *Cassia tora* had greater inhibition effect on peroxidation of linoleic acid than that of α-tocopherol. Higher roasting temperature and longer roasting periods reduced the antioxidant activity of *Cassia tora*. The extracts of unroasted *Cassia tora* also exhibited good antioxidant activity in the liposome peroxidation system induced by Fenton reaction as well as in the enzymatic microsome peroxidation system. Therefore, the oxidation of biological membrane *in vivo* may be inhibited. The extracts of *Cassia tora* also inhibited the oxidative DNA damage in human lymphocytes induced by hydrogen peroxide. Overall, the antioxidant activity of extracts from roasted *Cassia tora* decreased as compared with that of unroasted samples. This result might be caused by the reduction of phenolics and anthraquinones in *Cassia tora* as a result of roasting. Moreover, this may also be due to the degradation of Maillard reaction products under over-roasting.

Acknowledgment

This research work was supported in part by National Science Council, Republic of China, under drant NSC 88-2313-B005-005..

References

1. Stampfer, M. J.; Henneekens, C. H.; Manson, J. E.; Colditz, G. A.; Rosner, B.; Willet, W. C. *N. Engl. J. Med.* **1993,** *328,* 1444-1449.
2. Saija, A.; Scalese, M.; Lanza, M.; Marzullo, D.; Bonina, F.; Castelli, F. *Free Radical Bio. Med.* **1995**, *19,* 481-486.
3. Hertog, M. G. L.; Feskeens, E. J. M.; Hollman, C. H.; Katan, M. B.; Kromhout, D. *Lancet* **1993,** *342, 1007-1011.*

4. Hertog, M. G. L.; Hollman, P. C. H. *Eur. J. Clin. Nutr.* **1996,** *50,* 63-66.
5. Benavente-Garcisa, O.; Castillo, J.; Marin, F. R.; Ortuno, A.; Rio, J. A. D. *J. Agric. Food Chem.* **1997**, *45,* 4505-4515.
6. Huang, K. C. In *The Pharmacology of Chinese Herbs*; CRC Press: Boca Raton, FL, 1993; p 103.
7. Choi, J. S.; Lee, H. J.; Park, K. Y.; Ha, J. O.; Kang, S. S. *Planta Med.* **1997**, *63*, 11-14.
8. Wu, C. H.; Yen, G. C. *J. of the Chinese Agric. Chem. Soc.* **1999**, *37*, 263-275.
9. Su, J. -D. *Food Sci. (Chinese) **1992,** 19,* 12-24.
10. Kim, S. Y.; Kim, J. H.; Kim, S. K.; Oh, M. J.; Jung, M. Y. *J. Am. Oil Chem. Soc.* **1994,** *71,* 633-640.
11. Yen, G. C.; Chen, H. W.; Duh, P. D. *J. Agric. Food Chem.* **1998,** *46,* 820-824.
12. Zhang, Q.; Yin, J.; Zhang, J. *Chinese Herb* **1996,** *27,* 79-81.
13. Nicoli, M. C.; Anese, M.; Manzocco, L.; Lerici, C. R. *Lebensm.-Wiss. U.-Technol.* **1997,** *30,* 292-297.
14. Yen, G. C.; Chen, H.Y. *J. Food Prot.* **1994,** *57,* 54-58.
15. Halliwell, B.; Gutteridge, J. M. C. In *Free Radicals in Biology and Medicine*; Halliwell, B.; Gutteridge , J. M. C., Eds.; Clarendon Press: Oxford, 1989.
16. Salah, N.; Miller, N. J.; Paganga, G.; Tijburg, L.; Bolwell, G. P.; Rice-Evans, C. *Arch. Biochem. Biophys.* **1995,** *2,* 339-346.
17. Ratty, A. K.; Sunamoto, J.; Das, N. P. *Biochem. Pharmocol.* **1988,** *37,* 989-995.
18. Malterud, K. E.; Farbrot, T. L.; Huse, A. E.; Sund, R. B. *Pharmacology* **1993,** *47(suppl 1),* 77-85.
19. Duke, J. A. In *Handbook of Phytochemical Constituents of GRAS Herbs and Other Economic Plants*; CRC Press: Boca Raton, FL, 1992; pp 143-144.
20. Fairbairn D. W.; Olive P. L.; O'Neill K. L. *Mutat. Res.* **1995,** *339*, 37-59.
21. Duthie S. J.; Johnson W.; Dobson V. L. *Mutat. Res.* **1997,** *390,* 141-151.
22. Yen, G. C.; Chung, D. Y. *J. Agric. Food Chem.* **1999**, *47,* 1326-1332.
23. Hayase, F.; Hirashima, S.; Okamato, G.; Kato, H. *Agric. Biol. Chem.* **1990**, *54,* 855-862.
24. Yen, G. C.; Hsieh, P. P. *J. Sci. Food. Agric.* **1995**, *67,* 415-420.
25. Lingnert, H.; Ericksson, C. E. *Prog. Fd. Nutri. Sci.* **1981,** *5,* 453-466.
26. Sato, T.; Inaba, H.; Kawai, K.; Furukawa, H.; Hirono, I.; Miyazawa. T. *Mutat. Res.* **1991,** *251,* 91-97.
27. Namiki, M.; Oka, M.; Otsuka, M.; Miyazawa, T.; Fujimotol, K.; Namiki, K. *J. Agric. Food Chem.* **1993,** *41,* 1704-1709.
28. Chen, M. -F.; Mo, L. -R.; Lin, R. -C.; Kuo, J. -Y.; Chang, K. -K.; Liao, C.; Lu. F. -J. *Free Rad. Biol. & Med.* **1997,** *23,* 672-679.
29. Tsushida, T.; Suzuki, M.; Kurogi, M. *Nippon Shokuhin Kogyo Gakkaishi.* **1994,** *41,* 611-618.

30. Taga, M. S.; Miller, E. E.; Pratt, D. E. *J. Am. Oil. Chem. Soc.* **1984,** 61, 928-931.
31. Koshioka, M.; Takino, Y. *Chem. Pharm. Bull.* **1978**, *26*, 1343-1347.
32. Fairbirn, J. W.; Moss, M. J. R. *J. Pharm. Pharmcol.* **1970**, *32*, 584-593.
33. van den Berg, A. J. J.; Labadie, R. P. *J. Chromatogr.* **1985**, *329*, 311-314.
34. Wu, C. H. Master thesis. National Chung Hsing University, Taiwan, **1999**.
35. Yen, G. C.; Duh, P.D.; Chuang, D. Y. *Food Chem.* **2000**, *70*, 437-441.

Chapter 16

Free Radical and Oxidative Reactions of (–)-Epigallocatechin and (–)-Epigallocatechin Gallate, Two Major Polyphenols in Green Tea

Nanqun Zhu[1], Tzou-Chi Huang[2], Jen-Kun Lin[3], Chung S. Yang[4], and Chi-Tang Ho[1]

[1]Department of Food Science, Rutgers University, 65 Dudley Road, New Brunswick, NJ 08901-8520
[2]Department of Food Science, National Pingtung University of Science and Technology, Pingtung, Taiwan
[3]Institute of Biochemistry, College of Medicine, National Taiwan University, Taipei, Taiwan
[4]Laboratory for Cancer Research, College of Pharmacy, Rutgers University, Piscataway, NJ 08854-8020

(-)-Epigallocatechin gallate (EGCG) and (-) epigallocatechin (EGC) are two main antioxidative catechins of green tea. The specific mechanisms of antioxidant action of tea catechins remain unclear. In this study, the scavenging mechanisms of EGC and EGCG on three different oxidants are evaluated by the identification of reaction products. Although the principal site of antioxidant reactions on an EGC or an EGCG molecule is the trihydroxyphenyl B ring, this study indicates that the A ring of catechins may also serve as an antioxidant site.

Tea is one of the most popular beverages worldwide because of its attractive flavor, aroma and taste. Recent study showed that tea has broad activity against cancer (*1-4*) and coronary heart disease (5-*6*). The polyphenols are the most significant group of tea components, especially certain catechins. The major tea catechins are (-)-epigallocatechin gallate (EGCG), (-)-epigallocatechin (EGC), (-)-epicatechin gallate (ECG) and (-)-epicatechin (EC) (see Figure 1 for structures). Many biological functions of tea catechins have been studied, including antimutagenic effects (*7*) and anticarcinogenic effects (*8-10*), in several systems. Growing evidence suggests these activities could be due to the potent antioxidative activity of tea catechins; especially since they can act as radical scavengers in both in *vitro* and in *vivo* systems (*5,11-14*). For example, there are three suggested mechanisms of tea catechins as tumor inhibitors: The first is tea catechins are strong metal ion chelators because of their catechol structure. They can bind and thus decrease the level of free cellular ferric and ferrous ions, which are required for the generation of reactive oxygen radicals by the Fenton reaction. Secondly, tea catechins are strong scavengers against superoxide and hydroxyl radicals, which can damage DNA and other cellular molecules and initiate lipid peroxidation reactions. Finally, catechins can trap peroxyl radicals and thus suppress radical chain reactions and terminate lipid peroxidation (*1*). However, the specific mechanisms of catechin antioxidant reactions remain unclear. The identification of oxidation products formed by reactions of catechins with free radicals could provide some insight into those antioxidant reactions. Here, we report the structures of seven products formed by the reaction of EGCG and EGC with three kinds of oxidants. The possible mechanisms for the formation of these compounds are proposed.

EC	R_1 = H	R_2 = H
EGC	R_1 = H	R_2 = OH
ECG	R_1 = gallate	R_2 = H
EGCG	R_1 = gallate	R_2 = OH

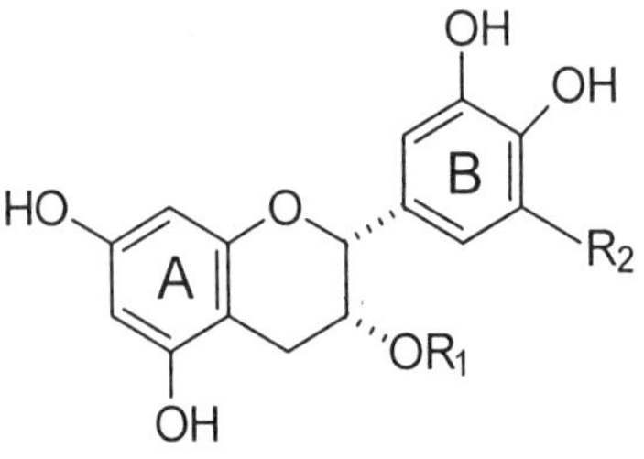

Figure 1: Structure of four major tea catechins.

Materials and Methods

Isolation of EGC and EGCG

Green tea polyphenol extract (10 g, Lipton Company) was dissolved in 95% ethanol solution and the solution loaded onto a Sephadex LH-20 column (38 mm i.d. × 45 mm). After being eluted with 95% ethanol and monitored by TLC [chloroform-methanol-water (3:1:0.2) as eluent], 700 mg EGC and 1300 mg EGCG were yielded.

Oxidation of EGC and EGCG with H_2O_2

Two hundred milligrams of EGCG or EGC in 5 mL of water were treated with 1 mL of H_2O_2 (50%). The reaction mixture was kept at room temperature for 48 hours until no EGCG or EGC was present as monitored by thin layer chromatography (TLC).

Oxidation of EGCG by Fenton Reaction

One hundred milligrams of EGCG in 5 mL of water were treated with 0.1 mL of H_2O_2 (50%) and 0.1 mg of $FeSO_4$. This mixture was kept at room temperature for 20 minutes. The same products as the reaction of EGCG with H_2O_2 (50%) without $FeSO_4$ were detected by TLC analysis.

Oxidation of EGC and EGCG with DPPH

Three hundred milligrams of EGCG or 250 mg of EGC were mixed with 550 mg or 450 mg of DPPH, respectively in 10 mL of acetonitrile. These mixtures were kept in darkness for two days.

Oxidation of EGC and EGCG with Peroxidase

Five hundred milligrams of EGCG or 350 mg of EGC in a mixture of acetone-water (5:8, 32.5 ml) were added to 1 mg of horseradish peroxidase. While being stirred, 1.5 mL of 3.14% H_2O_2 was added four times during 45 min.

At the end of 45 minutes another 0.5 mL of H_2O_2 was added and the stirring was continued for an additional 15 min.

Purification and Identification of the EGCG and EGC Oxidation Products

Using repeated column chromatography on silica gel and Sephadex LH-20, reaction products were isolated from the oxidation reactions. Spectrometric methods, particularly MS and NMR (^1H NMR, ^{13}C NMR, ^1H-^1H COSY, NOESY, HMQC and HMBC) techniques were used for structures determination of these products.

Results and Discussion

H_2O_2 Oxidation

Two compounds (compounds **1** and **2**) were isolated as the isolable oxidative products of EGCG (Figure 2) when it reacted with H_2O_2. When EGC was reacted with H_2O_2, only one product (compound **3**) was isolated (Figure 2). The structures of these three compounds were identified by their MS and NMR spectra. From their structures, it is clear that they were formed by the oxidative cleavage of the A-ring of the catechin skeleton. It is of great interest since it was previously considered that the ring-A of catechin molecules is very insensitive to oxidation. The observation that the ring-A can be oxidized to form carboxylic acid groups provides an unambiguous proof that oxidation can occur at the A-ring. It was reported recently that (-)-5-(3′,4′,5′-trihydroxyphenyl)-γ-valerolactone and (-)-5-(3′,4′-dihydroxyphenyl)-γ-valerolactone were formed as a major metabolite of catechins in the urine and plasma of human volunteers after ingestion of green tea (*15-16*). It was proposed that the large intestine, harboring a huge population of anaerobic bacterial, is likely to be the main site for the formation of these two A-ring cleavage metabolites (*16*). Although the exact mechanism for A-ring fission in the formation of compounds **1-3** is not known, we hypothesize that the hydroxyl radicals may be involved in the reaction. It is indeed when the EGCG was oxidized through Fenton reaction that compounds **1** and **2** were detected by TLC analysis. Thus, we have a proposed mechanism as shown in Figure 3. The key step is the attack of hydroxyl radical on the A-ring, which resulted in the break down of the A-ring.

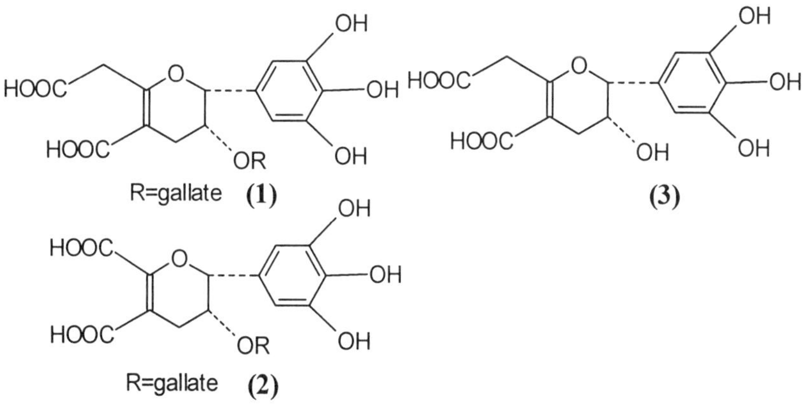

Figure 2. *Oxidation products by H_2O_2*

DPPH oxidation

The reaction of 300 mg of EGCG with DPPH radical produced 20 mg of compound **4** (6.7% yield). The other compounds formed were dark color polymeric material and not able to be purified by column chromatography. Similarly, the reaction of 250 mg of EGC with DPPH yielded only 8 mg of compound **5** (3.2%). The structure of compounds **4** and **5** (Figure 4) was confirmed by comparisons of their ^{1}H and ^{13}C NMR data with those in the literature (17) as theasinensin A and theasinensin C, respectively. Both compounds have previously been isolated from oolong tea (*17*). It is also of interest that theasinensin A was recently reported to be one of the two major metabolites during the incubation of EGCG with bile of rat (*18*). According to this report, 50% inhibitory concentration (IC_{50}, μm) for TBARS (thiobarbituic acid reactive substance) (using rat brain homogenates) of theasinensin A was even smaller than that of EGCG. Because *in vivo* antioxidant activity of tea might not be due only to the intact catechin itself, the oxidative metabolites of catechins may also have some contribution, our study may provide some insight for the evaluation of the antioxidant actions of tea catechins in biological systems. The proposed mechanism is shown in the Figure 5. The initial phenoxyl radicals were generated by DPPH and the resultant phenoxyl radical attacked at the same position of the galloyl moiety of the catechin molecule. Finally, the dimeric theasinensins can be obtained through keto-enol tautomerism.

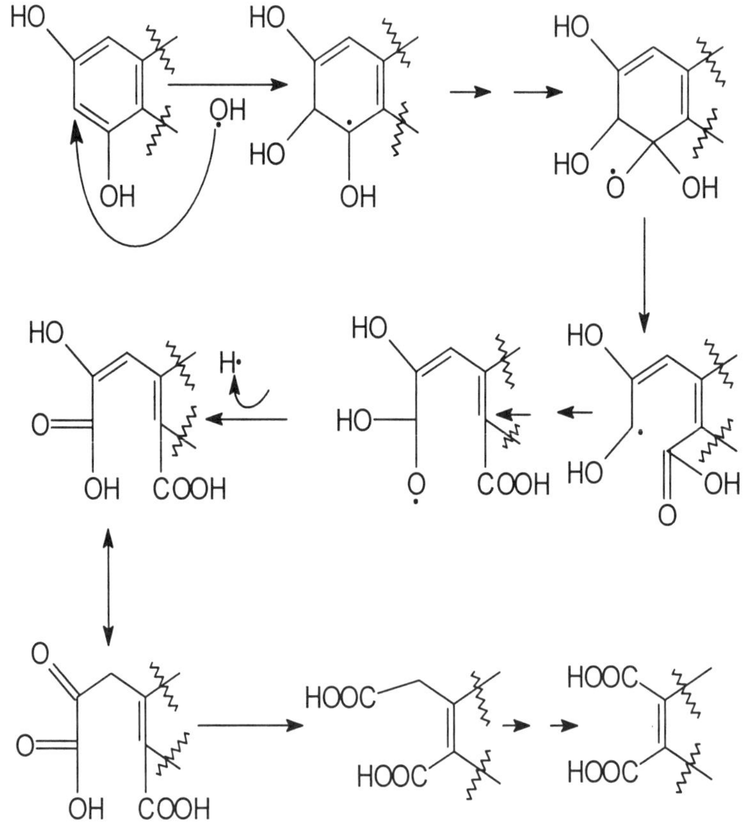

Figure 3. Proposed mechanism of for the oxidation of EGCG and EGC by H_2O_2.

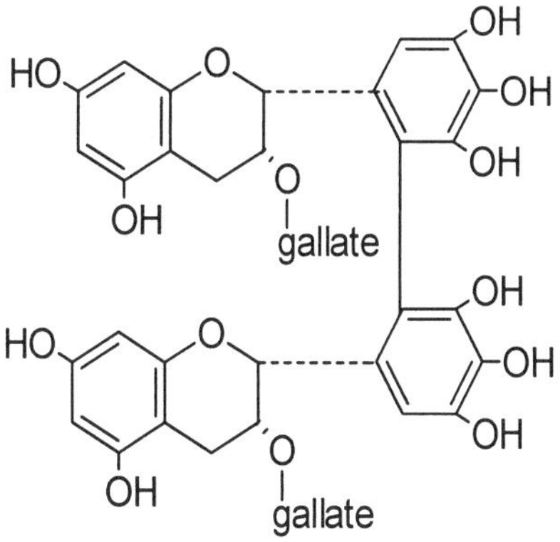

(4) 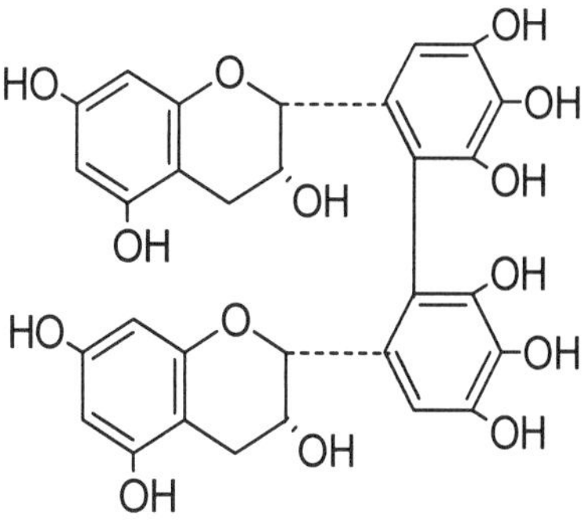 (5)

Figure 4. DPPH oxidation products of EGCG and EGC.

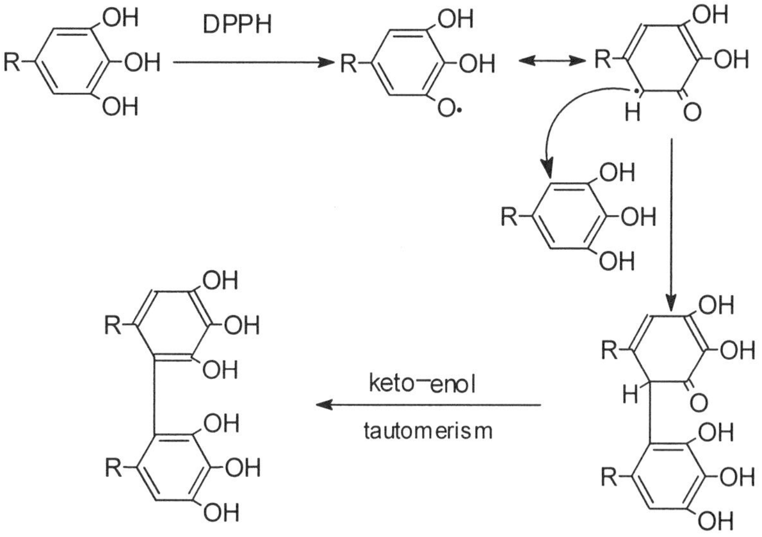

Figure 5. Proposed mechanism for the oxidation of EGCG and EGC by DPPH radical.

Peroxidase oxidation

Compounds **6** and **7** (Figure 6) were isolated when EGCG and EGC were catalytically oxidized by peroxidase. Compound **6** has recently been reported by Valcic et al. (*19*) as one of the peroxy radical reaction products of EGCG. It is therefore indicated that the peroxidase oxidation is very similar to the peroxyl radical reaction. It further confirms the view that the trihydroxyphenyl B ring, rather than the gallate moiety is the active site of antioxidant reaction in catechins. As the proposed mechanism shown in the Figure 7, the first step in this reaction is the formation of phenoxyl free radicals, which underwent a series of rearrangements and attacked another molecule of catechin to form a dimer radical. The resulting radical was then transformed into the symmetric dimer through another series of free radical rearrangements.

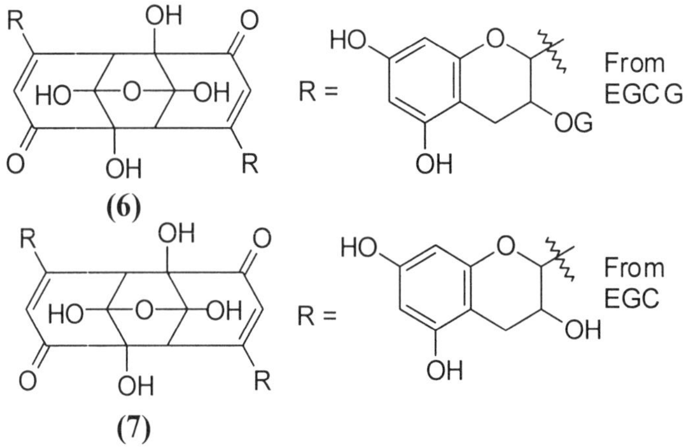

Figure 6. Peroxidase oxidation products of EGCG and EGC.

From the result of our present studies as well as the reports of Hirose et al. (*20*), Wan et al. (*21*), Valcic et al. (*19*) and Nanjo et al. (*22*), it can be concluded that the use of different oxidants can result in distinctively different oxidation products from catechins and that the main site of antioxidant action of catechins seems to depend on the oxidant used through the trihydroxyphenyl B ring as the active site rather than the gallate moiety.

Acknowledgement

This work was supported in part by NIH grants CA56673 and the Tea Trade Health Association.

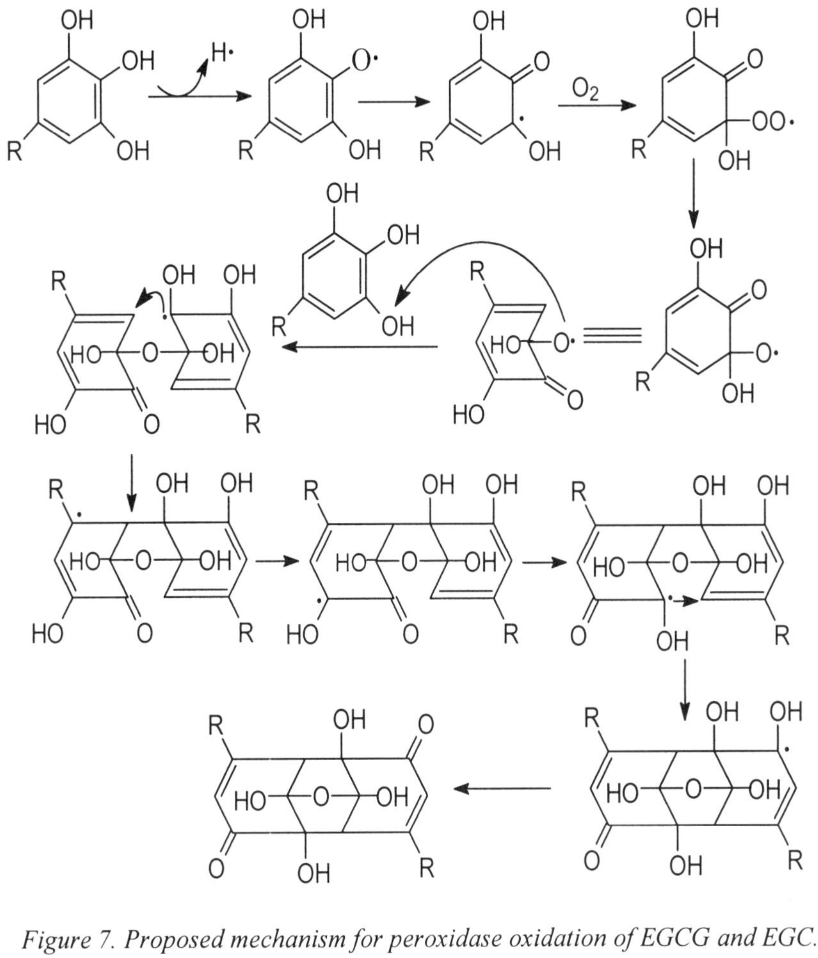

Figure 7. Proposed mechanism for peroxidase oxidation of EGCG and EGC.

References

1. Yang, C. S.; Wang, Z. Y. *J. Natl. Cancer Inst.* **1993**, *85*, 1038-1049.
2. Shim, J. S.; Kang, M. H.; Kim, Y. H.; Roh, J. K.; Robert, C.; Lee, I. P. *Cancer Epidem. Biomarkers Prev.* **1995**, *4*, 387-391.
3. Yang, C. S.; Kim, S.; Yang, G-Y.; Lee, M-L.; Liao, J.; Chung, J.; Ho, C.-T. *Proc. Soc. Exptl. Biol. Med.* **1999**, *220 (No. 4)*, 213-217.
4. Dreosti, I. E.; Wargovich, M. J.; Yang, C. S. *Crit. Rev. Food Sci. Nutr.* **1997**, *37*, 761-770.
5. Wiseman, S. A.; Balentine, D. A.; Frei, B. *Food Sci. Nutr.* **1997**, *37*, 705-718.
6. Ishikawa, T.; Suzukawa, M.; Ito, T.; Yoshida, H.; Ayaori, M.; Nishiwaki, M.; Yonemura, A.; Hara, Y.; Nakamura, H. Amer. *J. Clin. Nutr.* **1997**, 66, 261-266.
7. Wang, Z. Y.; Cheng, S. J.; Zhou, Z. C.; Athar, M.; Khan, W. A.; Bickers, D. R.; Mutkhtar, H. *Mutat. Res.* **1989**, *223*, 273-285.
8. Huang, M. T.; Ho, C.-T.; Wang, Z. Y.; Ferrano, T.; Finnegan-Olive, T.; Lou, Y. R.; Mitchell, J. M.; Laskin, J. D.; Newmark, H.; Yang, C. S.; Conney, A. H. *Carcinogenesis* **1992**, *13*, 947-954.
9. Xu, Y.; Ho, C.-T.; Amin, S. G.; Han, C.; Chung, F. L. *Cancer Res.* **1992**, *52*, 3875-3879.
10. Katiyar, S. K.; Agarwal. R.; Zaim. M. T.; Mukhtar, H. *Carcinogenesis* **1993**, *14*, 849-855.
11. Ho, C.-T.; Chen, Q.; Shi, H.; Zhang, K. Q.; Rosen, R. T. *Prev. Med.* **1992**, *21*, 520-525.
12. Yoshino, K.; Hara, Y.; Sano, M.; Tomita, I. *Biol. Pharm. Bull.* **1994**, *17*, 146-149.
13. Yen, G. C.; Chen, H. Y. *J. Agric. Food Chem.* **1995**, *43*, 27-32.
14. Lunder, T. L.. In *phenolic Compounds in Food and Their Effects on Health II*; Huang M. T.; Ho C.-T.; Lee, C. Y., Eds.; American Chemical Society, Washington, DC, 1992; pp 115-120.
15. Hollman, P. C.; Tijburg, L. B. M.; Yang, C. S. *Food Sci. Nutr.* **1997**, *37*, 719-738.
16. Li, C.; Lee, M.J.; Sheng, S.; Meng, X.; Prabhu, S.; Winnik, B.; Huang, B.; Chung, J. Y.; Yan, S.; Ho, C.-T.; Yang, C. S.; *Chem. Res. Toxicol.* **2000**, *13*, 177-184
17. Hashimoto, F.; Nonaka, G.; Nishioka, I.; *Chem. Pharm. Bull.* **1988**, *36*, 1676-1684.
18. Tomita, I.; Sano, M.; Sasaki, K.; Miyase, T. In *Functional Foods for Disease Prevention I;* Shibamoto, T.; Terao, T.; Osawa, T., Eds.; American Chemical Society: Washington, D.C., 1998; pp. 209-216.

19. Valcic, S.; Muders, A.; Jacobsen, N. E.; Liebler, D. C.; Timmermann, B. N. *Chem. Res. Toxicology*. **1999**, *12*, 382-386.
20. Hirose, Y.; Yamaoka, H.; Nakayama, M. *J. Amer. Oil Chem. Soc.* **1991**, *68*, 131-135.
21. Wan, X.; Nursten, H. E.; Cai Y.; Davis, A. L.; Wilkins, J. P. G.; Davies, A. P. A new type of tea pigment-from the chemical oxidation of epicatechin gallate and isolated from tea. *J. Sci. Agric.* **1997**, *74*, 401-408.
22. Nanjo, F.; Goto, K.; Seto, R.; Suzuki, M.; Hara, Y. Scavenging effects of tea catechins and their derivatives on 1, 1-diphenyl-2-picrylhydrazyl radical. *Free Radical Biol. Med.* **1996**, *21*, 895-902.

Chapter 17

Radical-Scavenging Activity of Green Tea Polyphenols

Takako Yokozawa and Erbo Dong

Institute of Natural Medicine, Toyama Medical and Pharmaceutical University, 2630 Sugitani, Toyama 930-0194, Japan

The effect of four types of tea on excessive free radicals were examined using spin trapping, 1,1-diphenyl-2-picrylhydrazyl radical, lipid peroxidation, and lactate dehydrogenase leakage from cultured cells. Green tea extract presented significant antiradical effects in these four assay systems; whereas oolong tea and black tea extracts showed a rather weak protective effect against free radicals. A more potent scavenger effect using cultured cells was found with green tea polyphenols. Similar to the effects of the green tea polyphenols, (-)-epigallocatechin 3-*O*-gallate, its main ingredient, had an inhibitory effect on oxidative stress-induced apoptosis. The activities of the antioxidation enzymes in rats after subtotal nephrectomy were increased, suggesting a protective action against oxidative stress. The increased levels of uremic toxins in the blood were also reduced in rats given (-)-epigallocatechin 3-*O*-gallate. These findings indicate that (-)-epigallocatechin 3-*O*-gallate helps to inhibit the progression of renal failure by scavenging radicals. On the other hand, green tea polyphenols (daily dose, 400 mg) administered for 6 months to 50 dialysis patients decreased blood levels of creatinine (Cr) and methylguanidine (MG) and also the MG/Cr ratio. A decrease in β_2-microglobulin and improved arthralgia were also noted in some patients. Based on the evidence available it appeared that green tea polyphenols ameliorate the state of enhanced oxidation in dialysis patients.

In the current era of overeating in developed countries, the importance of food now tends to be focused on health issues, particularly diseases, rather than on its role as a source of nutrients and energy. Furthermore, new aspects of the biological activity of food have been attracting attention. Along with recognition of the important role of active oxygen in such processes as aging, carcinogenesis, and the development of circulatory disorders, antioxidants in foodstuffs are now considered to be new edible and functional substances.

In previous studies, we investigated the antioxidant activity of various compounds chosen from Oriental medicine, paying particular attention to the correlation between molecular structure and activity. We have found that tannin and phenol compounds such as flavones possess such activity, and that tea, which contains these compounds as its major components, has high antioxidant activity (*1,2*).

Tea (*Camellia sinensis* L.), which originated in ancient China, is a widely used and enjoyed luxury grocery item. Recent studies have successively demonstrated the value of tea for the regulation of physiological function. Tea can be classified into three categories, i.e., unfermented, semi-fermented and fermented, depending on the degree of leaf fermentation. Each type of tea has a distinct aroma, color and flavor, which appeal to our senses of taste, smell and sight. The physiological actions of these three types of tea have rarely been compared, and there is limited information on the relationship between tea and free radicals, which mediate various diseases and pathological conditions.

Attention is currently being paid to free radical-forming systems in the body. Once produced, free radicals induce degeneration of membrane lipids and proteins, which causes damage to cell membranes or degeneration of DNA and enzymes, and this in turn leads to various pathological conditions. However, organisms possess various defense mechanisms against injury caused by radicals (*3*).

In light of this connection, the present study investigates the effect of tea and its components on injury due to excessive free radicals both in vitro and in vivo. In addition, we administered green tea polyphenols to dialysis patients who were in an enhanced oxidative condition, and we evaluated its usefulness in their treatment.

Tea

Fifty grams of commercially available green tea, oolong tea, or black tea were added to 1,000 mL of hot distilled water (70 °C) and shaken for 5 min. Each extract was then evaporated to dryness under reduced pressure. The yields of green tea, oolong tea, and black tea were 14.7%, 16.9% and 17.5%, respectively, relative to the weight of the starting materials. The green tea polyphenols studied were Sunphenon (Taiyo Kagaku Co., Yokkaichi, Japan), which was prepared from a hot-water extract of green tea. It was composed mainly of (-)-epigallocatechin 3-*O*-gallate (18.0%), (-)-gallocatechin 3-*O*-gallate (11.6%), (-)-epicatechin 3-*O*-gallate (4.6%), (-)-epigallocatechin (15.0%), (+)-gallocatechin (14.8%), (-)-epicatechin (7.0%) and (+)-catechin (3.5%). In addition, (-)-epigallocatechin 3-*O*-gallate was investigated; recycling HPLC was used to purify this compound from the mixture. Purification was carried out on a JAI-LC-908 HPLC (Japan Analytical Industry Co., Tokyo, Japan) equipped with JAI RI and JAI UV detectors, operating at 280 nm (*4*). The column was a PVA HP-GPC-column (JAIGEL GS-320, 50 x 2 cm i.d.). The eluent was methanol, and the flow rate was 3 mL/min. Purification was confirmed by GC-MS (JMS-DX 303, JEOL, Tokyo, Japan) and NMR (GSX-400, JEOL, Tokyo, Japan) analysis. Caffeine was purchased from Sigma Chemical Co., and theanine was obtained from Wako Pure Chemical Industries, Ltd. (Osaka, Japan).

In Vitro Experiments

Various substances in the body possess a guanidine group in their structures. Mori *et al.* (*5*) and Yokoi *et al.* (*6*) reported that an intraventricular injection of either α-guanidinoglutaric acid or homoarginine induces generalized seizures in rats, suggesting involvement of the formation of hydroperoxy ($\bullet O_2H$) and hydroxyl (•OH) radicals. On the other hand, as shown in Figure 1, the electron spin resonance (ESR) spectrum of homoarginine solution shows the 1:2:2:1 quartet pattern peculiar to 5,5-dimethyl-1-pyrroline-*N*-oxide (DMPO)-OH. There are many possible mechanisms by which •OH radicals are generated from an aqueous solution of a guanidine compound. In general, there is equilibrium between guanidine ions and electrons in a guanidine solution. These electrons are considered to react with oxygen molecules in solution to generate superoxide (O_2^-), resulting in formation of •OH through the Haber-Weiss reaction. However, in the presence of green tea extract at a concentration of 2 μg/mL, the production of •OH was markedly low. A similar •OH radical-eliminating action was noted for oolong tea and black tea, although the action in both cases was less potent than that of green tea (Figure 1).

The reactivity of •OH is particularly high among various different radicals. On the other hand, the 1,1-diphenyl-2-picrylhydrazyl (DPPH) radical is stable in ethanolic solution for more than 60 min (*7*). As a result of this, we used this system to assess the radical-scavenger activity of the three tea extracts. The green tea extract showed very significant scavenging activity on the DPPH radical, with 50% inhibitory activity observed at a low concentration of 4.14 μg/mL. The other tea extracts had a relatively weak effect, the IC_{50} value being 47.12 μg/mL for oolong tea and 27.02 μg/mL for black tea.

Lipid peroxides produced from unsaturated fatty acids by radicals have histotoxicity on their own, and also increase the production of free radicals *via* a chain reaction effect (*8*). On the basis of these findings, we added various concentrations of tea extracts to the incubation medium of an experimental system in which the Fenton reaction was induced in a kidney homogenate in the presence of hydrogen peroxide (H_2O_2) and Fe^{2+}. We then determined their IC_{50} values. Green tea extract showed 50% inhibition at a concentration of 7.22 μg/mL, whereas oolong tea and black tea extracts, with a large IC_{50}, showed a rather weak protective effect against lipid peroxidation.

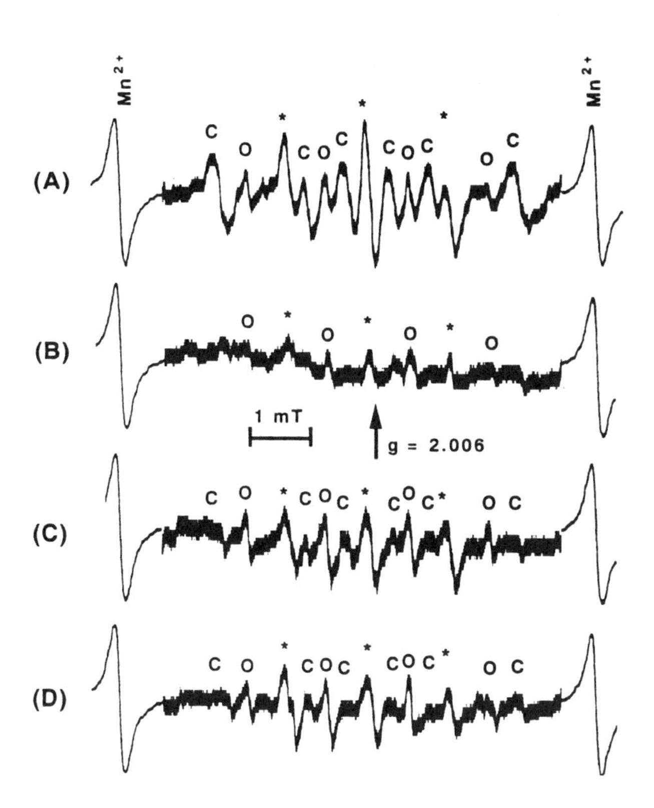

Figure 1. Influence of tea extract on the ESR spectra of a mixed solution of homoarginine and DMPO: (A) none; (B) green tea; (C) oolong tea; (D) black tea. C = DMPO-C, O = DMPO-OH, asterisk = DMPO-H. (Copyright 1998 American Chemical Society.)

Cell Culture Assay

Active oxygen species known to be produced in the body include O_2^-, H_2O_2, •OH, singlet oxygen ($^1O_2$) and LOOH. Although O_2^- does not have the strongest toxicity among these species, it plays an important role as an initiator of the generation process of various types of active oxygen (*9*). Therefore, in general, the crucial step is believed to be elimination of O_2^-, which is the origin of various other types of active oxygen in the body.

With regard to the source of active oxygen in ischemia-reperfusion injury, Saugstad and Aasen (*10*) reported that ATP molecules produced in ischemia are decomposed to adenosine and hypoxanthine, with conversion of xanthine dehydrogenase to xanthine oxidase. They further found that a large amount of O_2^- is produced in the presence of the increased hypoxanthine and xanthine oxidase when oxygen is supplied by blood flow in reperfusion. In the present studies, we used the swine kidney-derived cultured epithelial cell line LLC-PK_1 (*11*) to assess renal ischemic-reperfusion injury in vitro. These cells have features that resemble proximal uriniferous tubules: the site of severe injury in ischemic acute renal failure. When these cells were reoxygenated after incubating them under hypoxic conditions for 6 h, lactate dehydrogenase (LDH) leakage into the culture medium increased markedly and eventually reached a level of about 160 mIU/mL. On the other hand, enzyme leakage was significantly suppressed when green tea extract (1.25 μg/mL final concentration) was added to the medium prior to reoxygenation. Higher concentrations of green tea extract enhanced this effect. For black tea extract and oolong tea extract, significant suppression of LDH leakage into the medium was not obtained until double and 20-fold the corresponding concentration of green tea extract, respectively, was added to the medium (Table I). Unlike green tea, oolong tea is produced through a process of sun-drying and indoor drying, which includes turning the leaves over (stirring the tea levels while swinging them), which allows the serial steps of enzyme reaction to proceed slowly and steadily. This process also produces the necessary changes in the tea components. In the present study, oolong tea was shown to have the lowest activity among the three types of tea examined. The radical-scavenger activity of black tea, which is processed by fermentation after milling of the cells, was lower than that of green tea, but higher than that of oolong tea, demonstrating at the cellular level that the difference in the fermentation process affects the radical-scavenger activity of tea.

In all of the four experimental systems examined, the radical-scavenger activity of green tea extract was higher than that of black tea and oolong tea extracts. A similar, and more potent, scavenger effect using LLC-PK_1 cells was found with polyphenols, as shown in Table I. However, no such effect was found with caffeine, which is a component that has an analeptic action that is a characteristic property tea leaves, while theanine, a component that contributes to the taste of green tea, was associated with only a slight leakage of LDH.

Using LLC-PK_1 cells, Yonehana and Gemba (*12*) demonstrated that the intracellular antioxidant, glutathione, decreases significantly under hypoxic conditions. Snowdowne *et al.* (*13*) and Kribben *et al.* (*14*) also reported that hypoxia

Table I. Effect of Tea and Green Tea Component on LDH Leakage from $LLC-PK_1$ Cells Subjected to Hypoxia-Reoxygenation

Expt. No.	*Material*	*Concentration (µg/mL)*	*LDH activity (mIU/mL)*
1	Green tea	1.25	141.1 ± 5.4[b]
		2.5	134.2 ± 6.7[c]
		12.5	134.2 ± 7.3[c]
		25	128.9 ± 6.8[c]
		50	106.4 ± 79[c]
	Oolong tea	1.25	156.3 ± 8.3
		2.5	153.1 ± 7.9
		12.5	150.1 ± 7.3
		25	143.9 ± 6.8[a]
		50	138.6 ± 7.6[b]
	Black tea	1.25	147.6 ± 6.5
		2.5	144.5 ± 6.8[a]
		12.5	142.3 ± 7.2[b]
		25	137.9 ± 7.0[c]
		50	128.4 ± 6.8[c]
	Control		158.8 ± 5.3
2	Polyphenols	0.25	134.1 ± 8.4[b]
		0.5	112.4 ± 6.4[c]
		2.5	110.4 ± 5.0[c]
		5	92.9 ± 3.8[c]
		10	82.2 ± 2.4[c]
	Caffeine	0.25	146.8 ± 9.3
		0.5	147.5 ± 7.8
		2.5	140.0 ± 7.6
		5	133.2 ± 6.6
		10	135.6 ± 8.6
	Theanine	0.25	144.2 ± 6.5
		0.5	143.7 ± 8.4
		2.5	137.3 ± 7.2
		5	132.9 ± 6.8[a]
		10	129.3 ± 6.0[b]
	Control		147.4 ± 4.2

Statistical significance: [a]$p<0.05$, [b]$p<0.01$, [c]$p<0.001$ vs. value for none.
SOURCE: Copyright 1998 American Chemical Society.

causes the intracellular concentration of Ca to increase in another kidney-derived culture cell line, LLC-MK_2, or in isolated uriniferous tubules. On the other hand, in terms of cell injury due to reoxygenation, Paller *et al.* (*15*) proposed in 1984 a new theory whereby active oxygen is involved in the pathogenesis of renal ischemic-reperfusion injury. This theory has a considerable impact on this field of science. Since then, it has become gradually apparent that a close relationship exists between ischemic-reperfusion injury in various organs and diseases in these organs. In the proximal tubule-like LLC-PK_1 cells used in the present study, leakage of LDH was suppressed when dimethyl sulfoxide (DMSO), an •OH scavenger, was added to the medium prior to the start of culture (data not shown), suggesting that free radicals produced by renal epithelial cells through hypoxia and reoxygenation are responsible for the cell injury. This result is consistent with the finding of Paller and Neumann (*16*) that renal epithelial cells produce free radicals in primary renal cell culture. The fact that this type of cell injury was suppressed by green tea extract, and more potently by polyphenols, a component of green tea extract, is in vitro evidence for the direct effect of green tea polyphenols on renal cells. This supports our previous finding whereby oral administration of green tea polyphenols ameliorated renal failure in rats under oxidative stress (*17*). The viability of the renal cells, however, was not affected by this action.

DNA Fragmentation

In cells of the living body, H_2O_2 is produced by disproportionation of O_2^-, and the H_2O_2 produced is converted to a more potent oxidant, •OH. Significant attention is now being focused on the cytotoxicity of these active oxygen free radicals as the cause of various pathological conditions. Using thymocytes, Forrest *et al.* (*18*) revealed that active oxygen induces apoptosis, and that this apoptosis is in turn suppressed by a water-soluble vitamin derivative (Trolox). In addition, Ueda *et al.* (*19*) found that apoptosis is induced following activation of endonuclease when LLC-PK_1 cells are exposed to H_2O_2. In the present study using an experimental system in which LLC-PK_1 cells were exposed to •OH produced from H_2O_2 and $FeSO_4$ (Fenton reaction), the DNA of LLC-PK_1 cells was fragmented into lower-molecular-weight molecules even after a very limited exposure time, causing a dramatic increase in the proportion of free low-molecular-weight DNA per total amount of DNA (Table II). Analysis of the agarose gel electrophoretic pattern revealed a ladder, which was absent in cells that were not subjected to the Fenton reaction, indicating that oxidative stress induced apoptosis (Figure 2). However, when cells were subjected to the Fenton reaction in the presence of polyphenols at 30 μg/mL, the degree of DNA fragmentation was lowered. In the presence of the polyphenols at 60 μg/mL, DNA fragmentation was suppressed to a degree approximating that in cells that were not subjected to the Fenton reaction. When 2.5 μg/mL (-)-epigallocatechin 3-*O*-gallate, the major ingredient of the tannin mixture, was added to the culture medium, the degree of apoptosis was lowered, as confirmed by the rate of DNA

Table II. Effect of Green Tea Components on DNA Fragmentation

Addition	*Fragmentation rate (%)*
None	22.2 ± 2.9
$H_2O_2 + Fe^{2+}$	69.0 ± 5.8[b]
$H_2O_2 + Fe^{2+}$ + polyphenols (30 μg/ml)	45.1 ± 3.9[b,c]
$H_2O_2 + Fe^{2+}$ + polyphenols (60 μg/ml)	29.3 ± 3.4[c]
$H_2O_2 + Fe^{2+}$ + caffeine (30 μg/ml)	63.9 ± 5.4[b]
$H_2O_2 + Fe^{2+}$ + caffeine (60 μg/ml)	59.6 ± 6.6[b]
$H_2O_2 + Fe^{2+}$ + theanine (30 μg/ml)	59.8 ± 6.3[b]
$H_2O_2 + Fe^{2+}$ + theanine (60 μg/ml)	51.5 ± 6.1[b,c]
$H_2O_2 + Fe^{2+}$ + (-)-epigallocatechin 3-*O*-gallate (2.5 μg/ml)	40.1 ± 3.4[b,c]
$H_2O_2 + Fe^{2+}$ + (-)-epigallocatechin 3-*O*-gallate (5 μg/ml)	32.1 ± 2.5[a,c]

Statistical significance: [a]$p<0.05$, [b]$p<0.001$ vs value for none; [c]$p<0.001$ vs value for $H_2O_2 + Fe^{2+}$ only.
SOURCE: Copyright 1998 American Chemical Society.

fragmentation and the electrophoretic pattern (Table II and Figure 2). However, similar to the results under hypoxia/reoxygenation, caffeine showed no inhibitory effect like that observed with the polyphenols. Theanine caused significant inhibition of DNA fragmentation at a concentration of 60 μg/mL, as shown in Table II. There may be two mechanisms for such an effect: one is metal chelation, due to the properties of the phenol-hydroxy group; the other may be reduction of the double bonds in the aromatic ring.

Animal Experiments

To cope with oxidative stress, organisms have graded antioxidant mechanisms inside and outside the cell. The present results demonstrated that (-)-epigallocatechin 3-*O*-gallate influenced the activity of radical-scavenger enzymes in subtotally nephrectomized rats, leading to increased activity of superoxide dismutase (SOD), an enzyme which catalyzes the disproportionation of O_2^- into H_2O_2. Moreover, (-)-epigallocatechin 3-*O*-gallate induced an increase in the activity of catalase, an enzyme which specifically eliminates H_2O_2 and suppresses the formation of •OH and OCl^-. On the other hand, there was no change in glutathione peroxidase (GSH-Px), which like catalase, is an

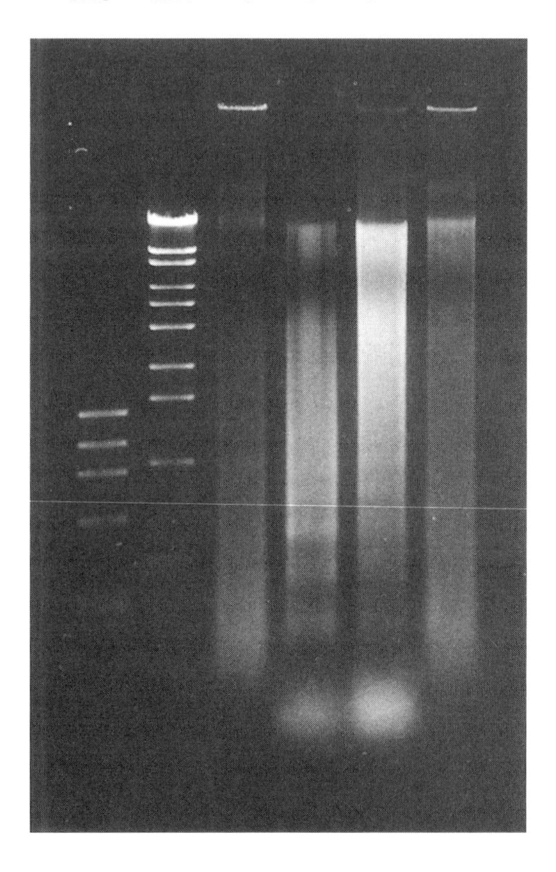

Figure 2. Agarose gel electrophoresis of DNA: Lane 1, none; lane2, H_2O_2 + Fe^{2+}; lane 3, H_2O_2 + Fe^{2+} + (-)-epigallocatechin 3-O-gallate (2.5 μg/ml); lane 4, H_2O_2 + Fe^{2+} + (-)-epigallocatechin 3-O-gallate (5 μM4, marker 4; M6, marker 6. (Copyright 1998 American Chemical Society.)

enzyme that helps eliminate H_2O_2 (Table III). GSH-Px is localized in the matrix of mitochondria. The above finding suggests that the site of action of (-)-epigallocatechin 3-O-gallate is the peroxisome. Peroxisomes in the kidney of the rat contain D-amino acid oxidase and flavin enzymes, such as α-hydroxyacetic acid oxidase, and Cu enzymes such as uricooxidase, that produce H_2O_2 through the oxidation of respective substrates. Catalase works to detoxify the H_2O_2 thus produced and to cleave long-chain fatty acids using H_2O_2 (a kind of β-oxidation). The fact that polyphenols, which are contained in a luxury grocery item such as tea, increase the activity of this enzyme, indicates the possible presence of a promising novel functional substance. In addition, creatinine (Cr), methylguanidine (MG), and guanidinosuccinic acid (GSA), which are known to accumulate in blood with the progression of renal failure, were also decreased in rats given (-)-epigallocatechin 3-*O*-gallate (Table IV), indicating that elimination of free radicals leads to relief of renal disorder. Thus, it is apparent that free radicals play an important role in the progression of renal failure.

The Role of Polyphenol in Dialysis Patients

In Japan, the number of patients on maintenance dialysis has been increasing year by year. Although this increase is likely mainly due to improved prognosis owing to advances in this form of treatment, it should also be recognized that the number of patients with terminal renal failure, for whom dialysis is necessary, is also on the rise.

Although many different diseases can induce chronic renal disease, which in turn may result in terminal renal failure, there is no established cure for this condition. Therefore, once a patient has developed chronic renal failure, control and prevention of disease progression and aggravation is critically important.

As routine dialysis treatment progresses, removal of β_2-microglobulin (β_2-MG) becomes less efficient, residual renal function deteriorates, and MG production increases (*20,21*). Prolonged dialysis may thus lead to various complications, such as a high incidence of malignant tumors, carpal canal syndrome, amyloid arthropathy, and immune deficiency. These conditions have been attributed to enhanced oxidative stress during prolonged dialysis. Recent advances in dialysis technology have enabled the efficient removal of β_2-MG, as well as, a certain degree of inhibition of MG production. However, even when improved dialysis methods are used, the patient's condition tends to worsen after a certain length of time, resulting in further oxidative stress. This highlights the importance of constant inhibition of oxidative stress and the need for antioxidant therapy.

We have previously isolated creatol from the urine of patients with chronic renal failure, and have found that the pathway of Cr metabolism to MG *via* creatol is a common one (*22*). We have further demonstrated, both in vitro and

Table III. Effect of (-)-Epigallocatechin 3-*O*-Gallate on the Activities of Reactive Oxygen Species-Scavenging Enzymes in Rats after Nephrectomy

Group	*Dose* *(mg/kg of BW/day)*	*SOD* *(U/mg protein)*	*Catalase* *(U/mg protein)*	*GSH-Px* *(U/mg protein)*
Nephrectomized rats				
Control		8.90 ± 0.32^{b}	135.6 ± 11.2^{b}	73.89 ± 4.91^{a}
(-)-Epigallocatechin 3-*O*-gallate	2.5	$11.86 \pm 0.54^{b,c}$	$172.1 \pm 10.3^{b,c}$	75.53 ± 4.28^{a}
(-)-Epigallocatechin 3-*O*-gallate	5	$14.49 \pm 0.70^{b,c}$	198.3 ± 9.4^{c}	77.98 ± 3.59
Normal rats		18.15 ± 1.23	210.0 ± 9.3	85.00 ± 3.02

Statistical significance: [a]$p < 0.01$, [b]$p < 0.001$ vs normal rats; [c]$p < 0.001$ vs nephrectomized control rats.

SOURCE: Copyright 1998 American Chemical Society.

Table IV. Effect of (-)-Epigallocatechin 3-*O*-Gallate on the Serum Guanidino Compounds

Group	*Dose (mg/kg of BW/day)*	*Cr (mg/dL)*	*MG (μg/dL)*	*GSA (μg/dL)*
Nephrectomized rats				
Control		2.16 ± 0.13^{a}	3.38 ± 0.18	51.02 ± 4.10
(-)-Epigallocatechin 3-*O*-gallate	2.5	$1.53 \pm 0.10^{a,b}$	2.68 ± 0.15^{b}	35.21 ± 3.16^{b}
(-)-Epigallocatechin 3-*O*-gallate	5	$1.20 \pm 0.09^{a,b}$	1.89 ± 0.13^{b}	29.82 ± 2.47^{b}
Normal rats		0.67 ± 0.04	ND	ND

ND, Not detectable; Statistical significance: [a]$p < 0.001$ vs normal rats; [b]$p < 0.001$ vs nephrectomized control rats.

in vivo, that under conditions in which •OH is generated, Cr is oxidized to creatol, and then to MG *via* the unstable intermediates, creatone A and creatone B (*23,24*). Since determination of these components is useful in evaluating the pathological features of renal failure, we measured changes in the blood levels of Cr and MG in chronic renal failure patients to assess the antioxidant activity of green tea polyphenols, which were administered to patients undergoing dialysis.

Cr is frequently used in the clinical setting as a renal function parameter, and is also a precursor in the conversion of creatol to MG. The blood level of this substance was found to be decreased significantly after 3 months of green tea polyphenol administration, and this effect was maintained until the end of the 6-month administration period. The decrease in the MG level preceded the decrease in Cr, and the decrease became significant after 1 month of green tea polyphenol administration. The level had fallen by 8-10 μg/dL at 5-6 months, as shown in Table V. This suggests, that in addition to decreasing the MG/Cr ratio, green tea polyphenols influence radicals involved in production of MG from Cr. Therefore, by determining the MG/Cr ratio, we found a significant reduction after 2 months of green tea polyphenols administration (Table V). Thus, the radical-scavenging activity of green tea polyphenols has been demonstrated, both in the present clinical study, as well as in previous animal experiments (*17,25*).

Green tea polyphenols had a strong effect on MG in patients with high baseline levels (>65 μg/dL). In contrast, there was no significant change in patients with low baseline MG levels (<50 μg/dL, about 35 μg/dL on average), and no increase occurred during the administration period. A similar effect was found for the MG/Cr ratio, the decrease being most prominent under the greatest oxidative stress.

Aggressive removal of β_2-MG is desirable in order to prevent the complications associated with prolonged dialysis, including amyloidosis. In this context, green tea polyphenols caused a significant decrease at every point of measurement during the 6-month administration period, except at 2 months after the start of administration, as shown in Table VI. When the suppressive effect was analyzed in three groups of patients classified according to their MG levels at the baseline (i.e., by the severity of oxidative stress), a significant fall in β_2-MG was found in the high-MG group during the green tea polyphenols administration period. It was notable that this decrease in β_2-MG occurred despite the use of non-high-performance dialysis, with which it is difficult to eliminate β_2-MG.

In 1977, Craddock *et al*. (*26*) reported that transient leukopenia occurred within 1 h of starting dialysis. They concluded that this was due to activation of the alternative complement pathway by the dialyzer and subsequent sequestration of polymorphonuclear cells by the pulmonary blood vessels caused by the chemotactic action of the activated complement. However, Haeffner-Cavaillon *et al*. (*27*) reported that complement promotes the production of interleukin 1 by

Table V. Effect of Green Tea Polyphenols on Serum Creatinine, Methylguanidine and the Methylguanidine/Creatinine Ratio in Patients Receiving Dialysis

Duration of treatment (month)	*Cr (mg/dL)*	*MG (μg/dL)*	*MG/Cr* ($x10^{-3}$)
0	13.51 ± 0.30	56.43 ± 2.67	4.12 ± 0.17
1	13.33 ± 0.27	53.65 ± 2.30[a]	3.99 ± 0.14
2	13.28 ± 0.22	51.92 ± 2.34[b]	3.86 ± 0.14[a]
3	12.81 ± 0.24[c]	48.66 ± 1.83[c]	3.78 ± 0.12[b]
4	12.65 ± 0.21[c]	49.12 ± 1.76[c]	3.87 ± 0.12[a]
5	12.37 ± 0.24[c]	45.06 ± 1.80[c]	3.62 ± 0.12[b]
6	12.43 ± 0.25[c]	48.41 ± 2.12[c]	3.85 ± 0.13[a]

Statistical significance: [a]$p<0.05$, [b]$p<0.01$, [c]$p<0.001$ vs value for pre-treatment.

Table VI. Effect of Green Tea Polyphenols on Serum β_2-Microglobulin in Patients Receiving Dialysis

Duration of treatment (month)	*β_2-MG (mg/dL)*
0	39.00 ± 1.27
1	34.95 ± 1.08[b]
2	37.46 ± 1.30
3	36.36 ± 1.13[a]
4	36.11 ± 1.03[b]
5	35.65 ± 1.20[a]
6	35.38 ± 1.11[a]

Statistical significance: [a]$p<0.01$, [b]$p<0.001$ vs value for pre-treatment.

monocytes. This in turn increases the level of acute-phase protein in the liver and acts on amyloid precursors to stimulate their conversion to amyloid. It has also been shown that oxygen radicals generated by polymorphonuclear cells or monocytes damage vascular endothelial cells (*21*). In this experiment, we demonstrated that administration of green tea polyphenols suppressed oxidative stress in dialysis patients and also ameliorated pain in the hip, cubitus, coxa, and finger in a certain percentage of patients. This suggests that green tea polyphenols inhibit the production of β_2-MG and its deposition in tissue. There were no significant changes in blood pressure, other general laboratory parameters, or subjective symptoms during the green tea polyphenols administration period.

Aggressive removal of β_2-MG and suppression of free radical activity are required in order to prevent the complications associated with prolonged dialysis, including amyloidosis. In this context, green tea polyphenols appear to represent a promising form of treatment. It may also allow the frequency of dialysis to be reduced in certain patients. However, research in this field is currently insufficient, and further investigations in a large number of patients are therefore necessary.

References

1. Yokozawa, T.; Chen, C.P.; Dong, E.; Tanaka, T.; Nonaka, G.; Nishioka, I. Study on the inhibitory effect of tannins and flavonoids against the 1,1-diphenyl-2-picrylhydrazyl radical. *Biochem. Pharmacol.* **1998,** *56*, 213-222.
2. Yokozawa, T.; Nakagawa, T.; Lee, K.I.; Cho, E.J.; Terasawa, K; Takeuchi, S. Effects of green tea tannin on cisplatin-induced nephropathy in LLC-PK_1 cells and rats. *J. Pharm. Pharmacol.* **1999,** *51*, 1325-1331.
3. Halliwell, B.; Gutteridge, M.C. Oxygen toxicity, oxygen radicals, transition metals and disease. *Biochem. J.* **1984,** *219*, 1-14.
4. Sakanaka, S.; Kim, M.; Taniguchi, M.; Yamamoto, T. Antibacterial substances in Japanese green tea extract against *Streptococcus mutans*, a cariogenic bacterium. *Agric. Biol. Chem.* **1989,** *53*, 2307-2311.
5. Mori, A.; Akagi, M.; Katayama, Y.; Watanabe, Y. α-Guanidinoglutaric acid cobalt-induced epileptogenic cerebral cortex of cat. *J. Neurochem.* **1980,** *35*, 603-605.
6. Yokoi, I.; Toma, J.; Mori, A. The effect of homoarginine on the EEG of rats. *Neurochem.* **1984,** *2*, 295-300.
7. Hatano, T.; Edamatsu, R.; Hiramatsu, M.; Mori, A.; Fujita, Y.; Yasuhara, T.; Yoshida, T.; Okuda, T. Effects of the interaction of tannins with co-existing substances. VI. Effects of tannins and related polyphenols on superoxide anion radical, and on 1,1-diphenyl-2-picrylhydrazyl radical. *Chem. Pharm. Bull.* **1989,** *37*, 2016-2021.

8. Fong, K.L.; McCay, P.B.; Poyer, J.L.; Keele, B.B.; Misra, H. Evidence that peroxidation of lysosomal membranes is initiated by hydroxyl free radicals produced during flavin enzyme activity. *J. Biol. Chem.* **1973,** *248*, 7792-7797.
9. McCord, J.M. Oxygen-derived free radicals in postischemic tissue injury. *N. Engl. J. Med.* **1985,** *312*, 159-163.
10. Saugstad, O.D.; Aasen, A.O. Plasma hypoxanthine concentration in pigs: a prognostic aid in hypoxia. *Eur. Surg. Res.* **1980,** *12*, 123-129.
11. Gastraunthaler, G.J.A. Epithelial cells in tissue culture. *Renal Physiol. Biochem.* **1988,** *11*, 1-42.
12. Yonehana, T.; Gemba, M. Cell injury due to hypoxia/reoxygenation in cultured renal epithelial cell line, LLC-PK_1. In *Kidney and Free Radical*; Koide, H., Ishida, M., Sugisaki, T., Tomino, Y., Owada, S., Ito, S., Eds.; Tokyo Igakusha: Tokyo, 1994; pp 139-141.
13. Snowdowne, K.W.; Freudenrich, C.C.; Borle, A.B. The effects of anoxia on cytosolic free calcium, calcium fluxes, and cellular ATP levels in cultured kidney cells. *J. Biol. Chem.* **1985,** *260*, 11619-11626.
14. Kribben, A.; Wieder, E.D.; Wetzels, J.F.M.; Yu, L.; Gengaro, P.E.; Burke, T.J.; Schrier, R.W. Evidence for role of cytosolic free calcium in hypoxia-induced proximal tubule injury. *J. Clin. Invest.* **1994,** *93*, 1922-1929.
15. Paller, M.S.; Hoidal, J.R.; Ferris, T.F. Oxygen free radicals in ischemic acute renal failure in the rat. *J. Clin. Invest.* **1984,** *74*, 1156-1164.
16. Paller, M.S.; Neumann, T.V. Reactive oxygen species and rat renal epithelial cells during hypoxia and reoxygenation. *Kidney Int.* **1991,** *40*, 1041-1049.
17. Yokozawa, T.; Chung, H.Y.; He, L.Q.; Oura, H. Effectiveness of green tea tannin on rats with chronic renal failure. *Biosci. Biotech. Biochem.* **1996,** *60*, 1000-1005.
18. Forrest, V.J.; Kang, Y.; McClain, D.E. Oxidative stress-induced apoptosis prevented by Trolox. *Free Rad. Biol. Med.* **1994,** *16*, 675-684.
19. Ueda, N.; Walker, P.D.; Hsu, S.M.; Shah, S.V. Activation of a 15-kDa endonuclease in hypoxia/reoxygenation injury without morphologic feature of apoptosis. *Proc. Natl. Acad. Sci. U.S.A.* **1995,** *92*, 7202-7206.
20. Gejyo, F.; Homma, N.; Suzuki, Y.; Arakawa, M. Serum levels of β_2-microglobulin as a new form of amyloid protein in patients undergoing long-term hemodialysis. *New Engl. J. Med.* **1986,** *314*, 585-586.
21. Ishizaki, M. Free radical activity in dialysis patients. *Renal Failure* **1989,** *1*, 81-87.
22. Oura, H.; Yokozawa, T.; Ienaga, K. Free radical and crude drug. In *Aging and Brain*; Oomura, Y., Oura, H., Eds.; Kyoritsu Shuppan: Tokyo, 1992; pp 66-90.

23. Ienaga, K.; Nakamura, K.; Yamakawa, M.; Toyomaki, Y.; Matsuura, H.; Yokozawa, T.; Oura, H.; Nakano, K. The use of ^{13}C-labelling to prove that creatinine is oxidized by mammals into creatol and 5-hydroxy-1-methylhydantoin. *J. Chem. Soc., Chem. Commun.* **1991,** 509-510.
24. Yokozawa, T.; Fujitsuka, N.; Oura, H.; Ienaga, K.; Nakamura, K. *In vivo* effect of hydroxyl radical scavenger on methylguanidine production from creatinine. *Nephron* **1997,** *75*, 103-105.
25. Yokozawa, T.; Oura, H.; Sakanaka, S.; Kim, M. Effects of tannins in green tea on the urinary methylguanidine excretion in rats indicating a possible radical scavenging action. *Biosci. Biotech. Biochem.* **1992,** *56*, 896-899.
26. Craddock, P.R.; Fehr, J.; Dalmasso, A.P.; Brigham, K.L.; Jacob, H.S. Hemodialysis leukopenia. *J. Clin. Invest.* **1977,** *59*, 879-888.
27. Haeffner-Cavaillon, N.; Fisher, E.; Bacle, F.; Carreno, M.P.; Maillet, F.; Cavaillon, J.M.; Kazatchkine, M.D. Complement activation and induction of interleukin-1 production during hemodialysis. *Contr. Nephrol.* **1988,** *62*, 86-98.

Chapter 18

Antioxidant Activity of Phytic Acid Hydrolysis Products on Iron Ion-Induced Oxidative Damage in Biological System

Sayuri Miyamoto[1], Kaeko Murota[1], Goro Kuwataz[2], Masatake Imai[2], Akihiko Nagao[3], and Juliji Terao[1]

[1]Department of Nutrition, School of Medicine, The University of Tokushima, Kuramoto-cho 3, Tokushima 770-8503, Japan
[2]Research Institute of Morinaga & Company Ltd., Simosueyoshi 2-1-1 Tsurumi-ku,, Yokohama 230-0012, Japan
[3]National Food Research Institute, Ministry of Agriculture, Forestry and Fisheries, Tsukuba 305-0856, Japan

Phytic acid (IP_6) is capable of chelating iron ion and thereby blocking the generation of reactive oxygen radicals via Fenton reactions. Some evidence suggests that the consumption of diets rich in phytic acid protect intestinal epithelial cells against iron ion-induced oxidative damage. During digestion. phytic acid is dephosphorylated yielding lower phosphorylated forms of inositol phosphate (IP_5 – IP_1). In this study, we evaluated the antioxidant properties of these hydrolysis products. Their ability to chelate iron ion decreased with the decrease in the number of phosphate groups on the inositol structure. However, IP_2 showed a unique ability to prevent iron ion-induced deoxyribose degradation, and similar to IP_6, IP_3 showed a strong inhibitory effect against iron ion-induced oxidation of large intestinal mucosa homogenate. Interestingly, addition of vitamin E into the liposomal suspension greatly increased the antioxidant activity of IP_3, suggesting a synergistic antioxidant effect. Moreover, oral administration of IP_6 protected large intestinal mucosa against iron ion-induced lipid peroxidation. These observations indicate an important antioxidant function for phytic acid hydrolysis products and suggest that their synergistic effect with vitamin E is an essential factor in the prevention of oxidative damage occurring in large intestinal mucosa.

Dietary iron remains largely unabsorbed in the small intestine, thus it is available to participate in the generation of reactive oxygen radicals in the large intestine (*1,2*). These radicals are capable of attacking all kind of molecules including protein, DNA, and lipids. In particular, radical attack on lipids constituting cellular membranes promotes lipid peroxidation, and thereby enhances oxidative tissue damage (*3*) (Figure 1). Recent studies have shown a positive correlation between high iron intake and colon cancer (*4-6*). On the other hand, epidemiological and animal model studies indicate that diets rich in phytic acid (*myo*-inositol hexaphosphate; IP_6) (Figure 2) may prevent colon carcinogenesis (*7-10*). The anticarcinogenic effect of IP_6 is probably related to its ability to chelate iron and block iron ion-catalyzed redox reactions (*11*). Since IP_6 is partly hydrolyzed during digestion (*12*), this study aimed to evaluate the antioxidant properties of phytic acid hydrolysis products. In the study, we evaluated the effectiveness of IP_5 – IP_2 against iron ion-induced oxidative reactions. It was demonstrated that IP_2 and IP_3 possess considerable antioxidant functions and oral administration of IP_6 was effective in protecting large intestinal mucosa against iron ion-induced lipid peroxidation.

Effect of Phytic Acid Hydrolysis Products on Deoxyribose Degradation Assay

Iron ion-binding ability of phytic acid (IP_6) and its hydrolysis products (IP_5, IP_4, IP_3, and IP_2) were assessed by the deoxyribose degradation assay (*13*). In the assay, the sugar 2-deoxyribose (2.8 mM) was degraded in the presence of $Fe(NO_3)_3$ (10 μM), ascorbic acid (100 μM), and H_2O_2 (1 mM), generating malonaldehyde-like fragments, which were determined by the TBA assay. IP_6 –

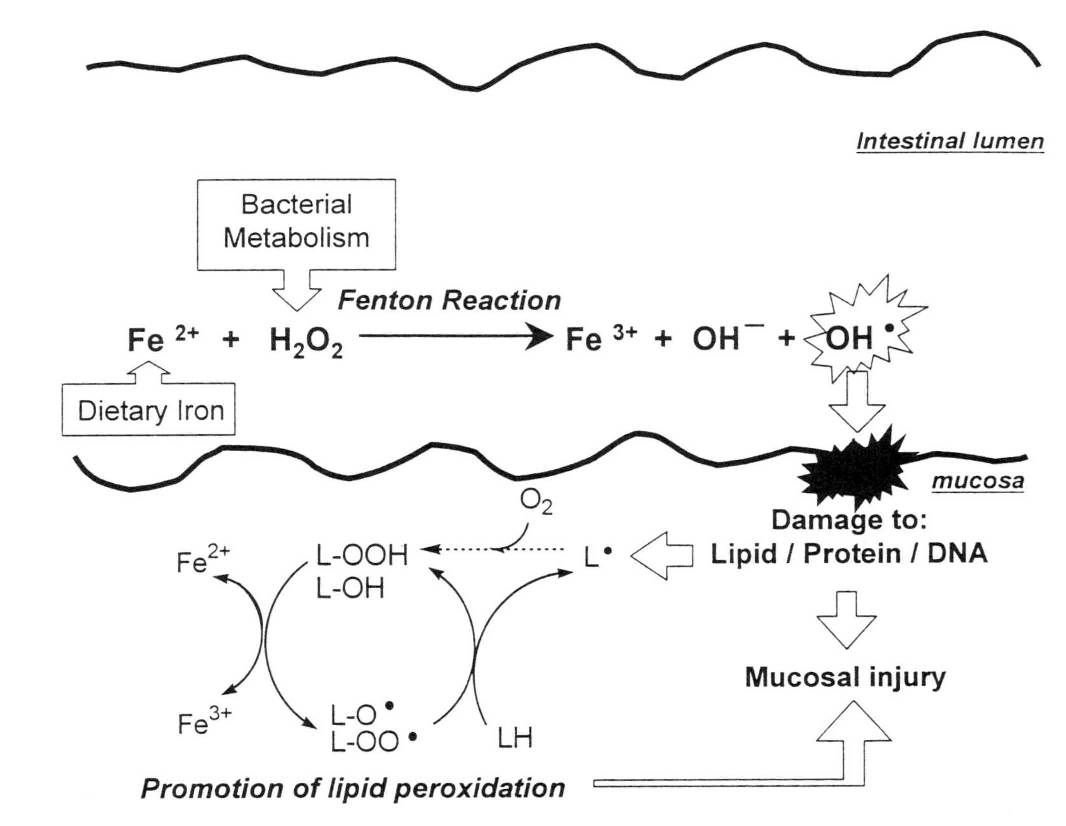

Figure 1. Possible mechanism of iron ion-induced oxidative damage in the large intestine.

IP_2 and Desferal (100 μM) were added to the reaction mixture and their inhibitory ratio was calculated (Figure 3). The strong iron chelator Desferal, used as a positive control, showed the highest inhibitory effect. IP_5 provided less inhibition than IP_6 and IP_4, and IP_3 showed no inhibitory effect, suggesting that dephosphorylation reduces the ability of IP_6 to inhibit Fenton reaction. However, IP_2 showed a unique, high ability to prevent deoxyribose degradation. This phenomenon can not be explained by its iron chelating activity alone. It has been suggested that deoxyribose degradation involves a series of reactions (*14*) and IP_2 probably has the ability to inhibit one of these reactions.

Synergistic Effect of IP_3 and Vitamin E on Iron Ion-induced Oxidation of Liposomal Membrane

Here, we focused in the antioxidant properties of IP_3 and tried to evaluate whether the addition of vitamin E has any effect on its antioxidant activity. Large unilamellar vesicle liposomes composed of phosphatidylcholine were prepared by the method previously described (*15*). The liposomal suspension was incubated with IP_3 (100 μM) with or without the addition of α-tocopherol (1 μM), and oxidation was initiated by addition of $Fe(NO_3)_3$ (10 μM) and ascorbic acid (100 μM). The reaction mixture was incubated at 37 °C, and accumulation of phosphatidylcholine hydroperoxides (PC-OOH) and the disappearance of α-tocopherol were determined by HPLC analysis (Figure 4). IP_3 slightly retarded the accumulation of PC-OOH, whereas α-tocopherol inhibited PC-OOH accumulation during the first 5 hours. However, the combination of IP_3 and α-tocopherol was much more effective in preventing the oxidation. Together they inhibited oxidation during approximately 15 hours. The disappearance of α-tocopherol was also retarded during this period. It is therefore likely that IP_3 and α-tocopherol have a synergistic antioxidant effect, and its combination may play an important role in the prevention of iron ion-induced oxidative damage.

Antioxidant Activity of Phytic Acid Hydrolysis Products in the Rat Large Intestine: in vitro and ex vivo Study

We examined the antioxidant activity of phytic acid hydrolysis products using rat large intestinal mucosa homogenate. Wistar male rats (6 weeks, 160-180 g) were used for the test. In the ex vivo study, 0.5 ml of an aqueous solution of phytic acid (100 mg) was administered intragastrically to the rats. The large intestinal mucosa was removed by scraping and then homogenized in Tris-HCl buffer (0.1 M, pH 7.4; containing 0.135 M KCl). The in vitro study was carried

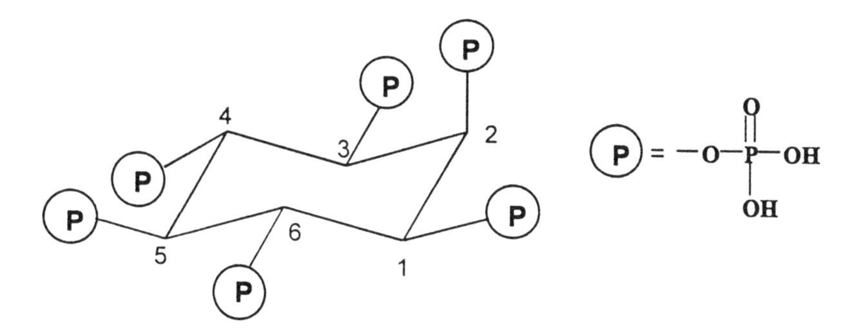

Figure 2. Structure of phytic acid.

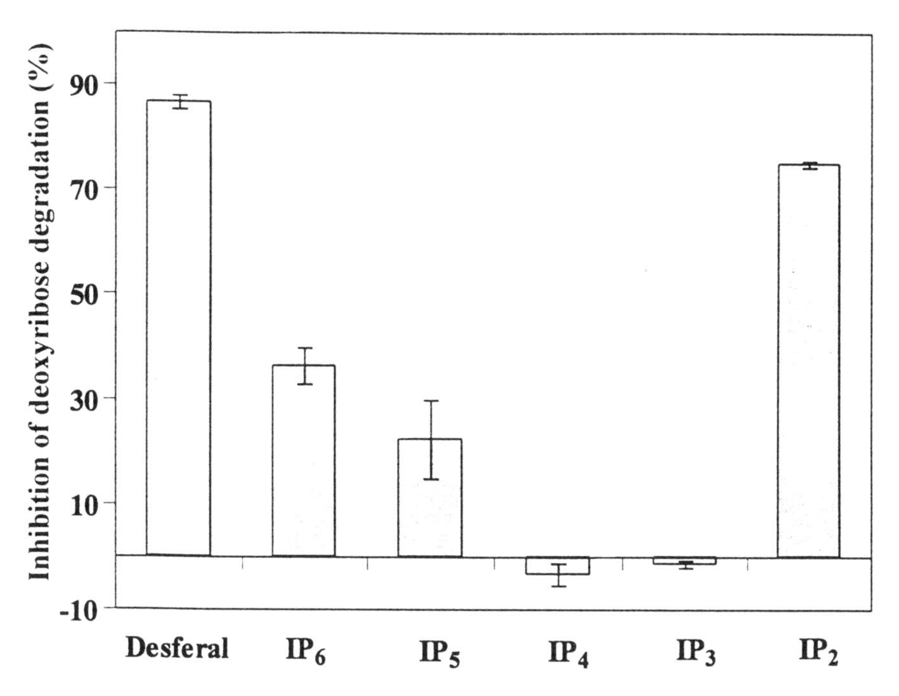

Figure 3. Effect of phytic acid hydrolysis products on iron ion-induced deoxyribose degradation.

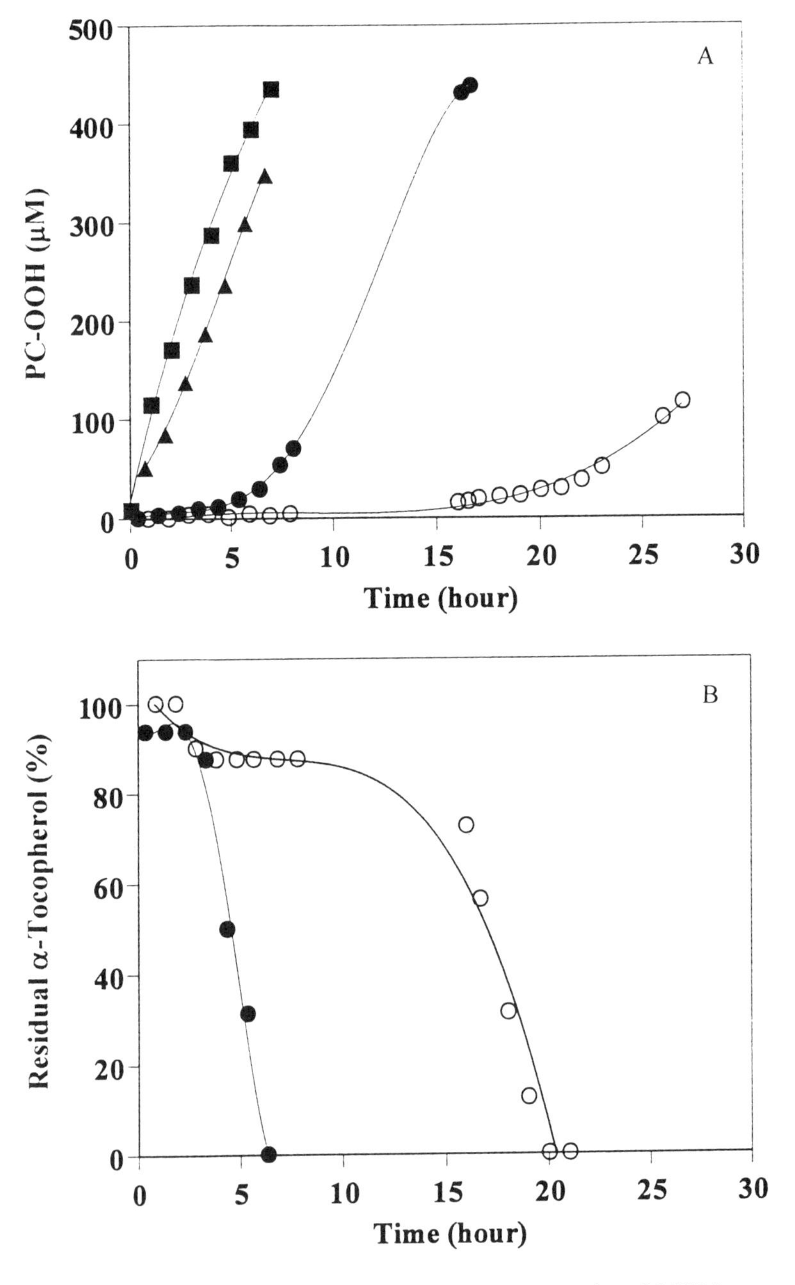

Figure 4. Effect of IP_3 and α-tocopherol on iron ion-induced LUV-liposome lipid peroxidation. (A) PC-OOH accumulation. (B) α-tocopherol disappearance. (■) Control, (▲) IP_3, (●) α-tocopherol, (○) α-tocopherol plus IP_3

out by the addition of IP_6, IP_3, or Desferal (1 mM) into the homogenate obtained from untreated animals. In both cases mucosal homogenate was oxidized by the addition of $Fe(NO_3)_3$ (10 μM) and ascorbic acid (100 μM). The incubation was carried out at 37 °C for 4 hours, and the oxidation level was analyzed by TBA assay (Table I). Similar to IP_6 and Desferal, IP_3 strongly inhibited formation of TBARS, suggesting that IP_3 is effective in preventing iron ion-induced lipid peroxidation in the large intestine. It is therefore expected that IP_6 can protect large intestinal mucosa against iron ion-induced lipid peroxidation even after digestion. Indeed in the ex vivo study, oral administration of IP_6 significantly protected large intestinal mucosa against iron ion-induced lipid peroxidation. Large intestinal mucosa from treated animals showed a significantly lower level of TBARS compared to results for untreated animals (Table I). Although the level of hydrolysis was not determined, phytic acid digestion products are suggested to exert an important antioxidant function in the large intestine. Further work is in progress to examine the ex vivo effects of IP_3 and IP_2.

Summary

Phytic acid is hydrolysed by the action of phytase during the digestive process (Figure 5). Dephosphorylation apparently decreases the ability of phytic acid hydrolysis products to inhibit hydroxyl radical generation via Fenton reaction. However, our results demonstrate that IP_2 and IP_3 still have important antioxidant functions. Indeed, administration of IP_6 protected large intestinal mucosa against iron ion-induced lipid peroxidation. Furthermore, the possible synergistic antioxidant effect of IP_3 with vitamin E implies that the combined effect of antioxidants exerts an essential role in the prevention of iron ion-induced oxidative damage.

References

1. Babbs, C.F. *Free Radic. Biol. Med.* **1990,** *8,* 191-200.
2. Lund E.K., Wharf S.G., Fairweather-Tait S.J., Johnson I.T. *Am. J. Clin.Nutr.* **1999,** *69,*250-5.
3. Halliwell, B.; Gutteridge, J.M.C. *Methods Enzymol.* **1990,** *186,* 1-85.
4. Nelson, R.L. *Free Rad. Biol. Med.* **1992,** *12,* 161-168.
5. Weinberg, E.D. *Biometals* 1994, 7, 211-216.
6. Lund E.K., Wharf S.G., Fairweather-Tait S.J., Johnson IT. *J. Nutr.* **1998,** *128.*175-179.
7. Graf, E.; Eaton J.W. *Cancer* **1985,** *56,* 717-718.
8. Graf E.; Eaton J.W. *Nutr. Cancer* **1993,** *19,* 11-19.

Table I. Effect of IP_6, IP_3 and Desferal on iron ion-induced large intestinal mucosa homogenate lipid peroxidation[a]

	TBARS (nmol/mg protein)			
	Control	*IP_6*	*IP_3*	*Desferal*
in vitro[b]	9.28 ± 1.42 [a]	1.25 ± 0.31 [b]	2.15 ± 0.77 [b]	0.84 ± 0.27 [b]
ex vivo[c]	8.89 ± 2.07 [a]	6.06 ± 1.59 [b]	n.d[d]	n.d[d]

Data are mean ±SD (n=4 or 5). Values in the same row with different superscripts are significantly different ($p<0.05$).

[a] Large intestinal mucosa homogenate was oxidized by the addition of $Fe(NO_3)_3$ and ascorbic acid solution (100 μM and 1 mM).

[b] IP_6, IP_3 and Desferal (1 mM) were added to large intestinal mucosa homogenate.

[c] IP_6 aqueous solution (100 mg) was administered orally to rats.

[d] n.d.= not determined

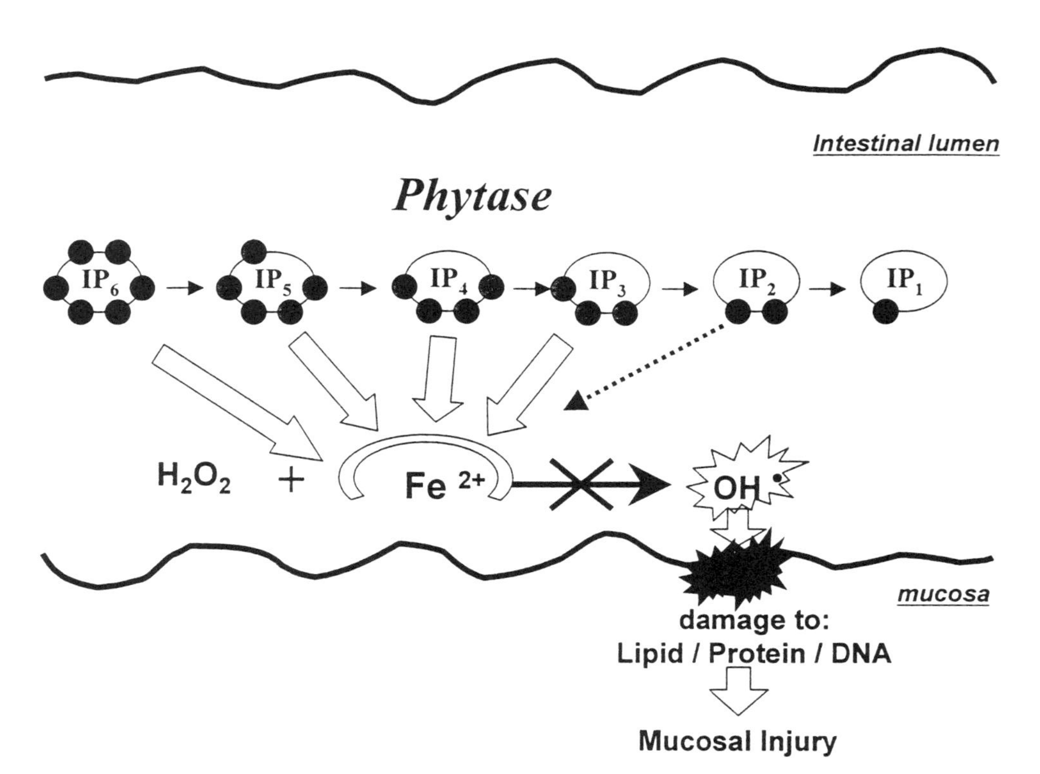

Figure 5. Proposed pathway for phytic acid hydrolysis and its effect on iron-ion induced oxidative damage of large intestinal mucosa.

9. Shamsuddin, A.M.: Elsayed, A.M.: Ullah, A. *Carcinogenesis* **1988,** *9,* 577-580.
10. Jenab, M.; Thompson, L.U. *Carcinogenesis* **1998,** *19,* 1087-1092.
11. Graf, E.; Empson, K.L. *J. Biol. Chem.* **1987,** *262,* 11647-11650.
12. Sandberg. A.-S; Andersson, H. *J. Nutr.* **1988,** *118,* 469-473.
13. Halliwell, B.. Gutteridge. J.M.C.; Aruorna, O.I. *Anal. Biochem.* **1987,** *165,* 215-219.
14. Winterbourn, C.C. *Free Rad. Biol. Med.* **1991,** *II,* 353-360.
15. Terao, J.; Piskula, M.; Yao, Q. *Arch. Biochem. Biophys.* **1994,** *308,* 278-284

Chapter 19

Bromination, Chlorination, and Nitration of Isoflavonoids

Brenda J. Boersma[1], Stephen Barnes[1,2], Rakesh P. Patel[3], Marion Kirk[2], Donald Muccio[4], and Victor M Darley-Usmar[3]

[1]Department of Pharmacology & Toxicology, [2]Comprehensive Cancer Center Mass Spectrometry Shared Facility, Departments of [3]Pathology and [4]Chemistry, University of Alabama at Birmingham, Birmingham, AL 35294

Isoflavones have been shown to be beneficial in several chronic diseases in which oxidants are involved. Isoflavones are weak antioxidants when tested in vitro in the context of scavenging lipid peroxyl radicals. However, at sites of inflammation proinflammatory oxidants such as peroxynitrite ($ONOO^-$), hypochlorous acid (HOCl), and hypobromous acid (HOBr) are formed that may also react with isoflavones. We present evidence herein, using reverse-phase HPLC-mass spectrometry and proton NMR, that at concentrations of these oxidants formed at these local sites, nitration, chlorination, and bromination of isoflavones occurs.

Introduction

There is an ever-growing interest in the study of the isoflavones, a group of plant polyphenols found in large amounts in soy and other tropical legumes used in the diet. Their importance in the prevention of chronic disease is gradually unfolding. The soy isoflavone genistein (5,7,4'-trihydroxyisoflavone) has been shown to inhibit the formation of mammary tumors in rodent models of breast cancer *(1,2)*. Isoflavones have an important role in lipid lowering in cardiovascular disease *(3)* and have direct effects on vascular function *(4)*. As has been observed for many polyphenols, isoflavones have antioxidant properties when tested using in vitro models of oxidation of low-density lipoprotein (LDL) and with specific peroxyl radical generating systems *(5-7)*. However, compared to the classical antioxidant butylated hydroxy toluene (BHT) which is an effective inhibitor of lipid oxidation in in vivo models of cardiovascular disease, the soy isoflavones genistein and daidzein (7,4'-dihydroxyisoflavone) have only weak antioxidant effects *(8)*. Because of their low blood concentrations (10-100 nM as the aglucone even on a soy-based diet) *(9)* and their weak antioxidant properties, it is unlikely that they significantly contribute to scavenging of oxidants such as lipid peroxyl radicals.

In atherosclerosis and certain other chronic diseases, specific inflammatory cells are recruited to the site of inflammation. These include macrophages, neutrophils, and eosinophils. These specialized cells emit large quantities of proinflammatory oxidants utilizing either superoxide anion or hydrogen peroxide to generate other more oxidizing compounds. Macrophages generate hydroxyl radicals (OH•) and peroxynitrite ($ONOO^-$) *(10)*. Neutrophils convert hydrogen peroxide to hypochlorous acid (HOCl) *(5)*, whereas eosinophils make hypobromous acid (HOBr) *(11)*. Each of these compounds reacts with tyrosyl residues on proteins. Nitrated, chlorinated, and brominated tyrosine residues have been observed in atherosclerotic lesions *(12-16)*.

The B-ring in isoflavones is phenolic and therefore similar to that in tyrosine (Figure 1). Accordingly, we hypothesize that isoflavones react with $ONOO^-$, HOCl, and HOBr, thereby forming new isoflavone derivatives at the site of inflammation. These modifications to the isoflavones may alter their properties at specific receptor sites such as the estrogen receptors and in signal transduction pathways.

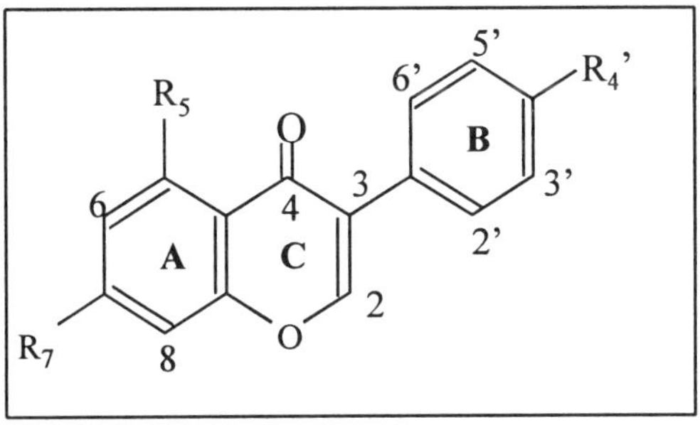

Figure 1. The structure of the isoflavones genistein (R_4'-OH, R_5-OH, R_7–OH), biochanin-A (R_4'-OCH_3, R_5-OH, R_7–OH), and daidzein (R_4'-OH and R_7–OH)

Materials & Methods

Materials

Genistein was extracted and purified as described previously *(17)*. Daidzein was obtained from LC labs (Worburn, MA). Biochanin A, HOCl, and bromine were obtained from Aldrich Chemicals (Milwaukee, WI). $ONOO^-$ was synthesized as reported previously *(18)* and quantified spectrophotometrically at 302 nm (pH =12, ε = 1670 $M^{-1}cm^{-1}$) in 1 M NaOH. HOCl concentrations were determined spectrophotometrically at 290 nm (pH =12, ε = 350 $M^{-1}cm^{-1}$). HOBr was prepared as described by Thomas et al. *(19)*. Briefly, 5 mM HOBr was prepared by reacting bromine (5 μL) with 900 μL cold 1.0 M sodium phosphate (NaPi) buffer, pH 7.0, and then diluting the reaction mixture with 17.4 mL cold water. The concentration of HOBr was confirmed by measuring the formation of mono-bromotaurine at 288 nm. HOBr (300 μL) was added to 700 μL of 0.1 M taurine in 1 M NaPi, pH 7.0 and the absorbance at 288 nm was measured and the concentration was determined using ε= 450 M^{-1} cm^{-1}. *Warning: bromine is very corrosive. May be fatal if swallowed or inhaled, and will cause severe burns to every area of contact.*

Reaction of Isoflavones with HOBr, HOCl, and $ONOO^-$

Varying amounts of HOBr (0-500 μM) were added with continuous mixing to the isoflavones (0-500 μM) in 100 mM sodium phosphate buffer, pH 7.0, to achieve the nominal concentration. Similar reactions were carried out with HOCl and $ONOO^-$ as described by Boersma et al. *(20)*. The reaction mixtures were then extracted prior to reverse-phase HPLC analysis as follows: 1 mL reaction mixture was added to 4 mL water and 2 mL diethyl ether with 100 μM internal standard. The samples were vortexed and centrifuged at 2,000 x g, whereupon the ethereal, top layer was removed. An additional 2 mL of ether was added and the same steps were repeated until a total volume of 10 mL of ether was added. Ether layers were combined and dried to evaporation under air. Prior to injection, 150 μL 80% methanol was added to redissolve the dried residues.

HPLC Analysis of Reaction Products

The HPLC conditions were as follows: all analyses were carried out on a Beckman HPLC 125 solvent module, diode array Model 168 detector (Fullerton, CA) with an Aquapore octyl RP-300, C-8, 22 cm x 4.6 mm i. d., 7-μm column pre-equilibrated with 10% aqueous acetonitrile in 0.1% trifluoroacetic acid (TFA). The column was eluted at a flow rate of 1.5 mL/ min with the following mobile phase composition: 0-10 min, linear gradient (10-50%) of acetonitrile in 0.1% TFA; 10-12 min, linear gradient (50-90%) of acetonitrile in 0.1% TFA; and 12-15 min, isocratically with 90% aqueous acetonitrile in 0.1% TFA. The eluent was monitored at 262 nm by the diode array detector.

Mass Spectrometry Analysis of Reaction Products

Samples were also analyzed by LC-MS. Samples were extracted as described above. The reaction mixtures were separated by HPLC using a 10 cm x 4.6 mm i. d., C-8 Aquapore reversed-phase column pre-equilibrated with 10 mM ammonium acetate (NH_4OAc). The mobile phase composition was: 0-10 min, linear gradient (0-50%) of acetonitrile in 10 mM NH_4OAc; 10-12 min, isocratically with 50% aqueous acetonitrile in 10 mM NH_4OAc; 12-15 min, linear gradient (50-90%) of acetonitrile in 10 mM NH_4OAc; and 15-17 min, isocratically with 90% aqueous acetonitrile in 10 mM NH_4OAc. Mobile phase flow was 0.4 mL/min with 0.1 mL/min of the column eluent being passed into the HN-APCI interface of a PE-Sciex (Concord, Ontario, Canada) API III triple quadrupole mass spectrometer. The voltage on the corona discharge needle was ~8000 V and the orifice potential was set at ~60 V. Negative ion spectra were recorded over a m/z range of 200-500. Selected $[M-H]^-$ molecular ions were analyzed by collision-induced dissociation with 90% argon-10% nitrogen, and the daughter ion mass spectra recorded. Data were analyzed using software provided by the manufacturer on Macintosh Quadra 950 and PowerPC 9500 computers (Apple Computers, Cupertino, CA).

Results

Identification of Halogenated and Nitrated Isoflavones

In the first series of experiments genistein, daidzein, and biochanin-A were reacted with a range of HOBr concentrations and subjected to analysis using

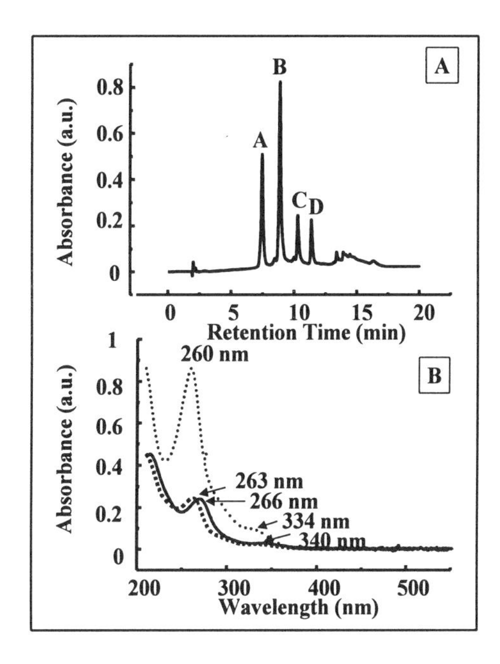

Figure 2. A) Reverse-phase HPLC chromatogram of genistein (Peak B) reacted with HOBr with daidzein as an internal standard (Peak A). Two novel peaks are shown (C and D). B) The UV-visible absorbance spectra of genistein (dotted line), peak C (dashed line), and peak D (bold line).

reverse-phase HPLC. Figure 2A shows the chromatogram of genistein reacted with HOBr, indicating there are four peaks. Peak A corresponds to the internal standard, daidzein, and peak B corresponds to genistein. The formation of peaks C and D are indicative of novel product formation. Figure 2B shows the UV-visible absorbance spectra of the new products resulting from the interaction of genistein with HOBr at physiologic concentrations of the oxidant. A small shift of 3 nm and 6 nm is seen in Peak C when compared to genistein at 260 and 344 nm, respectively. Similarly, Peak D also shifted 6 nm compared to genistein at 260 nm. Similar results were seen with daidzein and biochanin-A when reacted with HOBr (Table I). Furthermore, the reaction of the isoflavones with HOCl showed the same pattern on the HPLC chromatogram *(20)*. When the isoflavones were reacted with $ONOO^-$, only one novel product peak was detected for genistein and daidzein, whereas biochanin-A did not react *(20)*.

Table I. Values of Isoflavones Reacted with HOBr

Isoflavone	*Retention time HPLC (min)*	λ_{max} *(nm)*	*M/z [M-H]⁻*
Genistein	8.88	260/334	269
Product 1	10.31	263/340	347
Product 2	11.34	266/340	427
Daidzein	7.68	245/304	253
Product 1	9.32	248/307	331
Product 2	10.54	260/ND	411
Biochanin-A	11.40	257	283
Product 1	12.61	266	361
Product 2	13.37	260	441

HPLC-mass spectrometry was used to identify the novel products. Peak C and peak D had molecular ions of *m/z* 347 and 427, respectively (Figure 3, Table I). For peak C, this is an increase of *m/z* 78 over genistein, corresponding to monobromination of genistein (Figure 3A). Peak D had a *m/z* of 427, which is an increase of *m/z* 158 compared to genistein or *m/z* 80 over bromogenistein, confirming the addition of two bromine molecules to genistein (Figure 3B). The increase of *m/z* 80 as opposed to *m/z* 78 is a consequence of the approximately equal abundance of ^{79}Br and ^{81}Br. For dibromogenistein, the most abundant species contained ^{79}Br and ^{81}Br, i.e., an increase of *m/z* 158 compared with genistein.

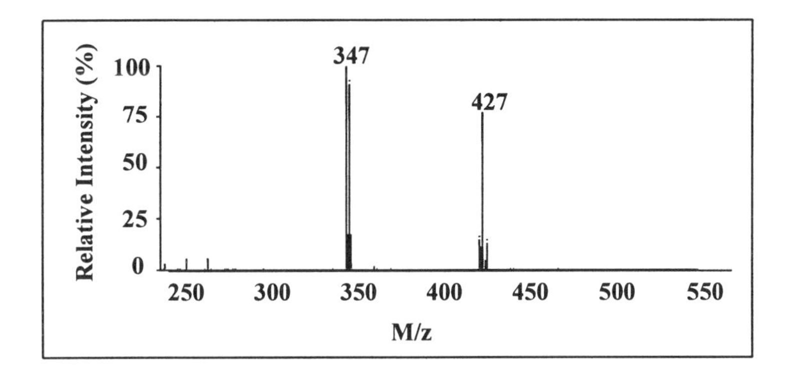

Figure 3. Reverse-phase LC- MS negative ion mass spectra selected ion chromatograms of m/z 347 (monobromogenistein), and m/z 427(dibromogenistein).

Selected ion chromatograms showed that in the neutral mobile phase the brominated products (Figure 4B and 4C) eluted more rapidly than genistein (Figure 4A). This was in contrast to the order observed in the acidic mobile phase (Figure 2A).. MS-MS fragmentation patterns of the mono- and dibromogenistein products were also determined (Figures 5A and B, respectively). In these fragmentation patterns, the *m/z* 79 ion is apparent, which is evidence for bromine. The novel products of the reaction of biochanin-A and daidzein showed similar results. Both isoflavones were mono- and dibrominated when in the presence of HOBr (Table I).

When genistein and biochanin-A react with physiologic concentrations of HOCl, mono-and dichlorinated products can be detected. Daidzein forms only a monochlorinated metabolite. When genistein and daidzein react with $ONOO^-$, mononitrated derivatives were detected. However, biochanin-A did not react with $ONOO^-$ *(20)*.

Discussion

In the present study we have shown that the isoflavones can react with proinflammatory oxidants to form novel derivatives. By using LC-MS we have confirmed that isoflavones react with HOBr to form mono- and dibrominated derivatives. These results correlate to the chlorination and nitration of the isoflavones our laboratory previously reported *(20)*. Mono- and dichlorinated isoflavones were formed when reacted with HOCl. Nitration of the isoflavones also occurs when reacted with $ONOO^-$. Nitration occurs at the C3' site of B-ring of the molecule.

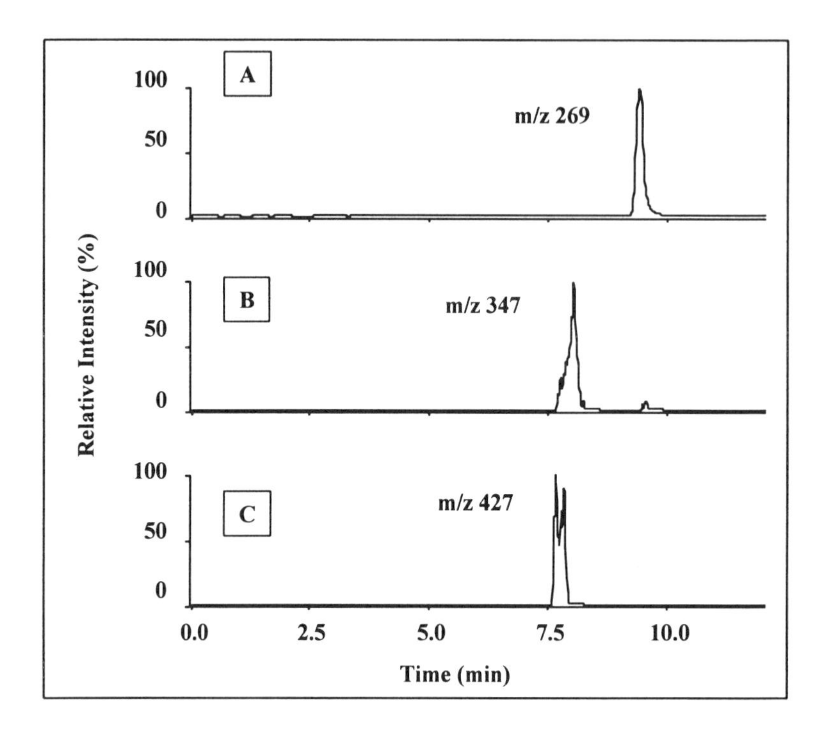

*Figure 4. Reverse-phase LC-MS selected ion chromatograms of (**A**) m/z 269 (genistein), (**B**) m/z 347 (monobromogenistein), and (**C**) m/z 427(dibromogenistein).*

Proinflammatory oxidants are formed at local sites in the body in the micromolar concentration range. HOBr is formed by the reaction of H_2O_2 with Br^- and is catalyzed by the enzyme eosinophil peroxidase (EPO), which is secreted by eosinophils, or by myeloperoxidase (MPO), secreted by neutrophils. EPO has a 1000-fold preference for Br^- ion as a substrate over Cl^- *(11,21)*. HOCl is formed by the reaction of H_2O_2 with Cl^- catalyzed by MPO or EPO.

In our previous study *(20)*, sites of chlorination were determined by proton NMR. It is likely that the reaction between the isoflavones and HOBr results in the formation of multiple brominated isomers. Given the similar reaction, we propose the sites of bromination are similar to that of chlorination, namely at the C6, C8, or C3' sites (Figure 1).

These reactions may have additional complexity. We have shown that chlorinated and nitrated derivatives of genistein are formed in two different cellular models *(22)*. We are currently isolating these species to determine their biological effects.

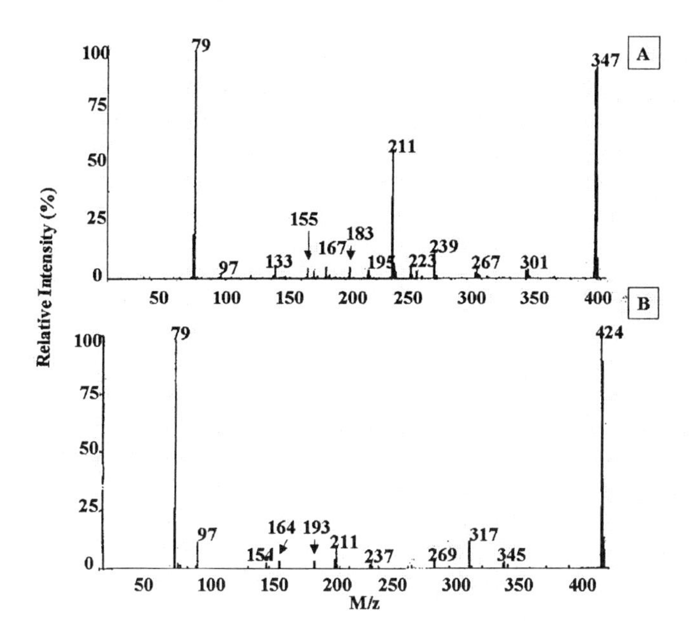

Figure 5. MS-MS spectra of m/z 347, monobromogenistein (A) and m/z 427, and dibromogenistein (B).

In summary, we have shown that the isoflavones can react with proinflammatory oxidants and form novel derivatives. Given the known biological targets of the isoflavones, such as estrogen receptors *(23)* and tyrosine kinases *(24)*, we have hypothesized that structural modifications to the isoflavones may alter these activities *(25)*. Indeed, Traxler et al. *(26)* have shown that from a pharmacophore analysis that 3'-chloro-5, 7-dihyroxyisoflavone is a tyrosine kinase inhibitor with a low nM K_i. Since a 3'-chloroisomer of genistein is formed from reaction with HOCl, it and other isoflavones may have potent biological activity in regulating tyrosine kinases in the vicinity of inflammatory cells such as macrophages, neutrophils, and eosinophils.

Acknowledgements

The authors would like to thank Wei-Wei Huang for her technical support. These studies were supported by a grant from the United Soybean Board (#7312) to SB and VDU. The mass spectrometer was purchased by funds from a NIH Instrumentation Grant (S10RR06487) and from this institution. Operation of the UAB Mass Spectrometry Shared Facility has been supported in part by a NCI Core Research Support Grant (P30 CA13148-27) to the University of Alabama at Birmingham Comprehensive Cancer Center.

References

1. Barnes, S. *Breast Canc. Res. Treat.* **1997,** 46, 169-179.
2. Fritz, W. A.; Coward, L.; Wang, J.; Lamartiniere, C. A. *Carcinogenesis* **1998,** 19, 1623-1629.
3. Anthony, M. S.; Clarkson, T. B.; Hughes, C. L. J.; Morgan, T. M.; Burke, G. L. *J. Nutr.* **1996,** 126, 43-50.
4. Williams, J. K.; Hodgin, J.; Smithies, O.; Korach, K. *J. Nutr.* **2000,** 130, 672S-673S.
5. Harrison, J. E.; Schultz, J. *J. Biol. Chem.* **1976,** 251, 1371-1374.
6. Kapiotis, S. M.; Hermann, M.; Held, I.; Seelos, C.; Ehringer, H.; Gmeiner, B. M. K. *Arterio. Thromb. Vas. Biol.* **1997,** 17, 2868-2874.
7. Tikkanen, M. J.; Wahala, K.; Ojala, S.; Vihma, V.; Aldercreutz, H. *Proc. Natl. Acad. Sci.* **1998,** 95, 3106-3110.
8. Patel, R. P.; Crawford, J.; Boersma, B.; Barnes, S.; Darley-Usmar, V. M. *J. Med. Foods* **1999,** 2, 163-166.
9. Coward, L.; Kirk, M.; Albin, N.; Barnes, S. *Clin. Chim. Acta* **1996,** 247, 121-142.
10. Beckman, J. S.; Beckman, T. W.; Chen, J.; Marshall, P. A.; Freeman, B. A. *Proc. Natl. Acad. Sci.* **1990,** 87, 1620-1624.
11. Weiss, S. J.; Test, S. T.; Eckmann, C. M.; Roos, D.; Regiani, S. *Science* **1986,** 234, 200-203.
12. Domigan, N, M.; Charlton, T. S.; Duncan, M. W.; Winterbourn C. C.; Kettle, A. J. *J. Biol. Chem.* **1995,** 270, 16542-16548.
13. Hazen, S. L.; Heinecke, J. W. *J. Clin. Invest.* **1997,** 99, 2075-2081.
14. Daugherty, A.; Dunn, J. L.; Rateri, D. L.; Heinecke, J. W. *J. Clin. Invest.* **1994,** 94, 437-444.
15. Kaur, H.; Halliwell, B. *FEBS Lett.* **1994,** 350, 9-12.

16. Beckman, J. S.; Ye, Y. Z.; Anderson, P. G.; Chen, J.; Accavitti, M. A.; Tarpey, M. M.; White, C. R. *Biol. Chem. Hoppe-Seyler* **1994,** 375, 81-88.
17. Peterson, T. G.; Barnes, S. *Biochem. Biophys. Res. Commun.* **1991,** 179, 661-667.
18. White, C. R.; Thomas, S.; Green, I; Crow, J. P.; Beckman, J. S.; Patel, R. P.; Darley-Usmar, V. M. *Methods in Molecular Biology, Nitric Oxide Protocols* Humana Press Inc.: Totowa, NJ, 1998; Vol. 100, pp 215-230.
19. Thomas, E. L.; Bozeman, P. M.; Jefferson, M. M.; King, C. C. *J. Biol. Chem.* **1995,** 270, 2906-2913.
20. Boersma, B. J.; Patel, R. P.; Kirk, M.; Jackson, P. L.; Muccio, D.; Darley-Usmar, V. M.; Barnes, S. *Arch. Biochem. Biophys.* **1999,** 368, 265-275.
21. Mayeno, A. N.; Curran, A. J.; Roberts, R. L.; Foote, C. S. *J. Biol. Chem.* **1989,** 264, 5660-5668.
22. Boersma, B. J.; Patel, R. P.; Kirk, M.; Darley-Usmar, V. M.; Barnes, S. *Free Rad. Biol. Med.* **1999,** 27, S32.
23. Pike, A. C. W.; Brzozowski, A. M.; Hubbard, R. E.; Bonn, T.; Thorsel, A. G.; Engstrom, O.; Ljunggren, J.; Gustafsson, J. K.; Carlquist, M. *EMBO J.* **1999,** 18, 4608-4618.
24. Akiyama, T.; Ishida, J.; Nakagawa, S.; Ogawara, H.; Watanabe, S.; Itoh, N. M.; Shibuya, M.; Fukami, Y. *J. Biol. Chem.* **1987,** 262, 5592-5595.
25. Barnes, S.; Kim, H.; Darley-Usmar, V.; Patel, R.; Xu, J.; Boersma, B.; Luo, M. *J. Nutr.* **2000,** 130, 656S-657S.
26. Traxler, P.; Green, J.; Mett, H.; Sequin, U.; Furet, P. *J. Med. Chem.* **1999,** 42, 1018-1026.

Nutritional Biochemistry and Health

Chapter 20

Inhibition of Xanthine Oxidase and NADPH Oxidase by Tea Polyphenols

Jen-Kun Lin[1], Ping-Chung Chen[1], Chi-Tang Ho[2], and Shoei-Yn Lin-Shian[3]

[1]Institute of Biochemistry, College of Medicine, National Taiwan University, Taipei, Taiwan
[2]Department of Food Science, Rutgers University, 65 Dudley Road, New Brunswick, NJ, 08901-8520
[3]Insitute of Toxicology and Pharmacology, College of Medicine, National Taiwan University, Taipei, Taiwan

Superoxide, hydroxyl radical and hydrogen peroxide are major reactive oxygen species (ROS) that have been implicated in the pathogenesis of many diseases including respiratory disease syndromes, rheumatoid arthritis, neurodegenerative disorders, cardiovascular diseases and cancer. Tumor promoters such as phorbol-12-myristate-13-acetate (PMA) enhance the generation of these ROS through protein kinase C pathway to activate NADPH oxidase and xanthine oxidase. Tea polyphenols such as (-)-epigallocatechin-3-gallate (EGCG) and theaflavins including theaflavin, theaflavin-3-gallate and theaflavin-3,3′-digallate are able to inhibit the formation of ROS through the competitive inhibition of xanthine oxidase or NADPH oxidase. These tea polyphenols are also able to scavenge extracellular hydrogen peroxide. It has been demonstrated that cellular ROS are important mitotic signals that modulate cell proliferation and apoptosis. Furthermore, ROS are pivotal factors that stimulate the process of tumor promotion. These findings suggest that tea polyphenols may suppress the development of tumor promotion and progression through both ROS scavenging and direct inhibition of xanthine oxidase and NADPH oxidase.

Reactive oxygen species (ROS) is a collective term often used by biological scientists to include not only hydrogen peroxide, hydroxyl radicals, peroxide anions, superoxide anion but also singlet oxygen, ozone, hypochlorous acid, peroxyl, alkoxyl, nitric oxide, and peroxynitrite. Normal loads of ROS are removed by enzymes such as superoxide dismutase (SOD), catalase, glutathione peroxidase and other endogenous reductants. Over-production of ROS could increase the oxidative stress in the body and accelerate membrane damage, DNA base oxidation, and DNA strand breaks and chromosome aberrations, most of which would be involved in the carcinogenesis process (*1*). It has been considered that both exogenous and endogenous ROS play important roles in carcinogenesis and chronic diseases (*2,3*). The levels of hydrogen peroxide and oxidized DNA bases in human neutrophils correlated well with the *in vivo* tumor-promoting potencies of the activating chemicals when stimulated with various tumor promoters (*4*).

ROS are increasingly being cited as agents of normal signal transduction, and, while their source is usually uncertain, xanthine oxidase XO is often a likely candidate. A phorbol ester tumor promoter such as PMA enhances the generation of ROS accumulation and declines the ROS detoxification enzymes in both epidermal and inflammatory cells. It triggers ROS accumulation through activation of xanthine oxidase (XO) or the stimulation of PMNs which cause NADPH oxidase activation.

Activation of NADPH oxidase results in an increase in phagocyte oxygen consumption unrelated to mitochondria respiration, which is called the respiration burst (*5*). Electrons are transferred from NADPH to molecular oxygen, producing superoxide anion (O_2^-), the direct product of NADPH oxidase. O_2^- is then rapidly converted to secondary toxic oxygen species such as hydrogen peroxide, hydroxyl radicals and hypochlorous acid. NADPH oxidase is a highly regulated bound complex that is composed of a number of cytosolic proteins (p47phox, p67phox, p40phox, and Rac1/2) and membrane-bound protiens (gp91phox, and p22phox). In resting cells, NADPH oxidase is dormant and its protein components are segregated into cytoplasmic and plasma membrane compartments. Because the active enzyme complex is located on the plasma membrane, the cytosolic proteins must translocate from the cytosol to the membrane during assembly of the functional enzyme complex (*6*).

NADPH oxidase is normally dormant but can be rapidly activated by a number of stimuli. These include opsonized particles and f-MetLeuPhe which interact with surface receptors and protein kinase C agonists like PMA. During NADPH oxidase activation, p47phox is phosphorylated on several serine residues by several protein kinases which are dependent on different stimuli pathways (*7,8*).

Protein kinase C is involved in NADPH oxidase activation and phosphorylation of p47phox (*5*). PKC is a family of phospholipid dependent serine/threonine kinases, which act in multiple signal translocation. The cofactor requirement differs between the various classes of PKC isotypes.

It has been shown that both green tea and black tea have antioxidant activity, and are able to inhibit tumor induction (*9*). In recent years many animal studies and several epidermiological studies suggest the anti-carcinogenic effects of tea (*10-12*). Extracts of green, black and other teas inhibited PMA-induced JB6 cell transformation (*13*). It has been proposed that the mechanisms of action by tea and its polyphenols on anti-carcinogenesis may be through signal transduction blockade (*14)*.

In the induction of skin carcinogenesis, the phorbol ester PMA was used as a potent tumor promoter in mouse skin. A NADPH oxidase system of neutrophil is dominant for O_2^- generating in TPA-treated mouse skin (*15*). ROS from leukocytes, including O_2^-, plays an important role for continuous and excessive production of chemotactic factors, leading to chronic inflammation and hyperplasia.

In this study, the effects of tea polyphenols (TF1, TF2, TF3, and EGCG) and gallates (gallic acid and propyl gallate) on the activity of ROS producing enzymes, NADPH oxidase and xanthine oxidase were investigated. The inhibitory effects of these six compounds on PMA stimulated superoxide production of HL60 and their scavenging activity to exogenous superoxide were compared and discussed.

Materials and Methods

Chemicals and Reagents

Theaflavin (TF-1), a mixture of theaflavin-3-gallate and theaflavin-3′-gallate (TF-2), and theaflavin-3,3′-digallate (TF-3) were isolated from black tea as described previously (*16)*. (-)-Epigallocatechin-3-gallate (EGCG) was purified from Chinese tea (Longjing tea, *Camellia sinensis*) as described in our previous report (*17)*. The chemical structures of these tea polyphenols are illustrated in Figure 1. Lucigenin, luminol, phorbol-12-myristate-13-acetate (PMA) or 13-*O*-tetradecanoyl-phorbol-13-acetate (ATP), scopoletin, horseradish peroxidase

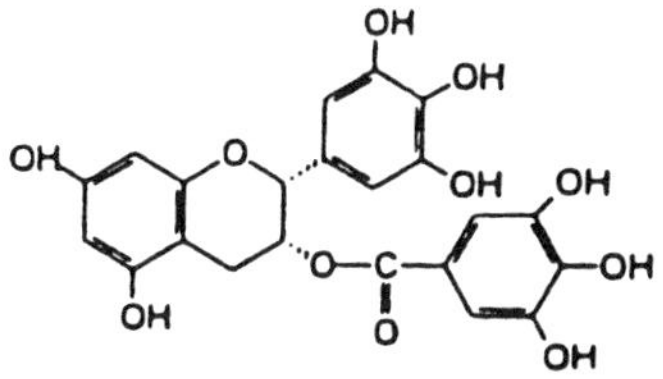

(-)-Epigallocatechin-3-gallate (EGCG)

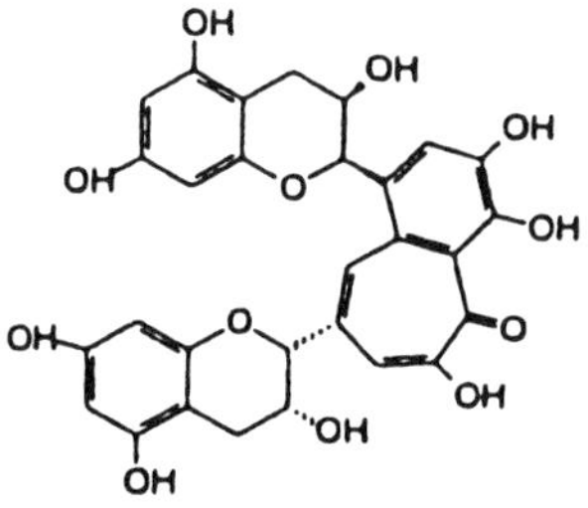

Theaflavin (TF-1)

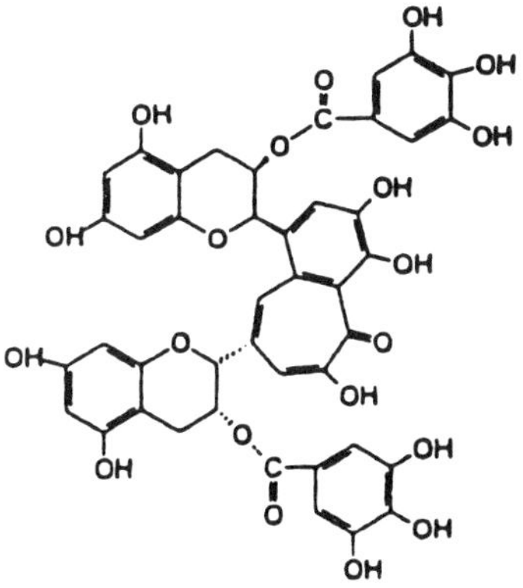

Theaflavin-3,3'-digallate (TF-3)

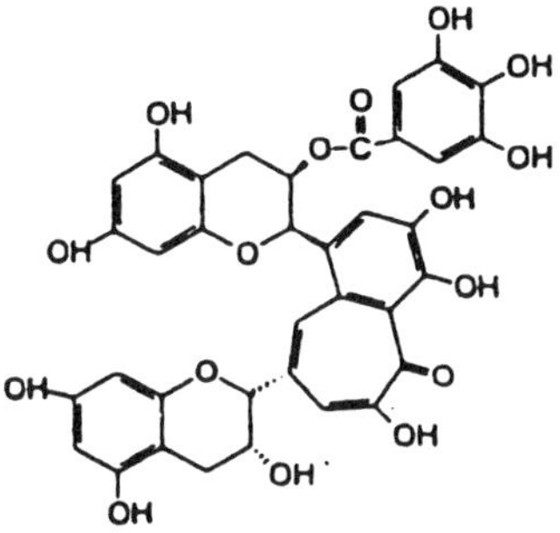

Theaflavin-3-gallate (TF-2a)

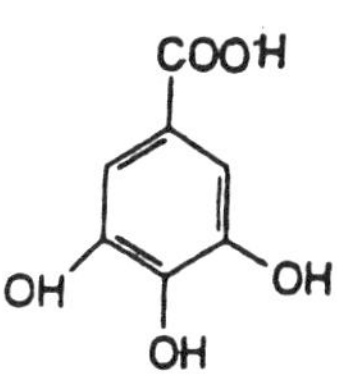

Gallic acid

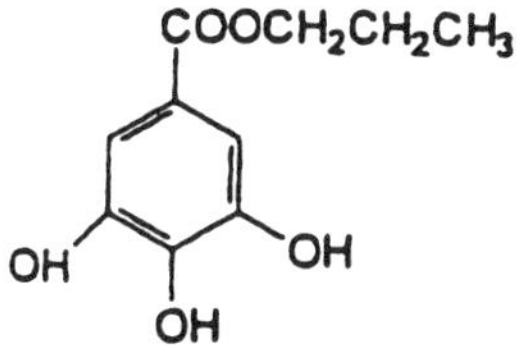

Propyl gallate

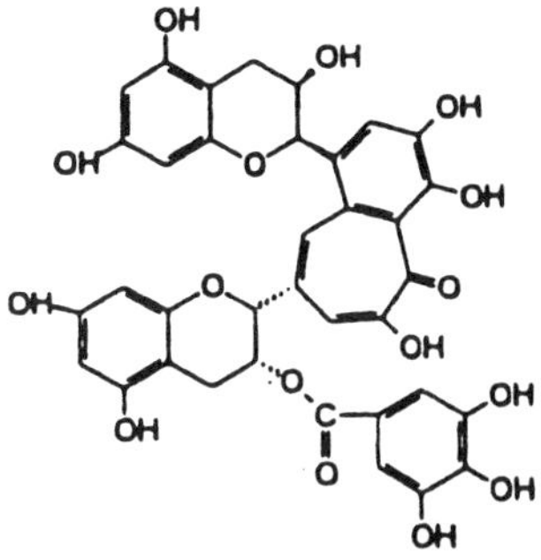

Theaflavin-3'-gallate (TF-2b)

Figure 1. Chemical structures of tea polyphenols and gallates.

(HRP), gallic acid, propyl gallate, allopurinol, xanthine oxidase, xanthine and uric acid were purchased from Sigma. 2',7'-Dichlorofluoresciein diacetate (DCF-DA) was purchased from Aldrich, and DMSO was purchased from Merck.

Cell Culture

Cells were cultured in a humidified atmospheric incubator at 37°C in 5% CO_2. Human promyelocytic leukemia cell line (HL-60) was cultured in RPMI 1640 supplemented with 15% heat-inactivated fetal bovine serum and 50 U/mL penicillin-streptomycin. To induced myeloid differentiation, cells were seeded at a density of 5×10^5 cells/mL and were cultivated for five days in RPMI 1640 containing 1.24% DMSO. The characteristics of mature cells were determined by smaller cell size, decreased nucleoli-to-cytoplasm ratio and pyknotic changes in nuclear chromatin. Cell numbers were counted using a hemocytometer, and cell viability was >98% as evidenced by trypan blue staining.

Estimation of ROS by Chemiluminescence Assay

The production of superoxide from PMA stimulated HL-60 cells was determined by lucigenin-amplified chemiluminescence (CL) and luminol dependent CL as described (*18*). CL was assessed by luminometers (models 1251, LKB, Bromma, Sweden) and recorded as mV (*19*). HL-60 cells (2.5×10^5) in HBSS with lucigenin 200 μM or luminol 20 μM were preincubated with various compounds for 5 minutes, readings were started by 100 ng/mL PMA and CL was measured and recorded for 120 min. The total amount of superoxide produced during the assay period was determined by integrating the area under the curve (in mV.min).

Superoxide Scavenging Activity Assay

The superoxide-generating system was constructed by mixing phenazine methosulfate, NADH and oxygen (air), and the formation of superoxide was quantitatively determined by the nitroblue tetrazonium method. The samples containing 10 μM phenazine methosulfate, 78 μM NADH and 25 μM nitroblue tetrazolium in 0.1 M phosphate buffer pH 7.4 were incubated for 2 min at room temperature and read at 560 nm against a blank containing PMS alone. Different concentrations of compounds were preincubated for 2 min before adding NADH.

Estimation of Xanthine Oxidase Activity

The enzyme activity was estimated spectrophotometrically at 295 nm by determining uric acid formation (U-3210 Hitachi) with xanthine as substrate (*20*). The enzyme reaction system consisted of a 1 mL reaction mixture containing 0.1 M potassium phosphate buffer pH 7.4, 0.004U XO and xanthine as substrate. All inhibitors were preincubated with enzyme for 5 min, then reaction was started by addition of xanthine. The reference cuvette was identical and only enzyme was absent. The IC50 and equations of dose-response curves were analyzed by computer with Sigma plot.

Translocation of Protein Kinase C to Cell Membrane

Cells ($3{\sim}5\times10^6$) in 1 mL HBSS with 1mM Ca^{2+}, Mg^{2+}. The compounds were incubated at 37°Cfor 30 min and stimulated with 100 ng/mL PMA. After 10 min the reaction was terminated by the addition of 4 vol ice-cold HBSS. The cells were pelleted at 400 g for 5 min at 4°C, suspended in 1 mL working buffer (100 mM KCl, 3mM NaCl, 3.5 mM $MgCl_2$, 10 mM PIPES, 1.25 mM EGTA) containing 1 mM PMSF and 10 μg/mL leupeptin and disrupted by homogenization. Unbroken cells and nuclear were pelleted at 800 g for 10 min at 4°C. Membrane preparation was made as described (*21*), 0.7 mL aliquots of the supernatants were loaded on top of a 2-mL layer of 15% sucrose (mass/vol in relaxtion buffer) and centrifuged (100,000 g 15 min) at 4°C. Supernatants were carefully removed and pellets were solubilized in SDS/PAGE sample buffer (Laemmli), and subjected to 10% SDS-PAGE and then blotted onto PVDF membrane .The upper part of the gel was revealed with anti-PKC (1:1000 dilution) and labeled by secondary antibodies (1/2000), followed by AP detection.

In Vivo Phosphorylation of p47phox

Cells were incubated overnight in phosphate-free medium, then transferred to fresh medium containing ^{32}P (0.5 mCi/mL) and incubated for 4 hr at 37℃. The compounds were incubated for 30 min. and then activated for 10 min with PMA (100 ng/mL 10^7 cells). The activation was terminated with 10 vol ice-cold buffer, then pelleted by centrifugation (400 × g for 10 min at 4℃) and resuspended at 1×10^8 cells/mL in ice-cold lysis buffer (20 mM Tris-HCL, pH 7.4, 150 mM NaCl, 5 mM EGTA, 15 μg/mL leupeptin, 10 μg/mL aprotinin, 1.5 mM phenylmethylsulfonyl fluoride, 0.5% Triton X-100, 25 mM NaF, 5 mM $NaVO_4$, 5 mM β-glycerophosphate, 1 mM *p*-nitrophenyl phosphate, 0.25 M sucrose, and 1 mg/mL DNase I) for 1 hr at 4 °C vortex every 10 min and centrifuge lysate 10 min at 3000 × g to remove nuclei.

Immunoprecipitation and Immunobloting

The supernatant obtained from the above procedure was incubated with p47phox antibody (1/200 dilution) or their respective IgG controls in the presence of 25 μL protein A/G plus-Agarose (Pharmacia Biotech, Inc.) overnight with gentle mixing. Then the beads were washed extensively with lysis buffer without DNase I. The immunoprecipitated proteins were eluted by boiling in electrophoresis sample buffer (62.5 mM Tris-HCl, pH6.8, 10% glycerol, 2.3% SDS, 2% 2-mercaptoethanol). The beads were pelleted by brief centrifugation, and the supernatant was subjected to 10% SDS-PAGE. Proteins were blotted onto PVDF membrane and detected by autoradiography or antibody labeling (p47phox 1/10,000 dilution), using colorimetric detection.

Results

Inhibition of XO by Tea Polyphenols

XO activity was determined by the formation of uric acid, which has maximum absorption at O.D. 295 nm providing the basis for a spectrophotometric assay.

The inhibitory effects of six polyphenols on XO were tested and the IC 50 values for these six compounds were as described in Table I. The IC50s of both gallic acid and propyl gallate on the inhibition of XO are more than 200μM . The inhibition ability of theaflavins and EGCG are: TF3 > TF2 > EGCG > TF1. Theaflavin-3,3′-digallate (TF3) exhibits the strongest inhibition with IC50 value of 4.5 μM. In the present system, the IC50 value for allopurinol is 0.68 μM. Upon further analysis of inhibition kinetics, it seems that TF3 acts on XO as a competitive inhibitior.

PMA Induced ROS Production of Differentiated HL-60

DMSO has been known to induce HL-60 differentiation which exhibits morphological alteration as well as functional maturity, e.g., the ability to generate O_2^-, H_2O_2, and increased myeloperoxidase and phagocytic activities

(*22*). Both HL-60 incubated with 1.24% DMSO for 0-5 days and HL-60 PMA stimulated ROS were assayed by DCF-DA and it was determined that production of PMA induced ROS is dependent on the time of DMSO treatment.

Table I. Inhibition of Xanthine Oxidase by Tea Polyphenols and Allopurinol

Compound[a]	*Inhibition of xanthine oxidase*	
	IC50 (μM)[b]	*Dose-response curve*[c]
Allopurinol	0.68	$y = -76.4x + 1.21$
EGCG	12.5	$y = -2.1x + 77.5$
Gallic acid	>200	$y = -0.1x + 101.6$
Propyl gallate	>200	$y = 0.16x + 96.5$
TF1	25.5	$y = 97.9\, e^{-0.023x}$
TF2	7.6	$y = 95.6\, e^{-0.0078x}$
TF3	4.5	$y = 95.6\, e^{-0.0077x}$

[a] The abbreviations are: EGCG, (-)-epigallocatechin-3-gallate; TF1, theaflavin; TF2, a mixture of theaflavin-3-gallate and theaflavin-3′-gallate, and TF3, theaflavin-3,3′-digallate.

[b] IC50, the concentration (in μM) of tested compound that induced 50% inhibition of xanthine oxidase under the described conditions.

[c] The equation of the dose-response curves were obtained by Sigma plot based on the experimental data.

Effects of Tea Polyphenols on PMA-stimulated ROS Production

The total ROS production was determined by luminol-derived chemiluminescence assay, PMA-induced ROS production was constitutively expressed at least at 50 min (Figure 2A). In the following experiments of chemiluminescence assay, chemiluminescence was recorded for 120 min after PMA treatment and the effects of compounds were compared with PMA alone.

The IC50 values of tested compounds on PMA-stimulated ROS production were less than 1 μM. DMSO (0.2 %) has inhibitory effects in this assay system (33.6%), but 10 μM of tested compounds have almost 90% inhibition (Figure 2B). It seems that ROS were scavenged by these compounds.

Experiments were performed to determined the effect of the selected tea polyphenols on O_2^- production by PMA-stimulated HL-60 cells over a wide concentration range (Figure 3). Gallic acid has no inhibitory effects on O_2^- production by PMA-stimulated HL-60 cells, but propyl gallate was the most potent inhibitor at 10 μM (data not shown). At 10 μM, the order of inhibition by

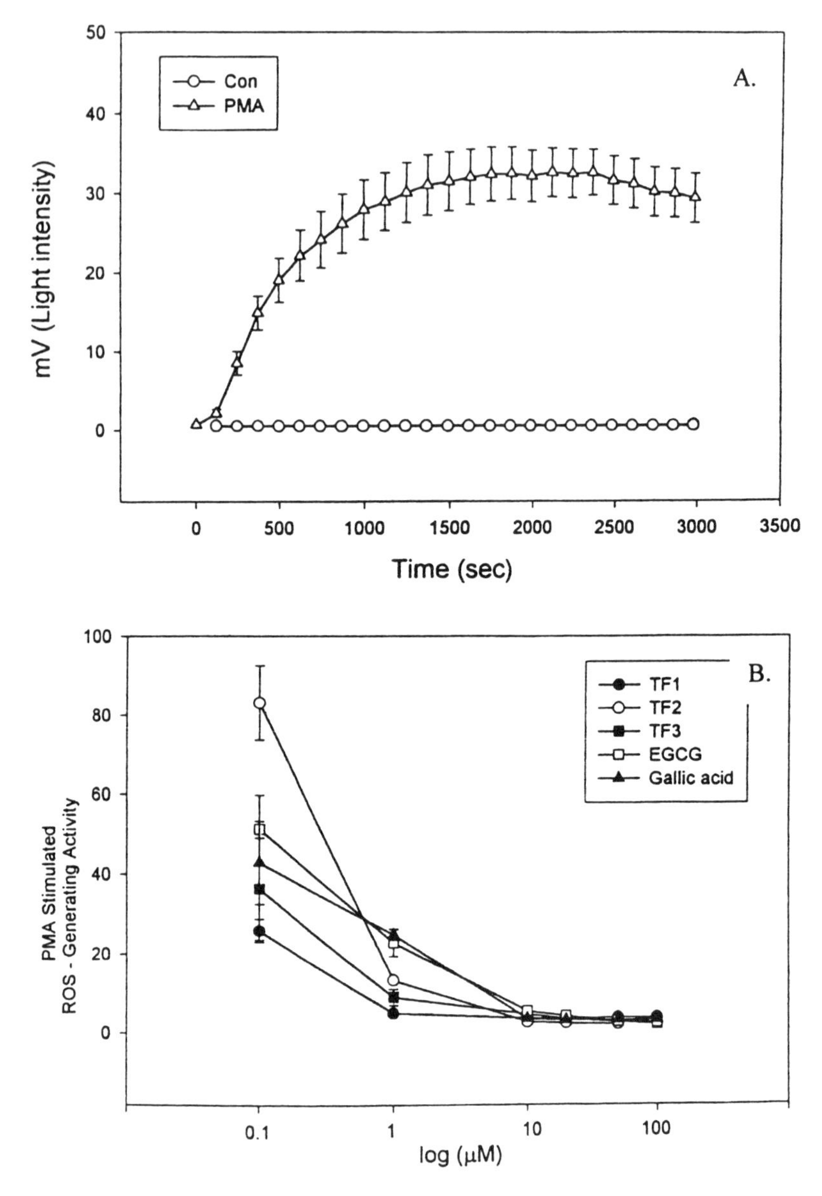

Figure 2. Inhibition of ROS-generation in PMA-stimulated HL-60 cells. (A) The total ROS-generation was determined by liminol-derived chemiluminescence (CL) as described in Materials and Methods. No increment in CL was detected in the control (Con) cells; while remarkable enhancement in CL was detected when the cells were treated with the tumor promoter PMA. (B) Dose-response curves describe the inhibition of tea polyphenols and gallic acid on the PMA-stimulated HL-60 cells . The concentrations of the tested compounds were in logarithmic scale.

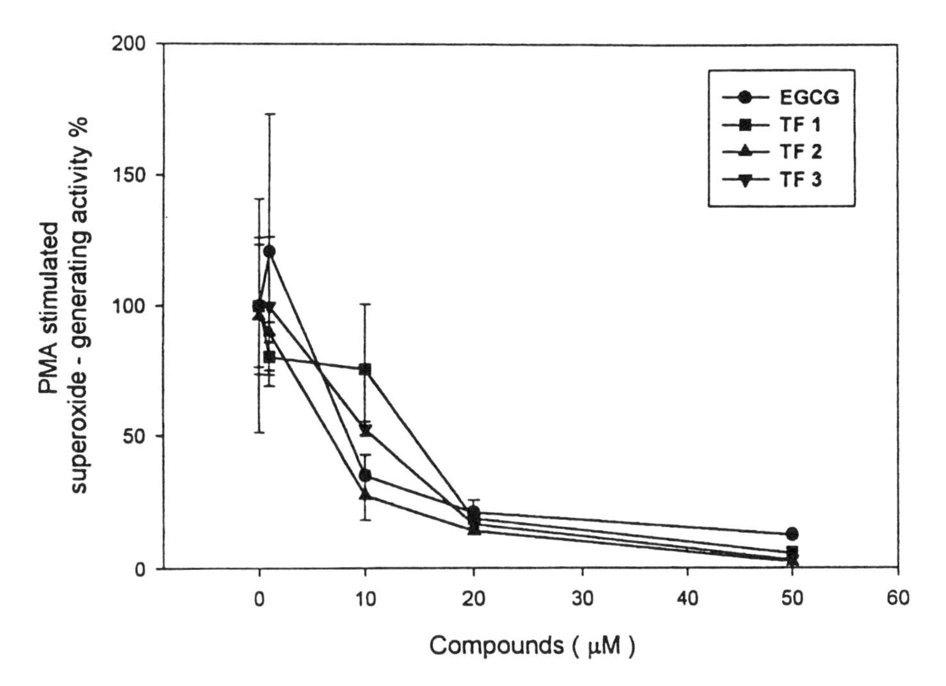

Figure 3. Inhibition of tea polyphenols on superoxide production in PMA-stimulated HL-60 cells. HL-60 cells were induced to differentiation for 5 days by 1.24% DMSO, then treated with various concentrations of tea polyphenols for 5 min and finally stimulated with PMA (100 ng/mL). The superoxide production was determined by lucigenin-derived CL as described in Materials and Methods.

tea polyphenols was: propyl gallate > EGCG > TF3 > TF1 > TF2. When the concentration is 20 μM or 50 μM, the inhibition ability is TF3 > TF1 > EGCG > TF2 > propyl gallate. The superoxide scavenging activity of polyphenols were estimated (Table II). The order of their inhibition was: EGCG > TF1 > TF2 > TF3 > GA > propyl gallate.

Table II. The Superoxide Scavenging Activity of Tea Polyphenols and Gallates

Compound[a]	*Superoxide scavenging activity*[b] *(IC50, μM)*[c]
EGCG	7.4
TF1	13.6
TF2	15.6
TF3	40.5
Gallic acid	42.5
Propyl gallate	50.4

[a] The abbreviations are described in Table I.

[b] The superoxide-generating system was constructed by mixing phenazine methosulfate, NADH and oxygen (air) and the formation of superoxide was determined as described in Materials and Methods.

[c] IC50: The concentration of tested compound (μM) for exhibiting 50% superoxide scavenging activity.

The inhibition of PMA-stimulated O_2^- production of HL-60 was comparable to the direct scavenging of O_2^- by the tested compounds. The results imply that EGCG, TF1 and TF2 play a major role in extracellular scavenging activity. The effects of TF3 and propyl gallate on O_2^- production by PMA-stimulated HL-60 cells may be involved in the signal transduction of PKC activation pathway that stimulated by PMA, since our previous study demonstrated that TF1, TF2, TF3 and EGCG have profound inhibitory effects on PKC and AP-1 binding activities *(23)*.

TF3 Inhibits PKC Membrane Translocation

PMA-stimulated production of O_2- , H_2O_2 and ROS in differentiated HL-60 cells may be through the activation of NADPH oxidase by PKC. To assess whether the inhibition of superoxide and hydrogen peroxide production by these compounds is through modulating PKC activation, we used Western blot

analysis to visualize the translocation of PKC from cytosol to membrane. The results showed that cytosolic PKC was translocated to membrane after PMA stimulation and the translocation was observed at 10 min. TF1, TF2 and TF3 were tested on the membrane translocation of PMA-stimulated HL-60 for 10 min. TF3 treatment showed remarkable blocking effects on its translocation (Figure 4). It seems that TF3 is a more potent inhibitor on PMA-induced PKC activation as compared with EGCG.

Effect of TF3 on p47phox Phosphorylation

The total protein lysates of neutrophil, HL-60, differentiated HL-60 (induced by DMSO for 5 days), macrophage RAW, colon 205, and HT29 were submitted to SDS-PAGE, and then blotted onto PVDF membrane and finally detected by immunobloting with anti-p47phox. It seems that only neutrophil and differentiated HL-60 showed significant expression of p47phox (Figure 5A).

Phosphorylation of p47phox in differentiated HL-60 was analyzed after ^{32}P labeling, p47phox immunoprecipitation, and by SDS-PAGE. It seems that PMA stimulation triggered the phosphorylation of p47phox in HL-60 cells (Fig. 5B). Based on our preliminary experimental data, TF3 (at 10, 20 and 50 μM) showed inhibitory effects on PMA-induced p47phox phosphorylation; the level of p47phox protein was also decreased by TF3 as demonstrated by immunoblot analysis (Figure 5B). Further investigation on the effects of tea polyphenols on the function of NADPH oxidase in cancer cell lines are now in progress.

Discussion

Cancer chemoprevention has a potential impact on the rates of cancer incidence through modulation of initiation and promotion stages (*24*). Epidemiological studies have suggested that sufficient uptake of tea polyphenols may reduce the risk of cancer through their role in the metabolism of carcinogens, antioxidant enzymatic activities and scavenging of free radicals (*14,25*). Several studies have shown that the differentiated HL-60 cells possess phagocytic properties and the capability of generating ROS upon stimulation (*21,26*). These phagocyte-generated oxidants are known to act as complete carcinogens and to cause harmful biologic effects via several pathways, i.e., damaging DNA, regulating gene expression and modulating signal transduction pathways (*3,27*). It is well known that generation of ROS is associated with tumor initiation and promotion (*14*). Green tea is widely ingested as a beverage in China, Japan and other Asian countries, whereas black tea is more popular in Western countries. In recent years, many animal studies and several

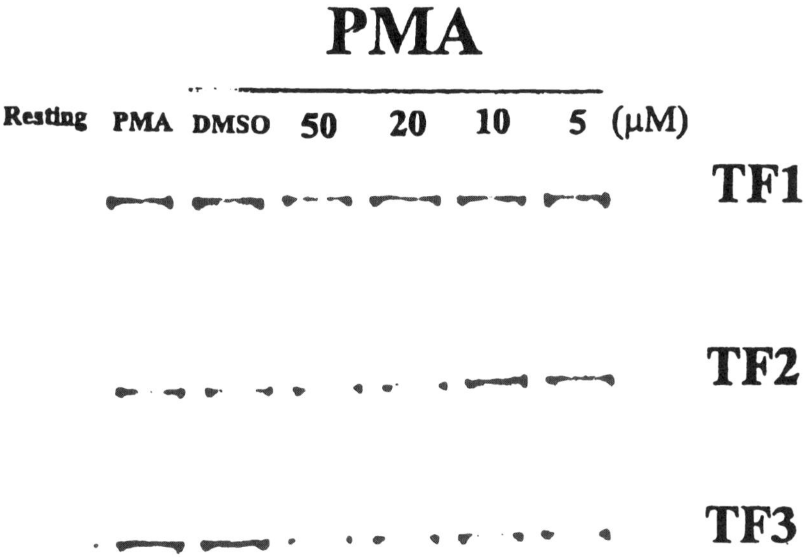

PMA

Resting PMA DMSO 50 20 10 5 (μM)

TF1

TF2

TF3

Figure 4. Effect of theaflavin-3,3'-digallate on PMA-stimulated PKC translocation. Differentiated HL-60 cells ($5x10^5$) were incubated with theaflavins (TF1, TF2 and TF3) for 30 min and the cells were stimulated with 100 ng/ mL PMA for 10 min. The membrane PKC was determined as described in Materials and Methods.

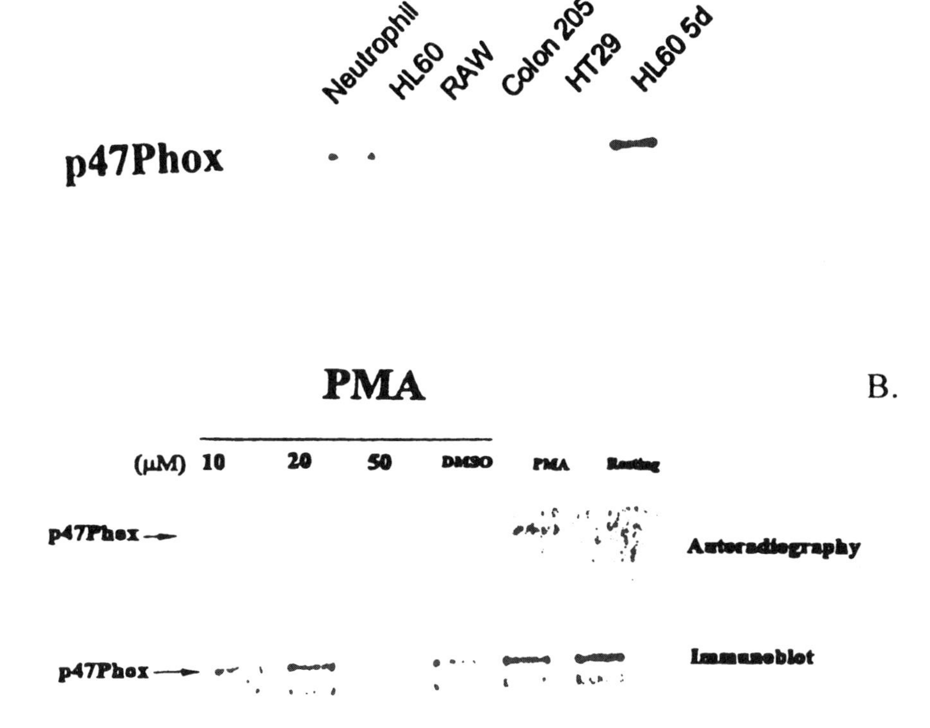

Figure 5. Effect of theaflavin-3,3'-gallate on p47phox phosphorylation. (A) Expression of p47phox protein in different cell lines. Only neutrophil and differentiated HL-60 cells (HL605d) are detectable for the presence of p47phox protein. (B) Theaflavin-3,3'-digallate (TF3) (10, 20, 50 μM) showed inhibitory effects on the PMA-stimulated p47phox phosphorylation (as detected by autoradigraphy) and expression (as detected by immunobloting). The experimental procedures were described in Materials and Methods.

epidemiological studies have suggested the anticarcinogenic effects of tea (*10*). Extracts of green, black and other teas inhibited PMA-induced JB6 cell transformation (*13*). However, the mechanisms of action by which these chemicals block carcinogenesis are not clear.

Because oxidative DNA damage is considered relevant to carcinogenic processes (*3,27)*, we have evaluated the possible anticarcinogenic effect by determining the effect of tea polyphenols on PMA-induced ROS generation and xanthine oxidase activity.

Enviromental carcinogenesis is known to proceed through multiple stages, such as initiation, promotion, and progression (*14*). Tumor promotion, which is a complex and important process in carcinogenesis, is generally studied using PMA as tumor promoter in mouse skin models (*28,29*). Both green tea and black tea have shown the ability to suppress PMA-mediated tumor-promoting effects on mouse skin (*9,10*).

The tumor promoter PMA is known to induce H_2O_2 production by phagocytic cells and epidermal cells (*15*), increase xanthine oxidase activity (*30*) and diminish antioxidant enzyme activities (*31*). Recent reports show that in PMA treated mouse skin, ROS from leukocytes, including superoxide, play an important role in conditions leading to chronic inflammation and hyperplasia, and that superoxide generation inhibitors are agent that act to inhibit this tumor promotion response. Tea polyphenols and propylgallate all have a potent inhibitory effect on PMA-stimulated superoxide production beytween 20~50 μM. Gallic acid (GA) has no effect on the inhibition of superoxide production in PMA stimulated HL-60 cells. At 10 μM, EGCG, TF3 and propyl gallate (PG) inhibit superoxide production more than 50%, and the order of inhibition was PG > EGCG > TF3; while the superoxide scavenging ability of these 6 compunds was: EGCG > TF1 > TF2 > TF3 > GA > PG (Figure 3). PG was the most potent inhibitor of PMA stimulated H_2O_2 production. H_2O_2 scavenging ability was TF2 >TF3 >TF1 > EGCG > PG > GA (Table II). From these data we suggest that inhibition of ROS production of PG and TF3 were not only by its scavenging ability but also through modulating the signal transduction of NADPH oxidase activation.

Based on previous investigations, ROS are increasingly being cited as agents of normal signal transduction (*32*), and, while their source is usually uncertain, xanthine oxidase (XO) is often a likely candidate. The intestinal mucosa and liver are the richest sources of XO. XO is the major source of ROS in the ischemic small intestine. Its candidature has been strengthened by recent analyses of promoter regions of human, mouse and rat enzymes, which suggest the presence of potential regulatory sites for cytokines known to stimulate generation of ROS. Four tea polyphenols and two gallates were tested in the XO activity assay. TF3 showed the most potent inhibition and acted as a competitive inhibitor. Although TF3 (IC50: 4.5 μM) is about 6-7 times less potent than allopurinol (IC50: 0.68 μM), it may still be a good inhibitor of XO *in vivo*. Several mechanisms have been proposed to explain the *in vitro* antimutagenic and the *in vivo* anti-cancer properties of tea. One of these mechanisms, i.e. antioxidant properties, can reduce oxidative stress by scavenging the ROS and therefore also hydroxyl radical formation. Tea polyphenols are particularly good *in vivo* antioxidants, due to their amphipathic properties. Inhibition of XO *in vivo* can be studied by measuring plasma concentrations of substrate (xanthine and hypoxanthine) and the product (uric acid). The lowering of uric acid in the urine can also be measured (*33*). Inhibition of XO can lower the ROS load of the body thereby making endogenous reducing equivalents available for other detoxification reactions. The present work indicates that tea polyphenols may have a novel mechanism, namely the limited inhibition of XO and thereby the reduction of ROS production in the body. Therefore, it is possible that tea polyphenols counteract the promotional effects of PMA through inhibiting oxidant formation by leukocyte and epidermal xanthine oxidase and by plasma membrane-bound NADPH oxidase. The exact mechanism for the inhibition of NADPH oxidase activity by tea polyphenols remains to be clarified. It should be noted that staurosporine, a microbial alkaloid protein kinase C inhibitor without antioxidative activity, inhibits O_2^- production in neutrophils (*34*). TF3 inhibit the membrane translocation of the PKC in differentiated HL-60 cells that are stimulated by PMA. It is possible that the inhibition of superoxide and hydrogen peroxide production of PMA-stimulated HL-60 was through the inhibition of the PKC pathway ability to activate NADPH oxidase. Phosporylation of p47phox , which is necessary in PMA, induces NADPH oxidase activation and P47phox translocation. TF3 showed an inhibitory effect on PMA stimulated p47phox phosphorylation at 10, and 20 μM. Furthermore, inhibition of protein kinase C by the tea polyphenols has been considered as possible mechanisms for the inhibition of tumor promotion (*22*).

Tea polyphenols may act, at an earlier stage than has previously been suspected, by inhibiting signal transduction for ROS production rather than only scavenging the already formed ROS. This could partly explain some of the unique beneficial properties of tea (*14*).

Acknowledgements

This study was supported by National Science Council NSC grants, NSC89-2316-B002-016 and by National Health Research Institute grant DOH88-HR-403.

References

1. Cerutti, P. A. *Science* **1985**, *227*, 375-380.
2. Breimer, L. H *Mol. Carcinogenesis* **1990**, *3*, 188-197.
3. Frenkel, K. *Phamacol. Ther.* **1992** , *53,* 127-166.
4. Frenkel, K.; Chrzan, K. *Carcinogenesis* **1987**, *8,* 455-460.
5. El Benna, J.; Fause, L. R. P.; Johnson, J. L.; Babior, B. M. *J. Biol. Chem.* **1994,** *269*, 23431-23436.
6. Clark, A. R.; Curnutte, J. T.; Rae, J. J.; Heyworth, P. G. *J. Clin. Invest.* **1990,** *85*, 714-721.
7. Ding, J.; Vlahos, C.J.; Liu, R.; Brown, R. F.; Badwey, J. A. *J. Biol. Chem.* **1995**, *270*, 11684-11691.
8. Park, J. W.; Hoyal, C. R.; Benna, J. E.; Babior, B. M. *J. Biol. Chem.* **1996**, *272,* 11035-11043.
9. Wang, Z. Y.; Huang, M. T.; Lou, Y. R.; Yie, J. G.; Reuhl, K.; Newmark, H. L.;Ho, C.-T.; Yang, C. S.; Conney, A. H. *Cancer Res.* **1994**, *54*, 3428-3435.
10. Katiyar, S. K.; Mukhtar, H. *Int. J. Oncol.* **1996**, *8*, 221-238.
11. Kelloff, G. J.; Boone, C. W.; Crowell, J. A.; Steele, V. E.; Lubert, R. A.; Doody, L. A.; Malone, W. F.; Hawk, E. T.; Sigman, C. C. *J. Cell. Biochem.* **1996**, *26s*, 1-28.
12. Bushman, J. L. *Nutr. Cancer* **1998**, *31*, 151-159.
13. Jain, A. K.; Shimoi, K.; Nakamura, Y.; Kada, T.; Hara, Y.; Tomita, I. *Mutat. Res.* **1989**, *210*, 1-8.
14. Lin, J. K.; Liang, Y. C.; Lin-Shiau, S. Y. *Biochem. Pharmacol.* **1999**, *58,* 911-915.
15. Robertson, F. M.; Beavis, A. J.; Oberysryn, T. M.; O'Connell, S.M.; Dokidos, A.; Laskin, D. L.; Laskin, J. D.; Reiners, J. J. Jr. *Cancer Res.* **1990** , *50*, 6062-6067.
16. Chen, C. W.; Ho, C.-T. *J. Food Lipids* **1995**, *2*, 35-46.
17. Lin, Y. L.; Juan, I. M.; Liang, Y. C.; Lin, J. K. *J. Agric. Food Chem.* **1996**, *44*, 1387-1394.
18. Gyllenhamer, H. *J. Immunol. Methods* **1987** ,*97*, 209-213.
19. Palmblad, J.; Gyllenhammar, H.; Lindgren, J. A.; Malmsten, C. L. *J. Immunol.* **1984,** *132*, 3041-3045.
20. Kalckar, H. M. *J. Biol. Chem* **1947***, 167*, 429-443.
21. Daphna, T.; Matya, H.; Daniela, M.; Yoel, K.; Irit, A. *Eur. J. Biochem.* **1996**, *242*, 529-536.
22. Nagy, K.; Pasti, G.; Bene, L.; Nagy, I. *Free Radical Res. Commun.* **1993**, *19*, 1-15.
23. Chen, Y. C.; Liang, Y. C.; Lin-Shiau, S. Y.; Ho, C.-T.; Lin, J. K. *J. Agric. Food Chem.* **1999**, *47*, 1416-1421.

24. Szarka, C. E.; Grana, G.; Engstrom, P. F. *Curr. Problems Cancer* **1994**, *18*, 16-79.
25. Lee, S. F.; Liang, Y. C.; Lin, J. K. *Chem. Biol. Interact.* **1995**, *98,* 283-301.
26. Wei, H., Frenkel, K. *Carcinogenesis* **1993**, *14*, 1195-1201.
27. Weitzman, S. A.; Gordon, C. I. *Blood* **1990** , *76*, 655-663.
28. Slaga, T. J. *Environ. Health Prospect.* **1983,** *50,* 3-14.
29. Lee, S. F.; Lin, J. K. *Nutr. Cancer* **1997**, *28*, 177-183.
30. Reiners, J. J. Jr.; Pence, B. C.; Barcus, M. C. S.; Cantu, A. R. *Cancer Res.* **1987**, *47*, 1775-1779.
31. Solanki, V.; Rana, R. S.; Slaga, T. J. *Carcinogenesis* **1981**, *2*, 1141-1146.
32. Burdon, R. H. *Free Rad. Biol. Med.* **1995**, *18,* 775-794.
33. Kojima, T.; Nishina, T.; Kitamura, M.; Hosova, Y.; Nishioka, K. *Clin. Chim. Acta.* **1984**, *137*, 189-198.
34. Yamamoto, S.; Kiyoto, I.; Aizu, E.; Nakadate, T.; Hosoda, Y.; Kato, R. *Carcinogenesis* (Lond.) **1989**, *10*, 1315-1322.

Chapter 21

Generation of Lipid Peroxyl Radicals from Oxidized Edible Oils and Heme-Iron: Suppression of DNA Damage by Unrefined Oils and Vegetable Extracts

Ayako Kanazawa[1], Tomohiro Sawa[2], Takaaki Akaike[2], and Hiroshi Maeda[2]

[1]Faculty of Education, Kumamoto University, Kumamoto, Japan
[2]Department of Microbiology, School of Medicine, Kumamoto University, Kumamoto 860-0811 Japan

A diet high in both fat and iron is known as a risk factor in cancer epidemiology. We found that lipid peroxyl radicals are generated from oxidized edible oils in the presence of heme-iron, which is present in high concentration in red meat, and they induce DNA damage including strand breakage and abasic site formation. The radicals are effectively scavenged by unpurified native vegetable oil containing flavonoids and other phenolic compounds. Although unpurified native vegetable oils contain a high amount of peroxyl radical scavengers, the conventional refining processes of edible oil seems to reduce the levels of such valuable components. The capability to suppress DNA breakage may indicate the preventive role of these components against fat-related carcinogenesis. In addition, peroxyl radical-scavenging activity parallels inhibitory effect on the tumor-promoting activity, as indicated by the “early antigen formation” in

Epstein-Barr virus harboring Raji cells. This effect dramatically increases in hot-water-extracts of vegetables i.e. soup as compared to cold-water-extracts of fresh vegetables.

These findings suggest that lipid peroxyl radicals, thus generated, may have a link between fat rich diet and colon carcinogenesis, involving DNA breakage, and tumor promoter effects, where red meat (heme) plays a crucial role. Therefore, the importance of peroxyl radical scavngers in crude plant oils and vegetable soup should be recognized in terms of cancer prevention and fat and aging-related diseases.

In cancer epidemiology, a high intake of dietary fat and/or meat appears to result in a high incidence of colorectal cancer (*1-3*). However, a question still remains as to how this dietary habit is associated with a high risk for cancer. We anticipate that endogenous genotoxic mutation is generated or facilitated by lipid peroxidation involving mutagenic substances formed endogenously (*4*). Polyunsaturated fatty acids, which abound in edible vegetable oils, are spontaneously oxidized in the presence of air and more so with light. The oxidation is facilitated much more effectively by free radicals such as hydroxyl radicals (•OH), and the resulting lipid hydroperoxide (LOOH) formation (Figure 1). Uniqueness of this formation is that the reaction is self propagating in the presence of all reacting components. LOOH has harmful and genotoxic potency generating radical species in the presence of radical initiator such as transition metal ion or heme-iron. Furthermore, free radical reacts with polyunsaturated fatty acid more effectively than protein or carbohydrates.

We found that the reaction of alkyl hydroperoxide (ROOH) (a model of oxidized lipid) with heme-iron generates alkyl peroxyl radical (ROO•) rather than alkoxyl radical (RO•) or alkyl radical (R•) (*6-8*). Instead the reaction with free Fe^{2+} generates predominantly RO•. Among the lipid related radical molecular species such as R•, RO•, and ROO•, ROO• appears to exert more profound biological effect because of its much longer half-life, which is estimated at longer than 30 min (*6*). ROO• may thereby have access to distant sites. On the contrary, much reactive free radicals such as •OH have extremely short half-life, and will react near the vicinity of its generation site. Consistent with this notion, we found that the incidence of colorectal cancer increased in rats fed oxidized oils and was further significantly enhanced by addition of heme-iron (*8*). In this chapter, we discuss the reactivity of ROO• as indicated by damage to DNA, and the suppression of this damage or biological effects by dietary antioxidants such as flavonoids and other phenolic compounds, and those found in hot water vegetable extracts.

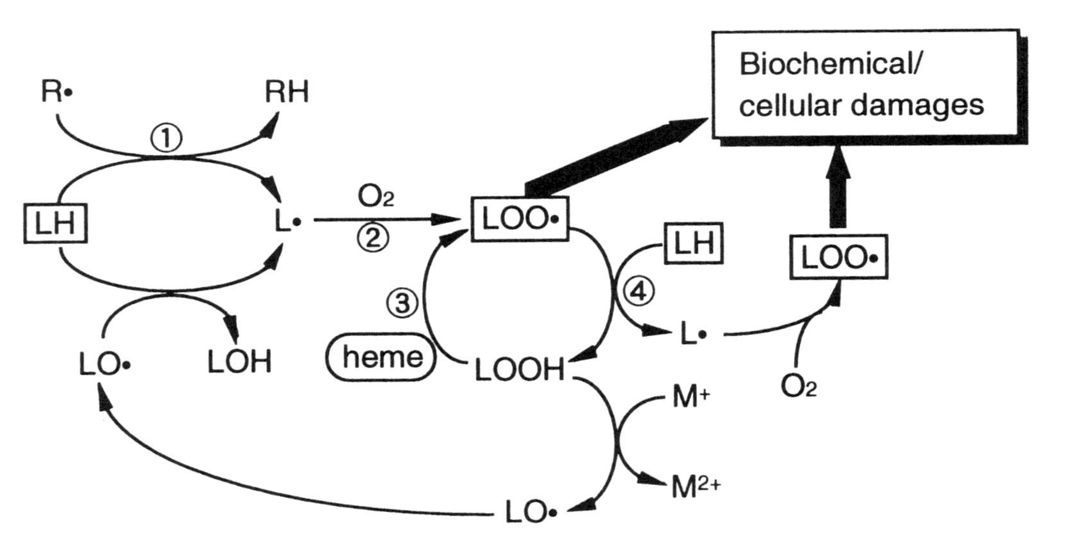

① Free radical dependent initiation: R•, free radical including •OH; LH, polyunsaturated fatty acid

② Peroxyl radical formation

③ Peroxyl radical formation

④ Lipid oxidation/radical propagation

Figure 1. *Free radical initiated propagating of lipid peroxidation. Reproduced with permission from reference 5. Copyright 2000 with IOS Press.*

Mechanism of Generation of Alkyl Peroxyl Radicals in the Presence of Heme Compounds

At least three types of radical species can be generated during decomposition of alkyl hydroperoxides catalyzed by free iron, nonheme, or heme-iron. The Fe^{2+}-catalyzed reaction to generate predominantly RO• from ROOH is,

$$ROOH + Fe^{2+} \rightarrow RO\bullet + OH^{-} + Fe^{3+}$$

whereas the reaction of ROOH and heme compounds generates ROO• (*5*): Namely,

$$ROOH + heme\ (Fe^{3+}) \rightarrow ROH + heme\ (Fe{=}O^{4+})\bullet + H^{+},$$

$$ROOH + heme\ (Fe{=}O^{4+})\bullet \rightarrow ROO\bullet + heme\ (Fe{=}O^{4+}) + H^{+}$$

These radical species were confirmed by use of electron paramagnetic resonance (EPR) spectroscopy using 5, 5-dimethyl-1-pyrroline-*N*-oxide (DMPO) as a spin trap agent (*6-8*). Figure 2 shows EPR spectra for the generation of ROO• from the reaction of ROOH plus heme-iron; *tert*-butyl hydroperoxide (*t*-BuOOH) and linoleic acid, oxidized for 3 days 60°C in the dark, were used as model compounds for ROOH. The spin-adduct of ROO• from the reaction of *t*-BuOOH and heme-iron was assigned hyperfine splitting constants as follows: $\alpha_N = 1.45$ mT, $\alpha_H{}^{\beta} = 1.04$ mT, and $\alpha_H{}^{\gamma} = 0.15$ mT. Signals assigned to the ROO• spin-adduct were also observed in the analysis of linoleic acid oxidized by air. The generation of ROO• continued to remain high at least for 30 min with maximum production at 5 min as estimated by the signal intensity of DMPO-OOR.

Bactericidal and DNA-Damaging Effects of Alkyl Peroxyl Radicals

We observed the bactericidal effect of various hydroperoxides in the presence of various heme compounds e.g., hemoglobin, myoglobin, hemin, and cytochrome c, but it was not observed in the reaction with diethylenetriamine-pentaacetic acid (DTPA)-Fe^{2+} (*6*). Furthermore, bactericidal potential was

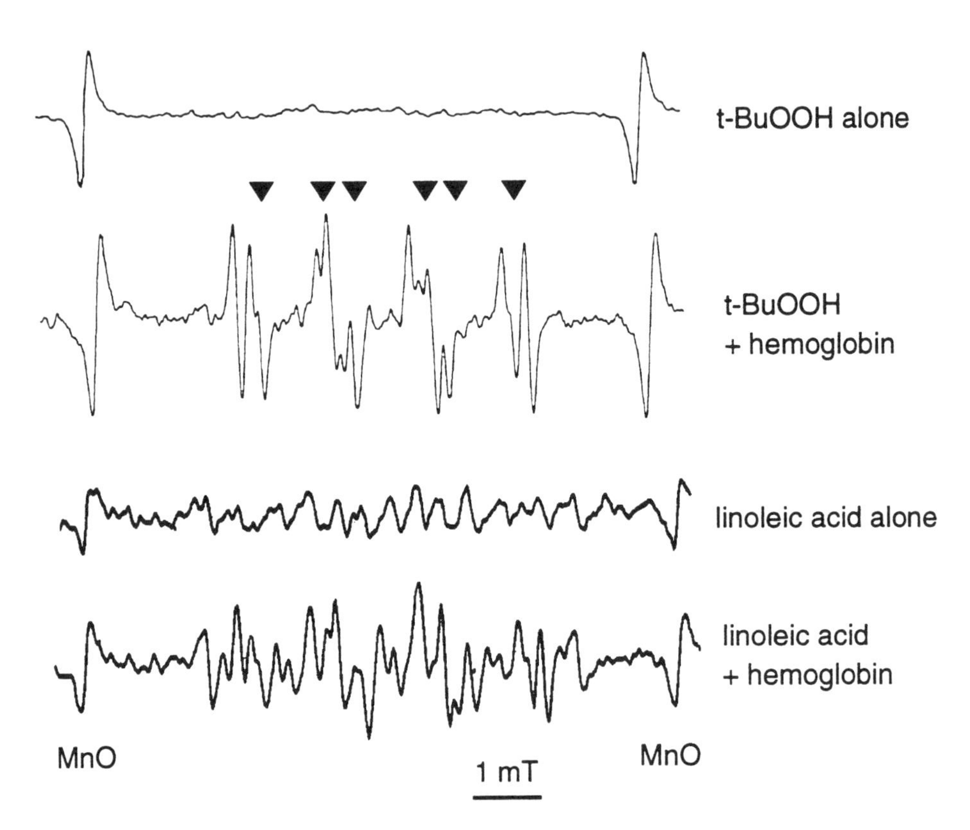

Figure 2. *Electron paramagnetic resonance spectra of DMPO spin adducts in the reaction systems of ROOH plus met hemoglobin. Data were reproduced from references 8 and 18.*

correlated with generation of ROO• as estimated by EPR spectra of the DMSO spin-adduct. When the reaction mixture containing hemoglobin, *t*-BuOOH, and gram-positive bacteria (*Staphylococcus aureus*) was added with ROO•-scavenging compounds (α-tocopherol, probucol, glutathione, L-cysteine, L-ascorbic acid, DMPO, etc.), strong inhibition against bactericidal action was observed. However, *N-t*-butyl-α-phenylnitrone (PBN) which trapped both R• and RO• but not ROO• did not exhibit this effect. Some antioxidants including well-known scavengers against hydroxyl radical (•OH) (e.g., dimethyl sulfoxide (DMSO), hyaluronic acid, L-methionine), superoxide anion radical ($O_2^{-•}$) (e.g., superoxide dismutase), or singlet oxygen (e.g., β-carotene, L-methionine) (*6, 9-14*) had no effect on this antibacterial action resulted from ROO• generation. These findings indicate that ROO• is a free radical species exhibiting bactericidal action and that the antioxidants scavenging ROO• can suppress the bactericidal activity.

Mutations resulting from DNA damage are thought to be a major cause of carcinogenesis. DNA damage by various free radical species involves strand breakage, depurination and depyrimidination, and base modification, such as deamination, nitration, hydroxylation, etc., and frequently leads to mutation. For instance, DNA replication errors that result from unpaired abasic sites and lead to carcinogenesis probably involve oncogenes (*15, 16*) and tumor suppressor gene such as p53. We also found that ROO• gives rise to abasic sites formation (*17*) and strand break (*8*). Further, we investigated abasic site formation by using a biotin-tagged derivative of *o*-(carboxymethyl)hydroxyamine, named aldehyde reactive probe (ARP®), and avidin-biotinylated peroxidase complex. Table I presents results from the formation of abasic sites on calf thymus DNA

Table I. Abasic Site Formation Induced by the Reaction of *t*-BuOOH with Hemoglobin[a] or EDTA/Fe^{2+}

t-BuOOH (mM)	*Abasic sites/10^4 nucleotides*			
	Hemoglobin		*EDTA/Fe^{2+}*	
	12.5μM	*25μM*	*50μM*	*100 μM*
1	0.18 ± 0.05	0.19 ± 0.05	nd[c]	nd[c]
10	0.61 ± 0.08 [b]	1.02 ± 0.08 [b]	nd	nd
50	0.59 ± 0.12 [b]	0.80 ± 0.09 [b]	nd	nd
100	0.62 ± 0.11 [b]	0.78 ± 0.04 [b]	nd	nd

[a] Data represent the mean ± SE of five independent experiments. [b] $P < 0.01$, as compared to 1 mM *t*-BuOOH. [c] nd, not detected.

exposed to the reaction of ROOH plus hemoglobin or EDTA-Fe^{2+}. In the presence of *t*-BuOOH (10 mM) and hemoglobin, 0.6-1.0 abasic sites/10^4 nucleotides were formed; however, they were not detected after replacing hemoglobin with nonheme-iron, e.g., EDTA/Fe^{2+}, which is known to initiate the production of R• and RO•. ROOH alone showed no appreciable abasic site formation (data not shown). Figure 3 shows the dose response relationship of abasic site formation in respect to the concentration of hemoglobin, in the presence of 100 mM *t*-BuOOH. Furthermore, we observed all four bases, C>T≥A>G, in the lyophilized filtrates of the reaction mixture, which were analyzed by HPLC using Asahipak GS-320 column and spectrophotometric detector.

Single-strand break (SSB) of plasmid DNA (pUC19) was examined by monitoring the structural transition from supercoiled to open circular form DNA by agarose gel electrophoresis (*8*). The cleaved DNA (open circular form) was separated from supercoiled DNA by electrophoresis, and then the DNA bands were stained with ethidium bromide, visualized under UV illuminator, and photographed. Figure 4 shows the effect of doses of *t*-BuOOH and hematin. The band of cleaved DNA was detected at the concentration of *t*-BuOOH more than 1 mM. The distinct conversion to generate open circular form at more than 10 mM *t*-BuOOH coincides with the enhanced formation of abasic sites (Table I). Scavengers for $O_2^{-\bullet}$ (SOD) and •OH (mannitol and DMSO) failed to suppress the generation of DNA breakage induced by ROO•. Therefore, •OH or $O_2^{-\bullet}$ do not appear to be contributors to the DNA breakage in the present ROO•-generating system. The present findings suggest that ROO• may have a genotoxic potential through unique reactions, including depurination and depyrimidination accompanied by DNA strand break.

Peroxyl Radical-Scavenging Activity of Flavonoids, Other Phenolic Compounds, and Unrefined Seed Oils

Antioxidants which scavenge ROO• will prevent the harmful biological and biochemical action. Our experimental results showed that flavonoids or phenolic compounds inhibited the induction of single strand DNA break, as shown in Figure 5A. It was envisaged that plant seeds rich in lipids contains high levels of antioxidant components. In this context, unpurified plant oil would also retain this antioxidative capacity. As shown in Figure 5B, suppression of DNA cleavage by oxidized oils (60 °C for 7 days) plus hematin become less efficient

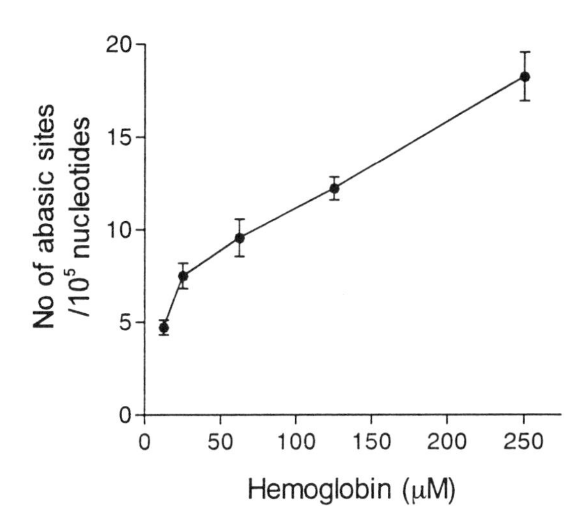

Figure 3. *Abasic site formation of the reaction of t-BuOOH with hemoglobin. Reproduced with permission from reference 17. Copyright 2000 Elsevier Science Ireland Ltd.*

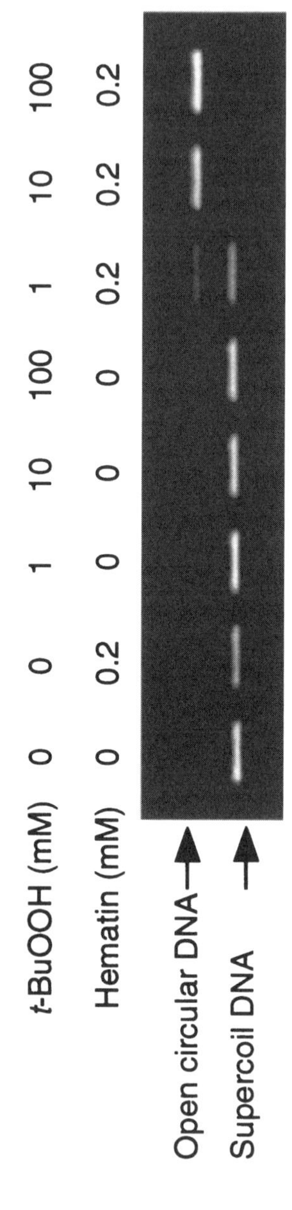

Figure 4. *DNA strand break by the reaction of t-BuOOH plus hematin. Data was reproduced from reference 8.*

as the purification/refining process proceed: i.e., more open circular DNA is seen as purification progress. For instance, safflower oil yielded a clear band of open circular form of DNA, which was not observed in the assay of crude native rapeseed oil. Figure 6A shows ROO•-scavenging activity of vegetable oils and refining steps estimated by chemiluminescence intensity and expressed as equivalent values to Torolox, a water soluble analogue of tocopherol. The intensity of luminol-enhanced chemiluminescence of the model radical component (*t*-BuOO•) generated was quantified in the presence of various edible oils or Trolox. The potential of ROO•-scavenging activity was highest in the extra virgin olive oil among different oils obtained from the market, where highly-purified commercial oils are scanty in ROO•-scavenging capacity. In addition, ROO•-scavenging capacity of antioxidants (flavonoids and phenolic components) was represented by the value equivalent to quercetin, as estimated by the bioassay based on the bactericidal effect of ROO• described previously (*7, 18*) (Figure 6B). The growth of bacteria was permitted by addition of various phenolic compounds to the *t*-BuOO•-generating system, where the minimal effective doses were estimated. Furthermore, ROO• generation, assessed by the use of a chemiluminescence analyzer, was more prominent as more purification proceeds, that is less suppression of the ROO• generation is seen. Table II shows the suppression of ROO• by various oils. Suppression was determined as the amount of oils required to quench 50% of the ROO• and is expressed as the IPOX50 value: the amount of oil required to reduce *t*-BuOO• to 50% as measured by the luminol enhanced chemiluminescence method.

Table II. Anti-Peroxyl Radical Activity of Various Vegetable Oils at Different Processing Steps

	IPOX50 (mg/ml)[a]			
	Soybean	*Rapeseed*	*Corn*	*Safflower*
Crude/original	8.99 ± 0.3	1.69 ± 0.04	7.31 ± 0.76	14.8 ± 0.7
Deacidefied	73.5 ± 4.3	17.8 ± 0.6	20.7 ± 0.7	172 ± 14
Decolorized	219 ± 3	208 ± 8	286 ± 11	198 ± 7
Deodorized	>300	>300	>300	>300

[a] IPOX 50: The amount of oil required to reduce *t*-BuOO• to 50%, as measured by luminol enhanced chemiluminescence method, which was used as an index. Adapted with permission from reference 5. Copyright 2000 IOS Press.

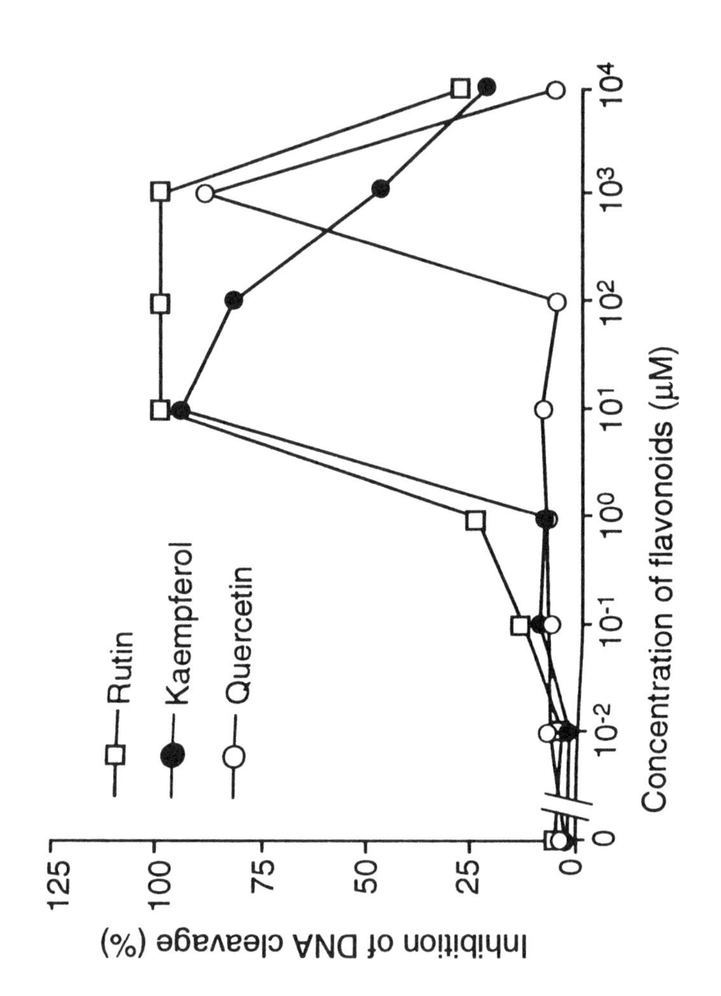

Rutin
Kaempferol
Quercetin
Inhibition of DNA cleavage (%)
125
100
75
50
25
0
0
10^{-2}
10^{-1}
10^{0}
10^{1}
10^{2}
10^{3}
10^{4}
Concentration of flavonoids (μM)

A

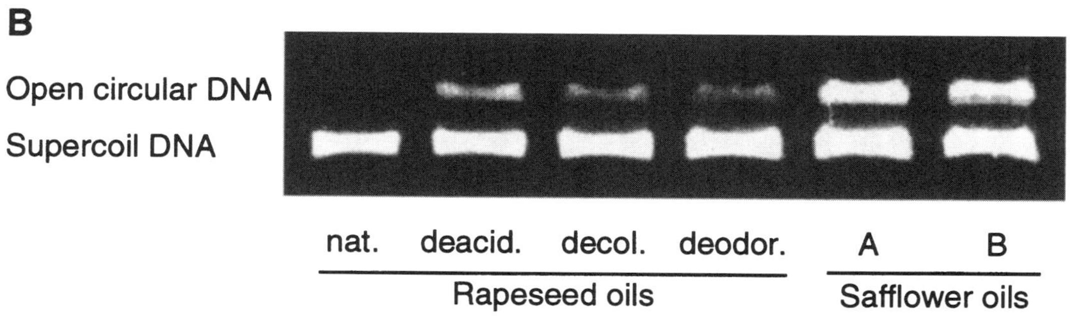

Figure 5. *Suppression of t-BuOO•-induced DNA strand breakage by flavonoids (A) and DNA strand breakage by the reaction of oxidized edible oil plus hematin (B). Data were reproduced from reference 8.*

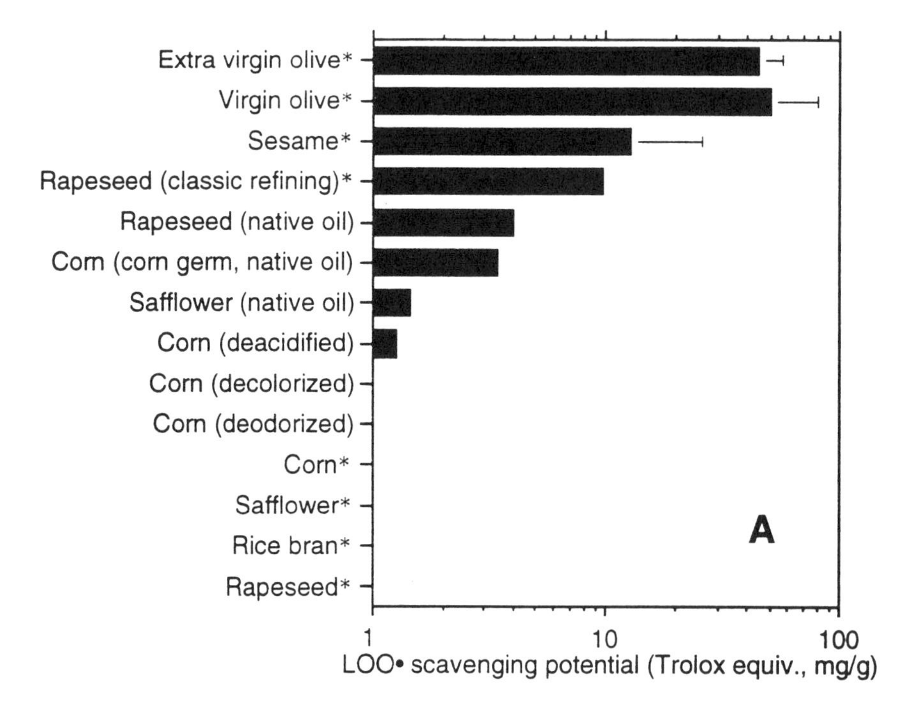

Figure 6. *The peroxyl radical-scavenging activities of various vegetable oils (A) and flavonoid and phenolic compounds (B). * Oils in markets. Adapted with permission from references 8 and 18. Copyrights 1999 American Association for Cancer Research (A) and ACS (B).*

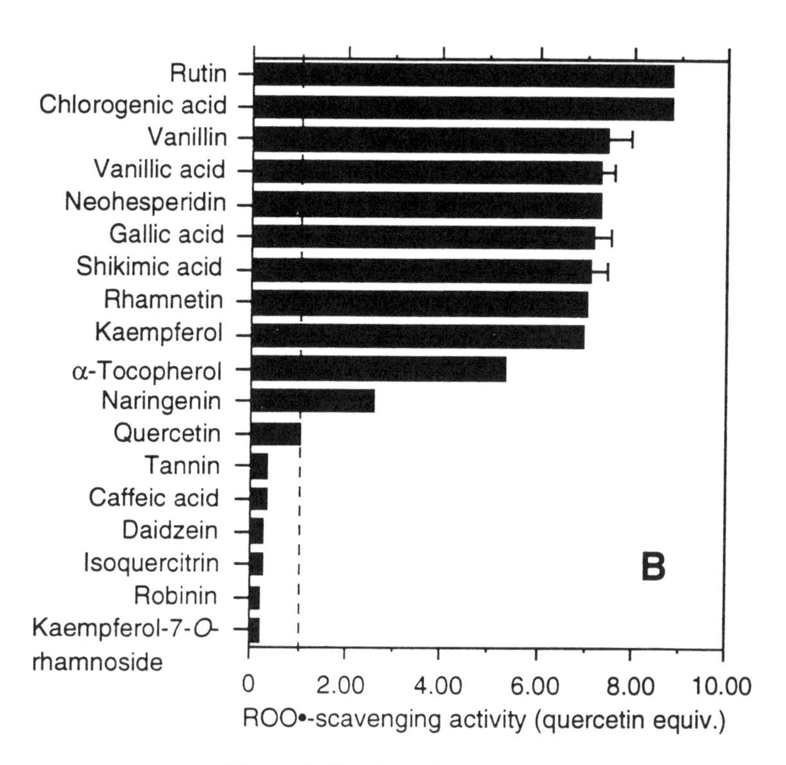

Figure 6. *Continued.*

Anti-Tumor-Promoter Effect and Alkyl Peroxyl Radical-Scavenging Activity of Boiled Vegetables Soup Compared with Cold-Water-Extracts of Fresh Vegetables

To reveal a correlation between tumor promoter effect induced by phorbol 12-myristate 13-acetate (PMA) and lipid peroxides or peroxyl radical thus formed, we examined the capacity to suppress the so-called "early antigen (EA)" formation, which is similar to transformation related antigen of human B-lymphocytes by PMA. This was carried out using Raji cells (B-lymphocytes), which harbor Epstein-Barr virus, and both cold- and hot-water-extracts of various vegetables. The ROO•-scavenging potential of the extracts was also examined (*19*). The anti-tumor-promoter effect of antioxidants was correlated to the anti-ROO• (*t*-BuOO•) radical activity. In this system (*20*), the tumor promoter PMA and *n*-butyric acid were added to Raji cells. When Raji cells are stimulated with PMA a proportion of EA positive cells increases; this phenomenon is regarded as a transformation-associated antigen as described above. Thus, if the suppression of EA formation correlated to the anti-ROO•

Table Ⅲ. The Activities of Scavenging Alkyl Peroxyl Radical and Suppressing EA Formation of Cold- and Hot-Water-Extracts of Vegetables.

Vegetable	*IPOX50*		*IEA50*	
	Fresh	*Boiled*	*Fresh*	*Boiled*
Greener				
Perilla (red)[a]	224	8100	233	3668
Perilla (green)[b]	218	5150	132	3120
Carrot , leaf[c]	71	1180	30	2112
Horwort (Mitsuba)[d]	77	1770	221	1500
Whitish				
Lettuce[e]	23	600	4	315
Celery[f]	26	316	33	314
Onion[g]	23	127	5	93
Cabbage, inside leaf[h]	11	24	30	55

See text for detail. EA, early antigen in EB virus harboring B-lymphocytes (Raji). Data were reproduced from reference 19.

activity obtained in vegetable extracts, either with hot or cold water, it is interpreted that anti-ROO• exhibits an anti-tumor-promoter effect. We were intrigued by the observation that suppression the EA formation was much higher in hot-water-extracts than cold-water-extracts (Table III) (*19*). Furthermore, the anti-tumor-promoting activity did parallel to the ROO•-scavenging activity of water extracts of vegetables (Figure 7, r = 0.82, n > 60) (*19*). Representive examples are given in Table III. In general, greener vegetables exhibited higher value than whitish vegetables (e.g., onion, inside leaf of cabbage, etc.). These findings suggest that ROO• has a tumor-promoter effect and antioxidant ROO•-scavenging would prevents tumor-promotion. The epidemiological evidence linking a high intake of green vegetables with a lower incidence of colon cancer is consistent with the finding (*3, 21*). It should also be emphasized that hot-water-extracts are much higher in these preferable effects than cold/fresh vegetable extracts. This may be explained by the fact that heating probably disrupts the plant cell walls, so that various intracellular antioxidants such as flavonoids and other phenolic components will be accessible to water milieu. This notion was further supported by more detailed experiment using mung bean sprout: namely, we confirmed that flavonoids identified by HPLC (robinin, rutin, kaempferol, quercetin, isoquercitrin, and kaempferol-7-o-rhamnoside) were present in higher concentration in boiled hot-water-extracts than in cold-water-extracts (*18*). Vitamin C is known to be relatively stable in vegetable even on heating to near boiling.

Conclusion

The simultaneous intake of fat and heme-iron is considered to result in higher incidence of colon cancer. We found that ROO•, which is generated in the presence of heme-iron, can damage DNA very effectively. ROO• species may induce cell death and mutation resulting from depurination/depyrimidination, strand breakage, base-modification, and perhaps stimulation of cell proliferation, which finally leads to carcinogenesis. Seed-derived natural vegetable oils, hot-water vegetable extracts, as well as antioxidants such as flavonoids and other phenolic compounds can suppress damage to DNA. Crude vegetable oils are essentially rich in antioxidant and ROO•-scavenging activities; however, the conventional process of oil-refining removes most of such radical-scavenging components not to mention antioxidant components. This indicates that an improved process of refining oil should be recommended because of the known adverse effect of ROO• which possesses a long in vivo half-life, propagating characteristics, and potential genotoxic activity. Furthermore, boiled vegetables, perhaps being much higher in bio-availability, possess much

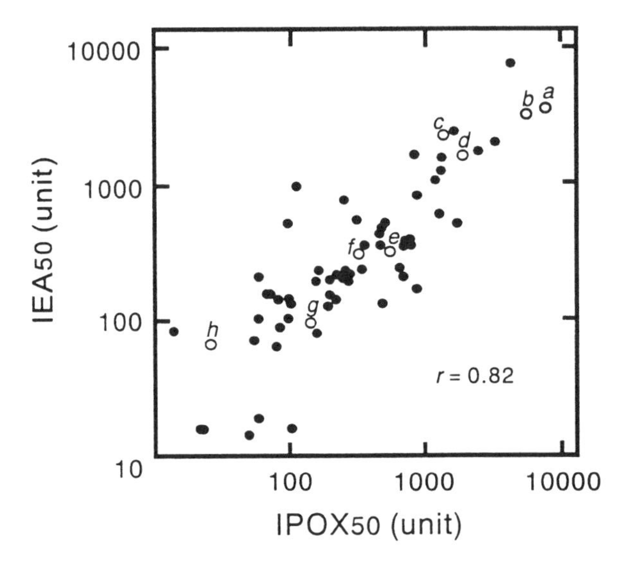

Figure 7. *Correlation between alkyl peroxyl radical-scavenging effect (IPOX50) and anti-tumor promoter effect (IEA50). Data were reproduced from reference 19.*

higher anti-ROO• activity as well as anti-tumor promoter effect than fresh vegetable extracts. The importance of heat cooking is thus obvious.

References

1. Willet, W. C.; Stampfer, M. J.; Colditz, G. A.; Rosner, B. A.; Speizer, F. E. *N. Engl. J. Med.* **1990**, *323*, 1664-1672.
2. Willett W. C. *Semin. Cancer Biol.* **1998**, *8*, 245-253.
3. Levi, F.; Pasche, C.; La Veccha, C.; Lucchini, F.; Franceschi, S. *Br. J. Cancer* **1999**, *79*, 1283-1287.
4. Burcham, P.C. *Mutagenesis* **1998**, *13*, 287-305.
5. Kanazawa, A.; Sawa, T.; Akaike, T.; Morimura, S.; Kida, K.; Maeda, H. *BioFactors* **2000**, *13*, 187-193.
6. Akaike, T.; Sato, K.; Ijiri, S.; Miyamoto, Y.; Kohno, M.; Ando M.; Maeda, H. *Arch. Biochem. Biophys.* **1992**, *294*, 55-63.
7. Akaike, T.; Ijiri, S.; Sato, K.; Katsuki, T.; Maeda, H. *J. Agric. Food Chem.* **1995**, *43*, 1864-1870.
8. Sawa, T.; Akaike, T.; Kida, K.; Fukushima, Y.; Takagi, K.; Maeda, H. *Cancer Epidemiol. Biomark. Prev.* **1998**, *7*, 1007-1012.
9. McCord, J. M.; Keele, B. B., Jr.; Fridovich, I. *Proc. Natl. Acad. Sci. USA* **1971**, *68*, 1024-1027.
10. Halliwell, B. *Free Radical Res. Commun.* **1990**, *9*, 1-32.
11. Dahl, T. A.; Midden, W. R.; Hartman, P. E. *J. Bacteriol.* **1989**, *171*, 2188-2194.
12. Akaike, T.; Ando, M.; Oda, T.; Doi, T.; Ijiri, S.; Araki, S.; Maeda, H. *J. Clin. Invest.* **1990**, *85*, 739-745.
13. Sato, K.; Akaike, T.; Kohno, M.; Ando, M; Maeda, H. *J. Biol. Chem.* **1992**, *267*, 25371-25377.
14. Van der Zee, J.; Van Steveninck, J.; Koster, J. F.; Dubblman, T. M. *Biochim. Biophys. Acta* **1989**, *980*, 175-180.
15. Kamiya, H.; Suzuki, M.; Komatsu, Y.; Miura, H.; Kikuchi, K.; Sakaguchi, T.; Murata, N.; Masutani, C.; Hanaoka, F.; Ohtsuka, E. *Nucleic Acids Res.* **1992**, *20*, 4409-4415.
16. Chakravarti, D.; Pelling, J. C.; Cavalieri, E. L.; Rogan E. G. *Proc. Natl. Acad. Sci. USA* **1995**, *92*, 10422-10426.
17. Kanazawa, A.; Sawa, T.; Akaike, T.; Maeda, H. *Cancer Lett.* **2000**, *156*, 51-55.
18. Sawa, T.; Nakao, M.; Akaike, T.: Ono, K.; Maeda, H. *J. Agric. Food Chem.* **1999**, *47*, 397-402.

19. Maeda, H.; Katsuki, T., Akaike, T; Yasutake, R. *Jpn. J. Cancer Res.* **1992**, *83*, 923-928.
20. Koshimizu, K; Ohigashi, H.; Tokuda, H.; Kondo, A.; Yamaguchi, K. *Cancer Lett*. **1988**, *39*, 247-257.
21. Franceschi, S.; Parpinel, M.; La Vecchia, C.; Favero, A.; Talamini, R.; Negri, E. *Epidemiology* **1998**, *9*, 338-341.

Chapter 22

Wasabi: A Traditional Japanese Food That Contains an Exceedingly Potent Glutathione *S*-Transferase Inducer for RL34 Cells

Yasujiro Morimitsu[1,3], Y. Nakamura[2,4], T. Osawa[1], and K. Uchida[1]

[1]Laboratory of Food and Biodynamics, Nagoya University Graduate School of Bioagricultural Sciences, Nagoya 464-8601, Japan
[2]Laboratory of Organic Chemistry in Life Science, Graduate School of Agricultural, Kyoto University, Kyoto 606-8502, Japan
[3]Current address: Laboratory of Food Chemistry Ochanomizu University, Otsuka 2-1-1, Bunkyo-ku, Tokyo 112-8610, Japan
[4]Current address: Laboratory of Food and Biodynamics, Nagoya University Graduate School of Bioagricultural Sciences, Nagoya 464-8601, Japan

In the course of our screening of edible plants on the induction of glutathione *S*-transferase (GST) activity in rat liver epithelial RL34 cells, 6-methylsulfinylhexyl isothiocyanate (MS-ITC) was isolated from wasabi (*Wasabia japonica*) as a potential inducer of GST. As a result from western blot analyses of GST expression induced by MS-ITC in RL34 cells, increasing of protein levels of both GST-Ya and GST-Yp were detected. And as a result from time dependent GSH level by MS-ITC, the GSH level in RL34 cells was initially decreased, then increased gradually to 1.8 times higher than its basal level after 12 hrs. By adding benzyl isothiocyanate (BITC) isolated from papaya instead of MS-ITC to the cells, the reactive oxygen intermediates (ROIs) detected by a fluorescence probe were increased immediately. From these data, we discuss the relation between levels of intracellular oxidation and the induction of GST.

Wasabi (*Wasabia japonica,* syn. *Eutrema wasabi*) is one of traditional foods in Japan, and is now worldwide used in the pungent spice of the "sushi". Wasabi belong to the family *Cruciferae* such as horseradish, mustard, cabbage and broccoli. When Cruciferous vegetables are damaged, glucosinolates are hydrolyzed by myrosinase, and characteristic isothiocyanates are formed immediately. Most of these Crucifer isothiocyanates (*cf.* allyl isothiocyanate) are well known for having antimicrobial, fungicidal, and pesticidal activities (*1,2*). For example, a major wasabi flavor compound 6-methylthiohexyl isothiocyanate, has been shown the antimicrobial activity (*3*).

On the other hand, epidemiological studies have demonstrated that consumption of cruciferous vegetables is associated with a lower incidence of cancers (*4-6*). Induction of phase II enzymes such as GST and/or quinone reductase (QR) has been demonstrated in broccoli (*7,8*), cabbage (*9*), and Brussel sprouts (*10,11*). Many natural isothiocyanates derived from cruciferous vegetables and some fruits have been shown to induction of phase II enzymes in cultured cells and rodents (*12-14*).

We have studied on physiological, pharmaceutical or therapeutic effects of various edible plants, especially the sulfur-containing compounds from onion (*15,16*). Our continuous screening for the induction of GST activity in RL34 cells resulted in isolation of MS-ITC from the ethyl acetate (EtOAc) extract of wasabi. In this study, we describe the isolation of MS-ITC from wasabi, and examined the protein levels of GST isozyme induced by MS-ITC. As a result from elucidation of the intracellular redox state of BITC instead of MS-ITC, the isothiocyanate moiety of these food factors plays an important role for the induction of phase II enzymes.

Experimental

Materials and Chemicals. Wasabi roots were kindly gifted from Kinjirushi Wasabi Co., Ltd (Aichi, Japan). All other vegetables and fruits were purchased from the market in Nagoya, Japan. Authentic MS-ITC and its derivatives were synthesized in our laboratory. BITC and all other chemicals were purchased from Wako Pure Chemical Industries (Osaka, Japan). Horseradish peroxidase-linked anti-rabbit IgG immunoglobulin and enhanced chemiluminescence (ECL) Western blotting detection reagents were obtained from Amersham Pharmacia Biotech. The protein concentration was measured using the BCA protein assay reagent obtained from Pierce. 2,7-Dichlorofluorescin diacetate (DCF-DA) was obtained from Molecular Probes (Leiden, the Netherlands).

Cell Cultures and Enzyme Assay. The induction of GST was measured in RL34 cells by using 1-chloro-2,4-dinitrobenzene as a substrate according to the reported method (*17*). The cells were grown as monolayer cultures in DMEM supplemented with 5% heat-inactivated fetal bovine serum, 100 units/ml penicillin, 100 μg/ml streptomycin, 0.3 mg/ml-glutamine, 0.11 mg/ml pyruvic acid, and 0.37% $NaHCO_3$ at 37°C in an atmosphere of 95% air and 5% CO_2. Postconfluency cells were exposed to samples for 24 hours in the medium containing 5% fetal bovine serum in 24-well or 96-well microtiter plates (*18,19*). Usually, 25 μg/ml and 2.5 μg/ml of the samples (as final concentrations in the medium) were tested. Assessment of cytotoxicity after treatment of the sample was measured by protein concentration in an aliquot of the digitonin cell lysate with the BCA protein assay reagent.

GSH Assay. Cellular GSH was measured by a fluorometric assay (*20*). Typically, cell lysates were diluted with 0.1M potassium phosphate buffer, pH 8.0, and proteins were precipitated with trichloroacetic acid. A 100 μl sample then mixed with 100 μl of *o*-phthaldialdehyde (10 mg in 10 ml of methanol) and 1.8 ml of 0.1M potassium phosphate buffer (pH 8.0) containing 5 mM EDTA. The solution was incubated at room temperature for 15 min, and formation of GSH-phthalaldehyde conjugate was then measured by a fluorescence spectrometer (Hitachi Model F-2000). The conjugate was excited at a wavelength of 350 nm, and the fluorescence emission was measured at 420 nm.

Western Blot Analysis. The sample treated and untreated cells were rinsed twice with PBS (pH 7.0) and lysed by incubation at 37°C for 10 min with a solution containing 0.8% digitonin and 2 mM EDTA (pH 7.8). Each whole-cell lysate was then treated with the Laemmuli sample buffer for 3 min at 100°C (*21*). The samples (20 μg) were run on a 12.5% SDS-PAGE slab gel. One gel was used for staining with Coomassie brilliant blue, and the other was transblotted on a nitrocellulose membrane with a semidry blotting cell (Trans-Blot SD; Bio-Rad), incubated with Block Ace (40 mg/ml) for blocking, washed, and treated with the antibody.

Intracellular Oxidative Products Determination. Intracellular oxidative products were detected by 2,7-dichlorofluorescin (DCF) as an intracellular fluorescence probe (*22,23*). The cells under confluency were treated with DCF-DA (50 μM) for 30 min at 37°C. After washing twice with PBS, BITC was added to the complete medium and incubated for another 30 min. A flow cytometer (CytoACE 150; JASCO, Tokyo, Japan) was used to detect DCF formed by the reaction of H_2DCF with the intracellular oxidative products.

Results and Discussion

Isolation of Inducers from Wasabi.

Our screening for induction of GST activity in RL34 cells was undergone on many extracts of various vegetables and fruits. In the course of our screening, the EtOAc extracts of cruciferous vegetables were found to show potent induction activities, especially wasabi. The EtOAc extract of wasabi showed the potent induction activity even though the adding amount of extract was diminished (2.5 μg/ml). After continuous fractionations of the EtOAc extract of wasabi by conventional, a potential inducer of GST in RL34 cells was isolated with yield of 70mg from 1.3 kg of wasabi root (Fig. 1A). And two homologous isothiocyanates were also isolated as minor inducible compounds. The major compound was identified as 6-methylsulfinylhexyl isothiocyanate (MS-ITC) and two minor compounds were also identified as 5-methylsulfinylpentyl isothiocyanate and 7-methylsulfinylheptyl isothiocyanate respectively by spectroscopic analyses (Fig. 1B). MS-ITC is a homologous compound of sulforaphane, which was isolated from broccoli as a potent inducer of phase II enzymes (*24,25*). MS-ITC was previously reported to isolate from certain wild plants and termed hesperin (*26*). And MS-ITC has been identified and quantified in wasabi root as well as 6-methylthiohexyl isothiocyanate, a major wasabi flavor compound (*27*).

Induction of GST in RL34 Cells.

Dose-dependent induction activity by MS-ITC in RL34 cells was examined (Fig. 2A). More than 10 μM, MS-ITC has begun to show cytotoxicity. As a result from western blot analyses of GST expression induced by MS-ITC in RL34 cells, increasing of protein levels of both GST-Ya and GST-Yp were detected (Fig. 2B). The GST-Yp isozyme could be one of the important determinants in cancer susceptibility, particularly in diseases where exposure to polycyclic aromatic hydrocarbons is involved. But a recent study using transgenic mice lacking the class π GST demonstrated that this class of GST was involved in the metabolism of carcinogens in mouse skin and had a profound effect on tumorigenicity (*28*). Additionally, the effect of MS-ITC on levels of phase II enzymes of ICR mice was also examined by gavage (Morimitsu *et al.*, prepared for submitting). As a result from determination of phase II enzyme activities, MS-ITC was also found to be a potential inducer of GST *in vivo*, owing to increasing protein levels of GST-Ya.

Intracellular Oxidation State Induced by Isothiocyanates.

As a result from time dependent GSH level by MS-ITC, the GSH level in RL34 cells was initially decreased, then increased gradually to 1.8 times higher

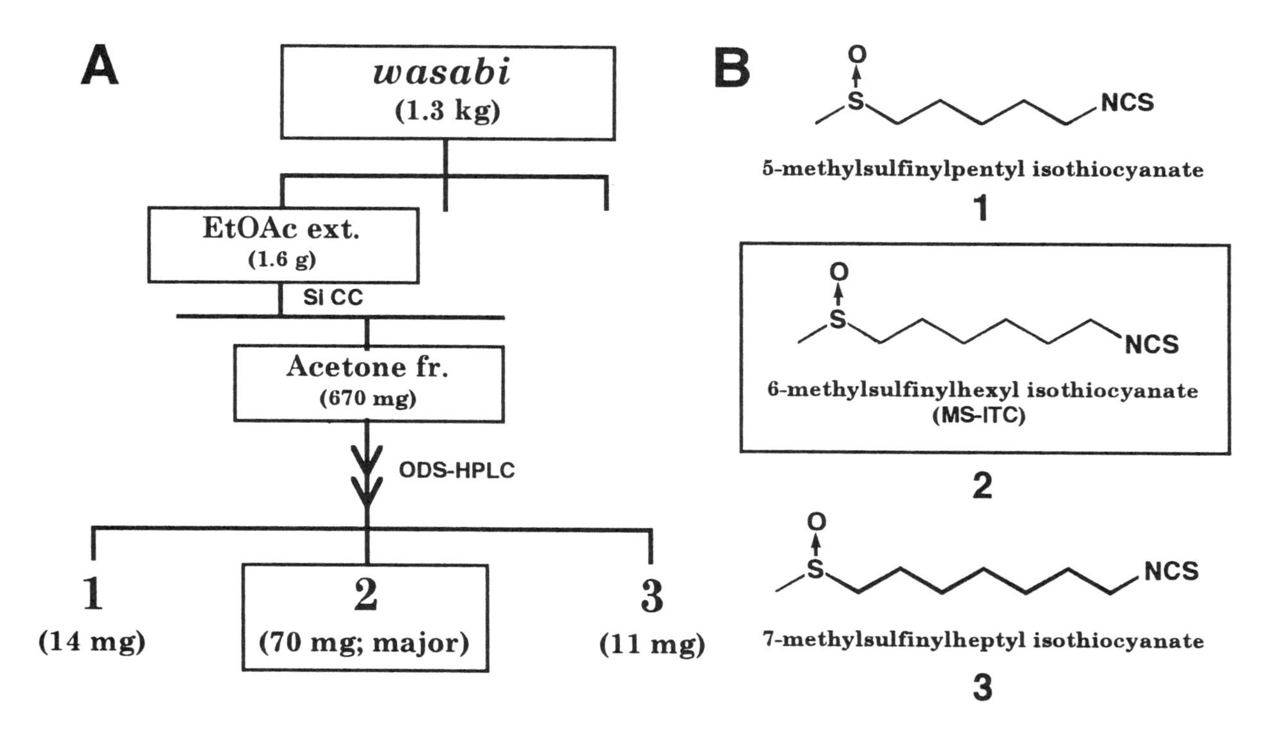

Figure 1. The isolation scheme (A) and the chemical structure (B) of potent GST inducers in wasabi.

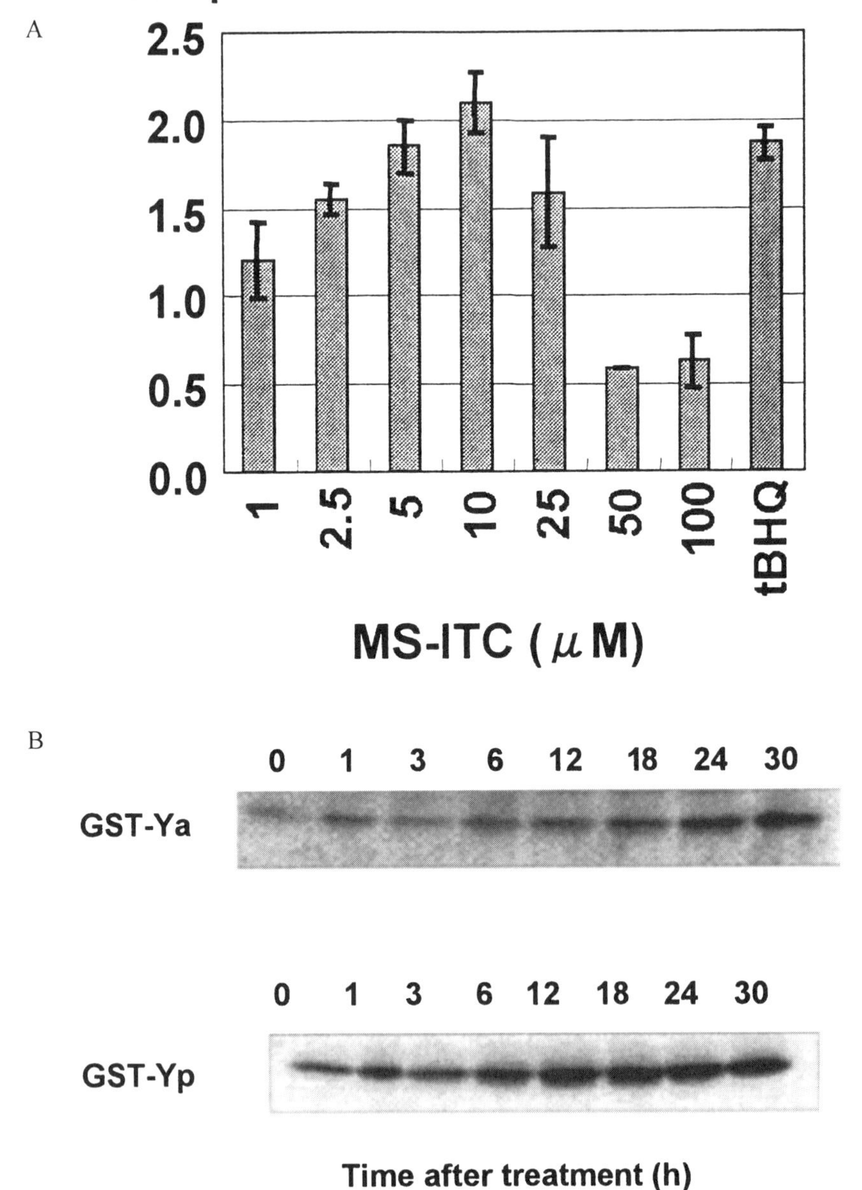

Figure 2. Dose-dependent effect of MS-ITC on the cellular GST (A), and time-dependent effect of MS-ITC on the cellular GST-Ya and GST-Yp proteins (B).

than its basal level after 12 hrs (Fig. 3A). In a recent study, we have isolated BITC from papaya as a potent GST inducer (*29*). In order to know intracellular oxidation degree, BITC in stead of MS-ITC was added to the cells. As shown in Fig. 3B, the reactive oxygen intermediates (ROIs; hydrogen peroxide, lipid peroxide, peroxinitrile, and so on) detected by a fluorescence probe were increased immediately. The level of ROIs induced by different concentrations of BITC was increased in a dose-dependent manner. The level at 10 μM BITC was near 50-fold higher than that of the control. This ROIs generation was mainly caused by the isothiocyanate's highly reactivity with nucleophiles such as GSH, sulfhydryl residues of proteins and so on (*25*). Concerning about the non-enzymatically conjugation reaction between isothiocyanate group and GSH in RL34 cells at initial step after the addition of MS-ITC resulted in slight GSH depletion. The Michael addition acceptors were reported to be able to induce phase II enzyme activities *in vitro* and perhaps even *in vivo* (*24,25*). Isothiocyanates is right classed in this category. BITC was also decreased the GSH level, then was induced oxidative stress in RL34 cells. It is considered that this GSH depletion comes to the generation of ROIs in the cells. We assume that this slight oxidative stress may lead to increase the protein levels of GST, after induction of signal transductions and gene expressions in cells (*30*).

Acknowledgements

The authors thank Kinjirushi Wasabi Co. Ltd. for providing us wasabi research samples. This research project was supported by the Program for Promotion of Basic Research Activities for Innovative Biosciences (PROBRAIN).

References

1. Soledade, M.; Pedras, C.; Sorensen, J. L. *Phytochemistry* **1998**, 49, 1959.
2. Tajima, H.; Kimoto, H.; Taketo, Y.; Taketo, A. *Biosci. Biotech. Biochem.* **1998**, 62, 491.
3. Ono, H.; Tesaki, S.; Tanabe, S.; Watanabe, M. *Biosci. Biotech. Biochem.* **1998**, 62, 363.
4. Lee, H. P.; Gourley, L.; Duffy, S. W.; Esteve, J.; Lee, J.; Day, N. E. *Int. J. Cancer* **1989**, 434, 1007.
5. Olsen, G. W.; Mandel, J. S.; Gibson, R. W.; Wattenberg, L. W.; Schuman, L. M. *Cancer Causes Contr.* **1991**, 2, 291.
6. Mehta, B. G.; Liu, J.; Constantinou, A.; Thomas, C. F.; Hawthorne, M.; You, M.; Gerhauser, C.; Pezzuto, J. M.; Moon, R. C.; Moriarty, R. M. *Carcinogenesis* **1995**, 16, 399.
7. Aspry, K. E.; Bjeldanes L. F. *Food Chem. Toxicol.* **1983**, 21, 133.

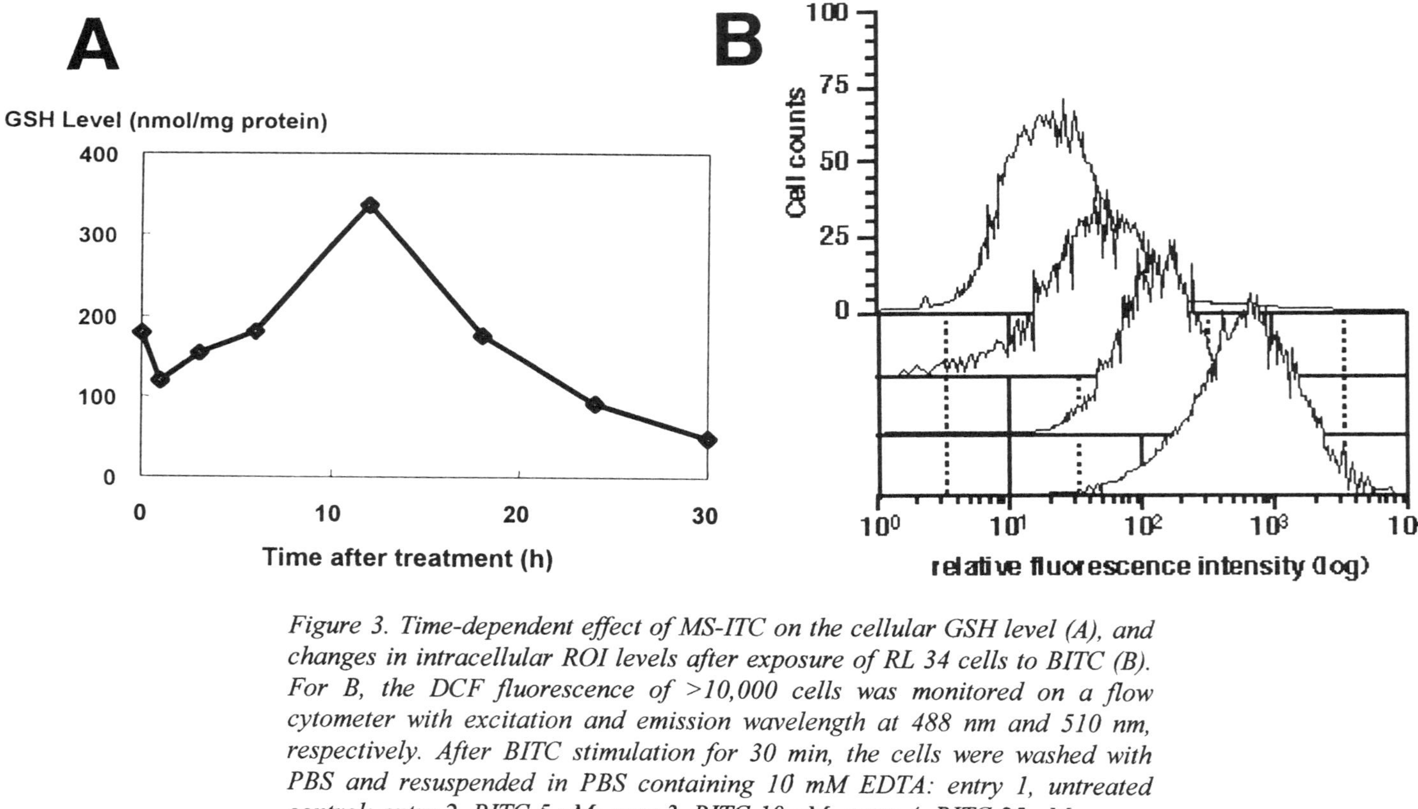

Figure 3. Time-dependent effect of MS-ITC on the cellular GSH level (A), and changes in intracellular ROI levels after exposure of RL 34 cells to BITC (B). For B, the DCF fluorescence of >10,000 cells was monitored on a flow cytometer with excitation and emission wavelength at 488 nm and 510 nm, respectively. After BITC stimulation for 30 min, the cells were washed with PBS and resuspended in PBS containing 10 mM EDTA: entry 1, untreated control; entry 2, BITC 5 μM; enry 3, BITC 10 μM; entry 4, BITC 25 μM.

8. Fahey, J. W.; Zhang, Y.; Talalay, P. *Proc. Natl. Acad. Sci. USA* **1997**, 94, 10367.
9. Whitty, J. P.; Bjeldanes, L. F. *Food Chem. Toxicol.* **1987**, 25, 581.
10. Nijhoff, W. A.; Groen, G. M.; Peters, W. H. M. *Int. J. Oncol.* **1993**, 3, 1131.
11. Wallig, M. A.; Kingston, S.; Staack, R.; Jeffery, E. H. *Food Chem. Toxicol.* **1998**, 36, 365.
12. Leonard, T. B.; Popp, J. A.; Graichen, M. E.; Dent, J. G. *Carcinogenesis* **1981**, 2, 473.
13. Yang, C. S.; Smith T. J.; Hong, J. Y. *Cancer Res.* **1994**, 54, 1982.
14. Barcelo, S.; Gardiner, J. M.; Gescher, A.; Chipman, J. K. *Carcinogenesis* **1996**, 17, 277.
15. Kawakishi, S.; Morimitsu, Y. *Lancet* **1988**, 1, 330.
16. Morimitsu, Y.; Kawakishi, S. *Phytochemistry* **1990**, 29, 3435.
17. Habig, W. H.; Pabst, M. J.; Jakoby, W. B. *J. Biol. Chem.*, **1974**, 249, 7130.
18. Fukuda, A.; Nakamura, Y.; Ohigashi, H.; Osawa, T.; Uchida, K. *Biochem. Biophys. Res. Commun.* **1997**, 236, 505.
19. Uchida, K.; Shiraishi, M.; Naito, Y.; Torii, Y.; Nakamura, Y.; Osawa, T. *J. Biol. Chem.* **1999**, 274, 2234.
20. Hissin, P. J.; Hilf, R. *Anal Biochem.* **1976**, 74, 214-226.
21. Laemmuli, U. K. *Nature (London)* **1970**, 227, 680.
22. Bass, D. A.; Parce, J. W.; Dechatelet, L. R.; Szejda, P.; Seeds, M. C.; Thomas, M. *J. Immnol.* **1983**, 130, 1910.
23. Nakamura, Y.; Murakami, A.; Ohto, Y.; Torikai, T.; Ohigashi, H. *Cancer Res.* **1998**, 58, 4832.
24. Zhang, Y.; Talalay, P.; Cho, C.-G.; Posner, G. H. *Proc. Natl. Acad. Sci. USA* **1992**, 89, 2399.
25. Zhang, Y.; Kensler, T. W.; Cho, C.-G.; Posner, G. H.; Talalay, P. *Proc. Natl. Acad. Sci. USA* **1994**, 91, 3147.
26. Daxenbichler, M. L.; Spencer, G. F.; Carlson, D. G.; Rose, G. B.; Brinker, A. M.; Powell, R. G. *Phytochemistry* **1991**, 30, 2623.
27. Etoh, H.; Nishimura, A.; Takasawa, R.; Yagi, A.; Saito, K.; Sakata, K.; Kishima, I.; Ina, K. *Agric. Biol. Chem.* **1990**, 54, 1587.
28. Henderson, C. J.; Smith, S. G.; Ure, J. Brown; K. Bacon; E. J.; Wolf, C. R. *Proc. Natl. Acad. Sci. USA* **1998**, 95, 5275.
29. Nakamura,Y.; Morimitsu, Y.; Uzu, H.; Ohigashi, H.; Murakami, A.; Naito, Y.; Nakagawa, Y.; Osawa, T.; Uchida, K. *submitted for publication.*
30. Nakamura, Y; Ohigashi, H.; Masuda, S.; Murakami, A.; Morimitsu, Y.; Kawamoto, Y.; Imagawa, M.; Uchida, K. *Cancer Res.* **2000**, 60, 219.

Chapter 23

Antioxidants and Antiatherosclerotic Effects of Chinese Medicinal Herb *Salvia Miltiorrhiza*

Ming-Shi Shiao[1], Yih-Jer Wu[2], Elaine Lin[3], Wan-Jong Kuo[1], and Bao-Wen Chang[1]

[1]**Department of Medical Research and Education, Veterans General Hospital, Taipei, Taiwan**
[2]**Department of Internal Medicine, Mackay Memorial Hospital, Taipei, Taiwan**
[3]**Institute of Biochemistry, National Yang Ming University, Taipei, Taiwan**

Recent studies have suggested that atherosclerosis is triggered by low-density lipoprotein (LDL) oxidation and endothelial dysfunction. *Salvia miltiorrhiza* (SM) is a Chinese medicinal herb widely used for the treatment of cardiovascular disorder. SM produces many water-soluble polyphenolic antioxidants. Among them, salvianolic acid B (Sal B) strongly inhibited human LDL oxidation *in vitro*. To evaluate the anti-atherosclerotic potential of SM, cholesterol-fed NZW rabbits and apoE(-) mice were used as the animal models. LDL samples obtained from treated animals were more resistant to oxidation *ex vivo*. The α-tocopherol contents of LDL were significantly higher in treated groups than that of control in NZW rabbits. Treatment of a water-soluble, Sal B-enriched fraction (SM-EW-1) of SM reduced atherosclerotic lesions in NZW rabbits and apoE(-) mice. Inhibition of atherosclerosis by SM relied heavily on the antioxidant effect. SM had an endothelial protective effect in rabbit model. This study concludes that SM, which contains Sal B as a major water-soluble antioxidant, increases α-tocopherol content in LDL, prevents LDL oxidation *in vitro* and *ex vivo*, and reduces atherosclerosis in these two atherosclerotic animal models.

Introduction

Many recent studies have indicated that oxidative modification of low-density lipoprotein (LDL) and endothelial dysfunction are involved in the early pathogenesis of atherosclerosis (*1-4*). Mildly oxidized LDL may occur in the plasma whereas extensively oxidized LDL (OxLDL) appears only in the intima of vessel wall (*2*). Uptake of OxLDL by the scavenger receptors of monocyte-derived macrophages causes the formation of foam cells and fibrous plaques. Antioxidants, which inhibit LDL oxidation, may reduce atherosclerosis (*5*). Vitamin E protects LDL from oxidative damage *in vitro* and decreases the morbidity of coronary heart disease in epidemiological studies. Probucol, a cholesterol-lowering drug and hydrophobic antioxidant, reduces atherosclerosis in WHHL and cholesterol-fed rabbits (*6*). N,N'-Diphenyl-phenylenediamine (DPPD) (*7,8*) and glabridin (from licorice) (Figure 1) have been demonstrated to reduce atherosclerosis in cholesterol-fed rabbits and apolipoprotein E-deficient (apoE(-)) mice (*9*), respectively. However, probucol does not reduce atherosclerosis in apoE(-) mice (*10*).

The potency of antioxidants to inhibit LDL oxidation *in vitro* does not correlate with their potency to reduce atherosclerosis *in vivo*. If functions of antioxidants are due to the protection of LDL from oxidation, water-soluble antioxidants, which are unable to adhere to LDL, may be disadvantageous to protect LDL from oxidation in the subendothelial space. However, epidemiological and animal studies have shown that many water-soluble antioxidants, such as flavonoids, reduce atherosclerosis (*5*). It is likely that water-soluble antioxidants reduce atherosclerosis by preventing the loss of endogenous lipophilic antioxidants in LDL (mainly α-tocopherol) (*11,12*). These antioxidants may also prevent endothelial injury causing by dyslipidemia and oxidative stress in the circulation.

Salvia miltiorrhiza Bunge (Labiatae) (abbreviated as SM) is a famous medicinal herb in the treatment of blood stasis in traditional Chinese medicine. The term blood stasis has no counterpart to a particular disease in western medicine. In a narrow and definable way, blood stasis has been described as poor microcirculation, hypertension, disturbances in platelet aggregation and thrombosis, dyslipidemia, and high oxidative stress (*13*). The above-mentioned symptoms are related to the pathogenesis of atherosclerosis. It is reasonable to speculate that Chinese medicinal herbs frequently used in the treatment of blood stasis may be effective in prevention and treatment of atherosclerosis-

α-Tocopherol

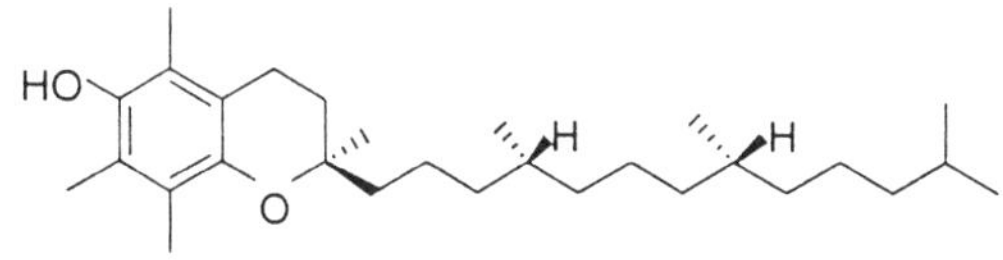

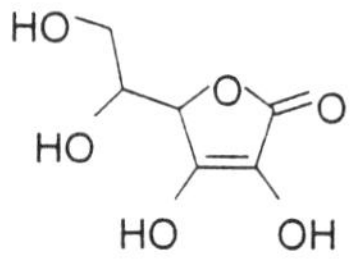

Ascorbic acid

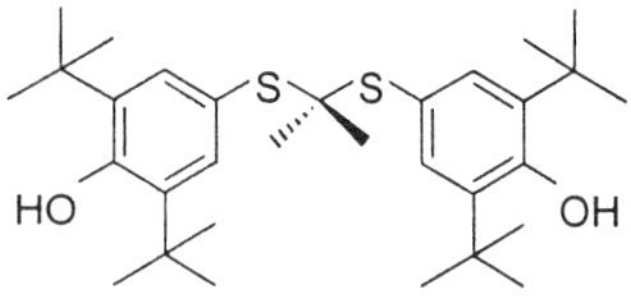

Probucol

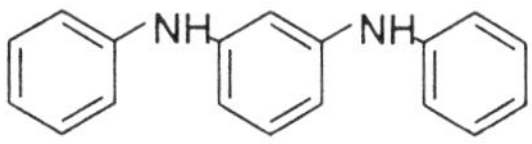

N,N'-Diphenyl-1,4-phenylenediamine (DPPD)

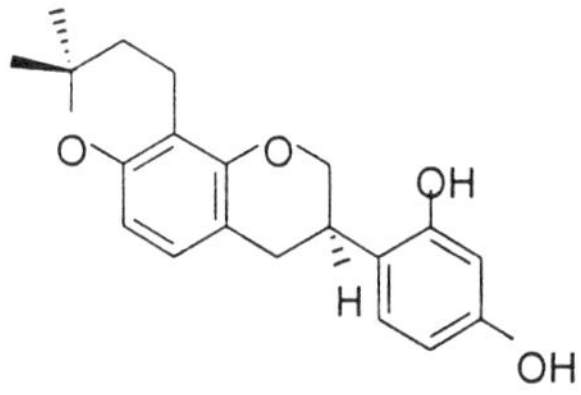

Glabridin (from Licorice)

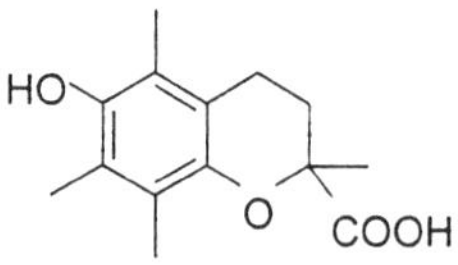

Trolox

Figure 1. Structures of antioxidants related to the inhibition of LDL oxidation.

related disorder. A summary of *Salvia* pharmacology based on *in vitro* and animal studies is shown below.

Pharmacology of *Salvia Miltiorrhiza*

Platelet Function

Many platelet aggregation inhibitors have been identified in SM. Among them, miltirone, Ro 096680, and salvinone exhibit the IC50 values of 5.76×10^{-6}, 2.14×10^{-6}, and 2.40×10^{-6} M, respectively (*14*). Compound 764-3 from SM inhibits platelet $[Ca^{2+}]i$ rise as well as aggregation evoked by collagen and thrombin in the presence of extracellular Ca^{2+}. After the removal of extracellular Ca^{2+} by EGTA that collagen and thrombin cause no aggregation, compound 764-3 still suppresses platelet $[Ca^{2+}]i$ rise. It inhibits platelet Ca^{2+} influx and Ca^{2+} mobilization with similar potency (*15*). Acetylsalvianolic acid A (ASAA), a semisynthetic analogue of salvianolic acid A (Figure 2), is also an effective inhibitor of platelet function. ASAA inhibits rat and rabbit platelet aggregation induced by ADP, collagen, arachidonic acid (AA), and thrombin. In *ex vivo* experiments with ADP, collagen, and AA as inducers, ASAA treatment inhibits platelet aggregation markedly. ASAA suppresses collagen induced 5-HT release while inhibiting platelet aggregation (*16*)

Blood Pressure

SM has been shown to produce hypotensive response in a dose-dependent manner in normo-tensive rats. The vasodepressor effect is not due to the presence of cations (K^+, Ca^{2+} and Mg^{2+}) in the extract. The hypotensive effect is not mediated via the alpha-, beta-adrenoceptors, histamine receptors, or autonomic ganglion; nor via direct vasodilation and diuresis. However, the vasodepressor effect is angiotensin- and bradykinin-related since captopril potentates the hypotensive effect. The vasodepressor effect can be accounted for by the positive inotropic and negative chronotropic effects (*17*). In SM treated rabbits, the concentrations of angiotensin II and atrial natriuretic polypeptide (ANP) are significantly lower in the plasma and slightly lower in brain and atrial. It indicates that the SM effect on blood vessels is related to changes of angiotensin II and ANP. The difference in SM effects between brain and plasma is likely due to the blood-brain barrier (*18*).

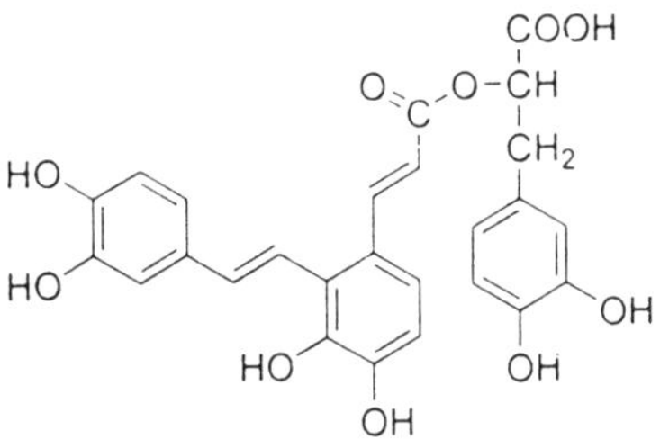

Salvianolic acid A (Sal A)

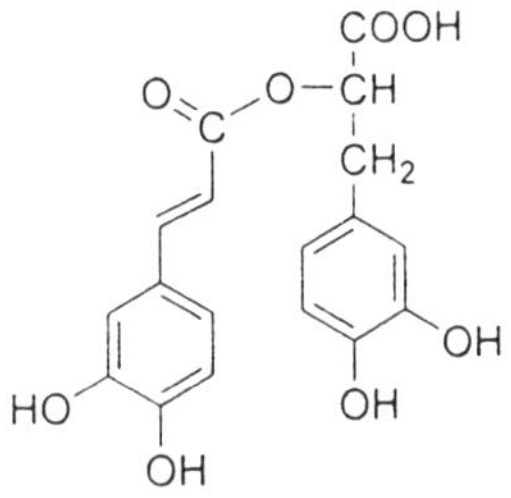

Rosmarinic acid

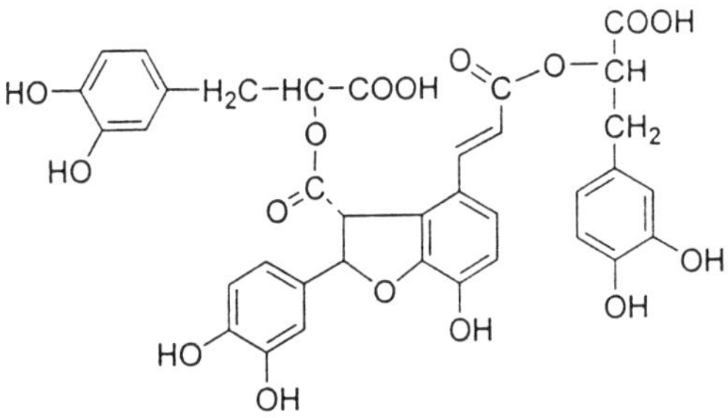

Salvianolic acid (Sal B)

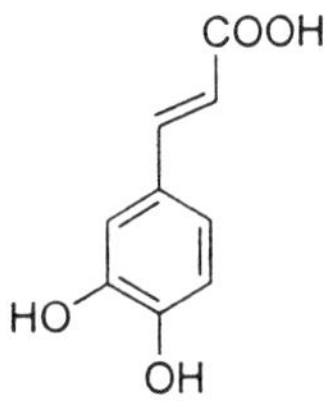

Caffeic acid

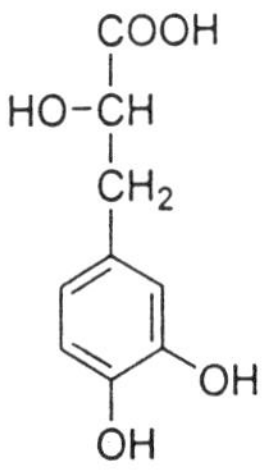

Danshensu

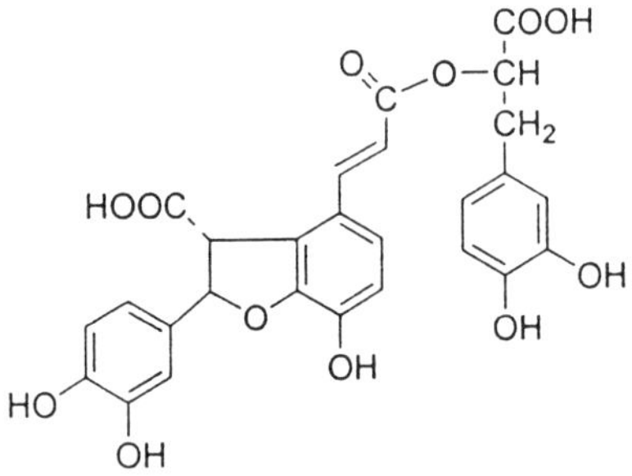

Lithospermic acid

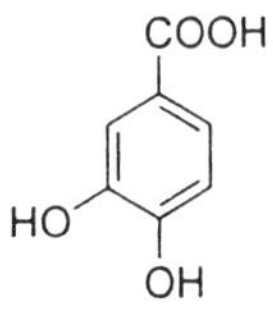

Protocatechuic acid

Figure 2. Phenolic natural products isolated from Salvia miltiorrhiza.

The cardiovascular action of SM to lower systemic blood pressure in the rats has been demonstrated. Langendorff cardiac preparation in guinea pig and four types of vasculature in dog, including coronary, renal, femoral, and mesenteric arteries are performed. SM induces dose-related hypotension without changing heart rate. Atropine, propranolol, and chlorpheniramine plus cimetidine antagonize the hypotensive effect. In the isolated whole-heart preparation, SM injection increases coronary blood flow and causes a positive inotropic action. SM relaxes all arteries at low concentration and contracts all, but the coronary artery, at high concentration (*19*).

In rats and rabbits, the suppressive effects of SM on systemic blood pressure are blocked or reversed by atropine, propranolol, and chlorpheniramine plus cimetidine. This reversed hypertension is also blocked by phenoxybenzamine indicating that SM has multiple effector sites in the cardiovascular system. One possibility is an increased utilization of extracellular Ca^{2+} since Ca^{2+} enhances the activity of SM on isolated blood vessels of rabbits. The aqueous extract and tanshinones from SM cause vasodilation of coronary arteries but induces vasodilation of renal, mesenteric, and femoral arteries. At higher concentrations, vasoconstriction is induced in these vessels. It has been suggested that SM is a useful antianginal agent since it dilates coronary vessels. However, its use in hypertension is questionable since SM induces vasodilation as well as vasoconstriction depending on dose and target vessel (*20*).

Protection of Myocardial Dysfunction and Oxidative Damage

The aqueous extract of SM (SM-H) protects chemically induced acute myocardial ischemia and arrhythmia in rats. SM-H protects the acute myocardial ischemia and arrhythmia induced by isoproterenol or $BaCl_2$ with the following observations: a. Injection (i.p.) of SM-H for 3 to 5 days or single i.v. injection reduces death rate, b. SM-H pretreatment increases the lethal dose of $BaCl_2$ infusion, c. SM-H (i.p.) decreases premature ventricular contraction and fibrillation, bradycardia, and mortality rate induced by $BaCl_2$, and d. i.p. injection of SM-H reduces ECG J-point displacement induced by isoproterenol (*21*).

Myocardial protection and cardioplegia by SM have also been reported. Myocardial ^{45}Ca sequestration in dogs during global ischemia and reperfusion, by using cardiopulmonary bypass (CPB) and myocardial function as markers, has been followed before and after CPB. SM effectively controls myocardial calcium sequestration during early reperfusion and reduces reperfusion injury. As to myocardial protection, cardioplegia with SM and verapamil are superior to hyperkalemic treatment alone (*22*).

The diterpenoids tanshinone I, cryptotanshinone, and tanshinone VI protect myocardium against ischemia-induced derangements. In isolated hearts subjected to hypoxic perfusion reoxygenation, tanshinone I, cryptotanshinone, and tanshinone VI enhance the recovery of the contractile force upon reoxygenation. The effects are likely due to the preservation of ATP metabolites in the myocardium and enhanced ATP restoration upon oxygen-replenishment (*23*). Sodium tanshinone IIA sulfonate (STS), a soluble derivative of tanshinone IIA, reduces myocardial infarct size in an ischemia-reperfusion rabbit model. Its effect is comparable to that of trolox, a water-soluble α-tocopherol analogue. Like trolox, STS does not inhibit oxygen uptake by xanthine oxidase. In contrast to trolox, STS prolongs the survival of cultured human saphenous vein endothelial cells but not ventricular myocytes exposed to xanthine oxidase-generated oxygen radicals. These findings suggest that STS is cardioprotective and may exert a beneficial effect on vascular endothelium (*24*).

SM injection scavenges reactive oxygen species (ROS) generated from ischemia-reperfusion injury in the myocardium as effectively as superoxide dismutase (SOD). Danshensu, one of its effective components, scavenges superoxide radicals generated from xanthine by xanthine oxidase. Free radicals generated from lipid peroxidation of myocardial mitochondrial membranes are scavenged by tanshinone. Danshensu protects the mitochondrial membrane fluidity and reduces lipid peroxidation from the ischemia-reperfusion injury (*25*). Seven phenolic natural products from SM including salvianolic acid A and salvianolic acid B (Figure 2) have been demonstrated to inhibit peroxidative damage to biomembrane (*26*). Salvianolic acid A also inhibits oxygen radical release and neutrophil function (*27*).

ROS, generated by xanthane-xanthane oxidase, inhibit the unitary current of the single K^+ channel activity in cardiac myocytes as demonstrated by using patch clamp technique. The inhibition is reversed by salvianolic acid A (*28*) (Figure 2). Lithospermic acid B is also active in preventing myocardial damage. When infused into the post-ischemic rabbit heart, lithospermic acid B reduces the myocardial damage in ischemia-reperfusion (*29*). Injection (i.p.) of the aqueous extract of SM prevents acute cardial ischemia induced by the ligation of the coronary artery in rats. The elevation of S-T segment in the electrocardiogram caused by ischemia is greatly reduced. The ischemia area in the left ventricle is reduced and animal survival rate is increased (*30*).

SM and polysaccharide sulphate (PSS) inhibit the adhesion of erythrocytes, from patients with cerebral thrombosis, to cultured human umbilical vein endothelial cells (HUVEC) (*25*). Treatment of SM, PSS, dextran 40 and mannitol improve the viscoelasticity properties of red blood cell suspension (*31*). SM inhibits the contraction of rabbit basilar artery rings evoked by $CaCl_2$

and KC1. The inhibition exhibits nonspecific antagonism in response to $CaCl_2$ dose. SM also inhibits the contraction of rabbit basilar artery rings evoked by KC1. However, it does not show myogenic activity of portal vein strips in rats (*32*)

SM reduces acute fatal ventricular fibrillation (VF) in male rats induced by injection of isoproterenol. VF occurs in all control animals resulting in 96% death with only 4% spontaneously reverted and survived. Pretreatment of animals with an aqueous extract of SM (SM-H) reduces J-point displacement and VF. Survival rate is significantly raised compared with the control. Immediate i.v. injection of SM-H to poisoned rats causes 71% of animals to recover temporarily their sinus rhythm. SM-H treatment also prolongs the survival time (*33*).

Microcirculation

SM treatment improves the microcirculation in the hamster cheek pouch (34). Improvement of mesenteric microcirculatory blood flow in dogs, followed by the Doppler effect of laser light, has also been reported (*35*).

Thrombosis

A middle cerebral artery thrombosis, generated by adding $FeCl_3$ to the surface of the right middle cerebral artery in rats, has been used to demonstrate the protective effects of acetylsalvianolic acid A (ASAA) on focal cerebral ischemic injury (*36*). ASAA significantly reduces the cerebral infarction and attenuate neurological deficits. Thrombosis in middle cerebral artery is reduced if animals are pretreated with ASAA. The effect of ASAA on focal cerebral ischemia is related to the anti-thrombotic activity.

Stenosis and Restenosis

In a coronary artery critical stenosis model produced by a micrometer constriction on left circumflex coronary artery in open-chest dogs, SM improves the left ventricular diastolic function (*37*). If SM is injected into left atrium after 15 min of stenosis, CBF, -dp/dt max, and -Vce are increased and T is decreased. The change of CBF is the earliest event responding to SM treatment. In air-injured carotid artery of rats, SM treatment reduces the intimal thickness of injured arteries indicating that SM prevents experimental restenosis in this animal model. SM also inhibits ^{3}H-thymidine uptake and proliferation of isolated rabbit arterial smooth muscle cells (SMC) in a dose-dependent manner. It suggests that SM is useful to prevent arterial restenosis after angioplasty (*38*).

Natural Product Chemistry of *Salvia miltiorrhiza*

Until now, over 50 natural products have been isolated from SM. The number does not include those common fatty acids, amino acids, and nucleosides. The following diterpenoids have been isolated from SM: 1,2,15,16-tetrahydro-tanshiquinone, tanshinaldehyde, Ro-090680, dihydro-isotanshone I, danshexinkun B, miltirone, nortanshinone, hydroxytanshinone II-A, tanshinone I, dihydrotanshinone I, tanshinone II-A, cryptotanshinone, methylenetanshiquinone, methyltanshinonate, spiroketallactone, neocrypto-tanshinone, isotanshinone IIB, danshexinkun A, danshenol A, and danshenol B (*39, 40*). SM is rich in water-soluble depsides, namely protocatechualdehyde, caffeic acid, methyl rosmarinate, rosmarinic acid, and salvianolic acids A, B and C (*41, 42*) (Figure 2).

The Effects of *Salvia miltiorrhiza* on Atherosclerosis

SM exhibits many biological and pharmacological activities, which are beneficial to the cardiovascular system. Until recently, direct evidence to correlate the antioxidant activity with anti-atherosclerotic potential in pathological animal models is still lacking. Previous studies have suggested that the active principles on the cardiovascular system are in the water-soluble fraction of SM in which danshensu, salvianolic acid A (Sal A), and lithospermic acid are the potential candidates (Figure 2). However, our recent study indicates that salvianolic acid B (Sal B) is the most abundant and key antioxidant in the water-soluble fraction of SM. Sal B strongly inhibits LDL oxidation *in vitro* and a Sal B-enriched fraction of SM reduces atherosclerosis in cholesterol-fed New Zealand white (NZW) rabbits and apoE(-) mice (*43*). These observations are summarized below.

Isolation of Sal B and Sal B-enriched fraction SM-EW-1

A Sal B-enriched fraction was obtained from the dried roots of plant material (3 kg) by using aqueous ethanolic extraction (ethanol : water = 1 : 2, 30 L). Depending on seasonal and geographical variations, the contents of Sal

B in the extracts were between 0.5-4.4% as determined by RP-HPLC (Figure 3). A Sal B-enriched fraction (SM-EW-1) was prepared for animal studies. There was no detectable α-tocopherol, ascorbic acid, or β-carotene in SM-EW-1 (*43*).

Based on the properties that Sal B is water-soluble and has pKa values similar to that of a carboxylic acid (pKa=4.5), the content of Sal B was further enriched by adjusting pH of concentrated SM-EW-1 to 7.5 and precipitating with acetone. After ion exchange and Sephadex LH-20 column chromatography, Sal B was obtained in higher than 98% purity.

Inhibition of LDL Oxidation *in vitro* by Sal B and SM

Two bioassays were used to evaluate the potency of Sal B and SM-EW-1 fraction to inhibit human LDL oxidation *in vitro*. The first assay was based on inhibition of malondialdehyde (MDA) formation, determined as TBARS by reacting with thiobarbituric acid (TBA), at fixed time (2-6 hr). The IC_{50} values indicated that Sal B (1.1 μM) was 4.0 times more potent than probucol (4.5 μM). The time course study of Sal B to inhibit conjugated diene formation in a standard LDL oxidation assay (LDL at 50 μg/mL was reacted with 5 μM Cu^{2+}) was also carried out in a microtiter plate (quartz) (*44,45*). Sal B prolonged the lag phase (ΔT_{lag}) of LDL oxidation 7.5 times longer than probucol both at 1.0 μM (Table 1) (Figure 4). Sal B accounted for up to 75% of antioxidant activity in SM-EW-1 fraction.

Table I. Lag Phase and Relative Potency of Antioxidants to Inhibit Human LDL Oxidation *in Vitro*

	T_{lag} *(min)*	ΔT_{lag} *(min)*	*Relative Potency*
(Control)	151	-	-
Probucol	302	151	1.00
Trolox	251	100	0.66
Quercetin	422	271	1.79
Genistein	216	65	0.43
Sal B	1286	1135	7.52

NOTES: Oxidation of LDL (50 μg/mL) was induced by Cu^{2+} (5.0 μM) at 25 °C. Conjugated diene formation was monitored continuously by the increase of absorption at 234 nm. The potency of an antioxidant is defined as its capability to prolong the lag phase. The relative potency of probucol was set as 1.00. Results are mean values of three determinations.

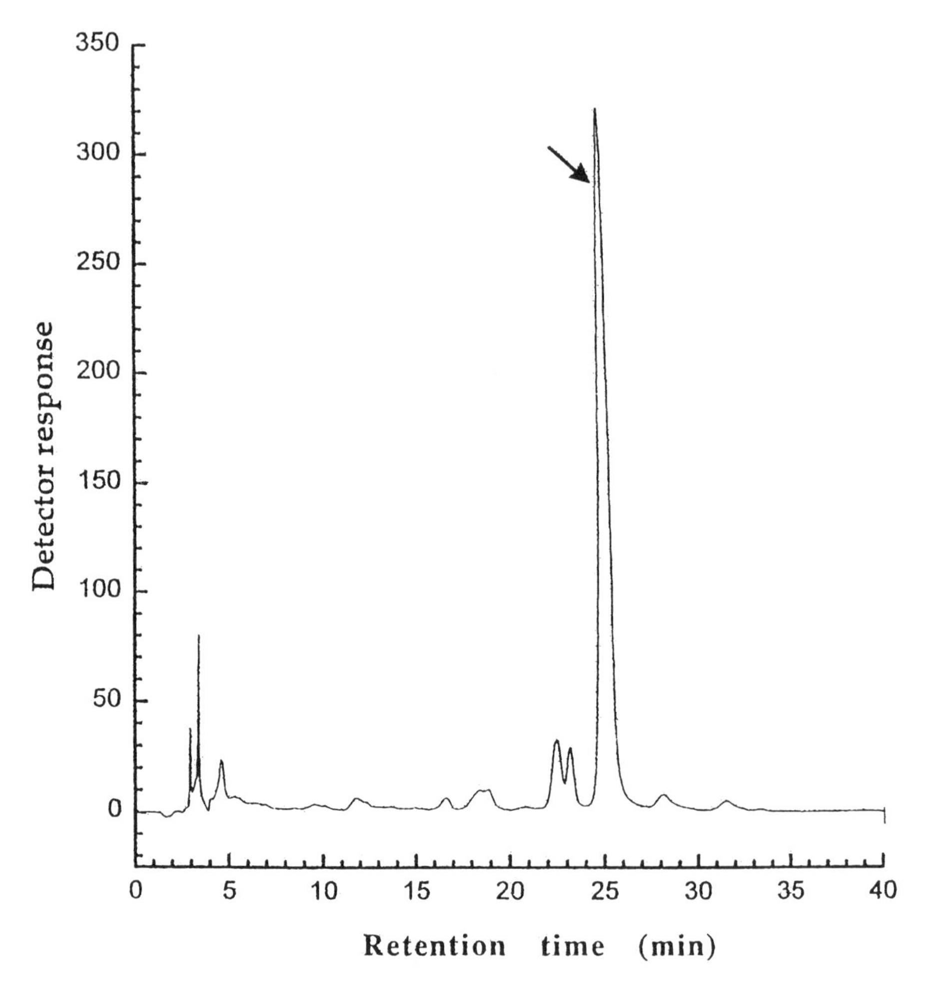

Figure 3. Reversed-phase HPLC profile of the aqueous ethanolic extract of S. miltiorrhiza. A pre-packed C18 column (25-cm) was used. A stepwise gradient elution of methanol-water binary system containing a constant volume % of acetic acid, from 25% to 40% methanol, was performed at 1.0 mL/min. The wavelength of UV detector was set at 290 nm. Peak corresponding to salvianolic acid B (Sal B) is marked by an arrow.

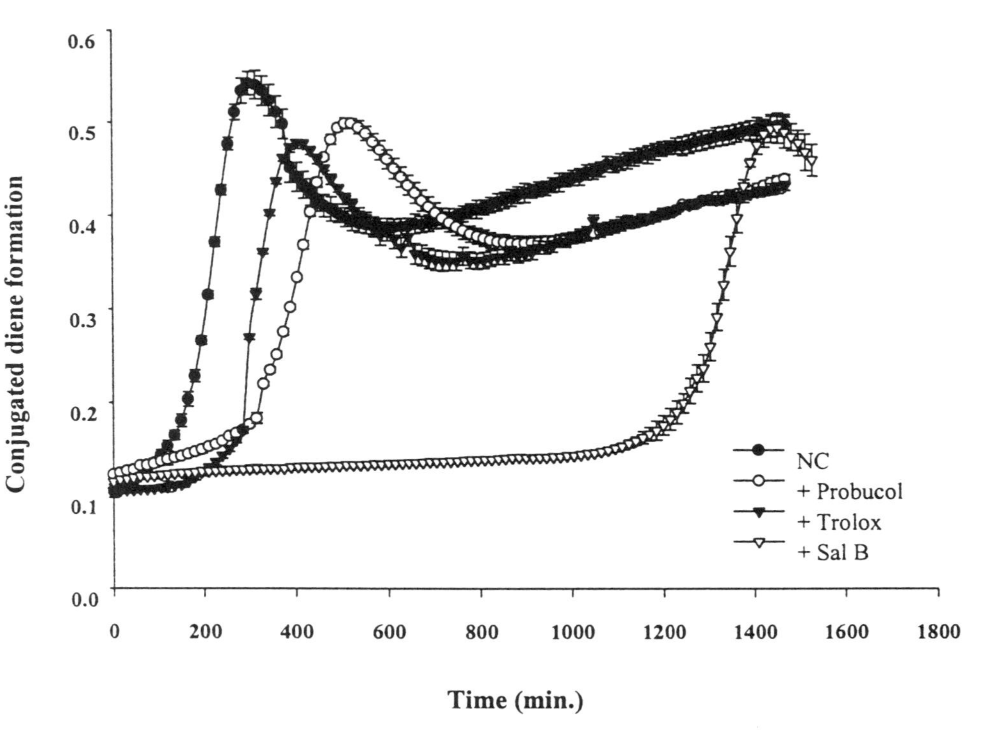

Figure 4. Time course of copper (II) induced LDL oxidation. Oxidation of LDL (50 μg/mL) was induced by Cu^{2+} (5.0 μM) at 25 °C. Conjugated diene formation was monitored continuously by the increase of absorption at 234 nm. Probucol and trolox were used as positive control.

Animal Treatment and Their Plasma Cholesterol and Lipoprotein

NZW Rabbits

Male NZW rabbits were fed with either a normal diet (ND group, n=5), a high cholesterol diet (1% cholesterol, w/w) (HC group, n=7), a high cholesterol diet containing 5% SM-EW-1 (SM group, n=7), or a high cholesterol diet containing 1% probucol (PB group, n=7). The feeding period was 12 weeks.

Plasma cholesterol levels were increased in rabbits fed with a high cholesterol diet. SM and PB groups exhibited less increase in plasma cholesterol (29% less in SM group and 44% less in PB) as determined by plasma cholesterol exposure. Reduction of plasma cholesterol by SM-EW-1 occurred predominantly in VLDL. Plasma HDL cholesterol was lowered in PB group. However, such a decrease did not occur in SM group (*43*).

ApoE(-) Mice

Three-month-old male ApoE(-) mice (Charles River Laboratory) (n=30) were randomly divided into three groups. Animals in the control group (CT group, n=10) were fed with a normal diet containing 0.15% cholesterol (w/w) (control diet) (8, 9). The test groups were either fed with a control diet plus 0.5% DPPD (w/w) (DPPD group as a positive control, n=10) or plus 5.0% SM-EW-1 fraction (w/w) (SaB group, n=10) (*46*). The dose of SM-EW-1 was chosen that the antioxidant potency was similar to that of DPPD. The content of Sal B in this fraction was 4.4% as determined by HPLC. The feeding period was three months.

After feeding experiment, plasma cholesterol concentration in CT group was 785 $\pm$ 151 mg/dL. There was no significant difference between CT and DPPD group (774 $\pm$ 185 mg/dL) or SaB group (743 $\pm$ 181 mg/dL). These results indicated that SM-EW-1 fraction did not lower plasma cholesterol in this animal model. The anti-atherosclerotic activity of SM-EW-1 was not related to cholesterol lowering.

Susceptibility to Oxidation *ex vivo* and α-Tocopherol Content of LDL

NZW Rabbits

LDL samples collected from SM and PB were more resistant to oxidation by Cu^{2+} than that from HC group. Time required to achieve half maximal TBARS production was defined as antioxidativity. The antioxidativity of SM

(6.2 ± 0.2 h) was significantly higher than that of high cholesterol diet (1.1 ± 0.1 h) (*42*). High cholesterol feeding decreased α-tocopherol content in LDL. In HC group, the content was 32% less than that in ND group. Probucol treatment (PB group) further decreases α-tocopherol content by 15%. The α-tocopherol content of LDL in SM group was 21.7 ±2.1 nmol/μmol cholesterol. This value was 2.3 folds higher than that in HC group.

ApoE(-) mice

The lipoproteins corresponding to VLDL and LDL in apoE(-) mice appeared as a smeared band in gel electrophoresis. A lipoprotein preparation of density range 1.006-1.063 was collected and designated as LDL. *In vitro* oxidation of the LDL by Cu^{2+} indicated that treatment of DPPD and SaB to animals reduced the susceptibility of their LDL to oxidation *ex vivo*. The lag phases of LDL from SaB group (380 ± 40 min) and DPPD group (361 ± 30 min) were significantly ($p<0.005$) longer that of CT group (145 ± 15 min).

Since the size of LDL in apoE(-) mice is heterogeneous, it is difficult to estimate the number of LDL particles in the corresponding LDL fraction (d, 1.006 to 1.063). The α-tocopherol content in LDL fraction was therefore expressed as number of α-tocopherol per 2000 cholesterol molecules. The value was 8.7 in CT group and 10.5 in SaB treated group ($p<0.05$ vs. CT). Adjusting by the total LDL-cholesterol, the difference was more pronounced. SM-EW-1 fed animals (SaB group) had 49.1% higher α-tocopherol content in LDL than that of CT group. Feeding with the hydrophobic antioxidant DPPD slightly decreased the α-tocopherol content in LDL by 10%.

Protection of Endothelial Damage

NZW Rabbits

The number of Evans blue-albumin (EBA) leaky foci in the thoracic aorta of HC group was significantly increased as compared with normal group (HC, 4.10 ± 0.95 foci/mm^2 vs. ND, 0.84 ± 0.03 foci/mm^2, $p<0.001$). Endothelial damage was reduced by 42% in SM group ($p<0.01$ vs. HC) and by 34% in PB group ($p<0.05$ vs. HC). It demonstrated that feeding with SM-EW-1 reduced endothelial damage.

ApoE(-) Mice

Limited by the body size, the endothelial damage experiment was not performed in apoE(-) mice.

Reduction of Atherosclerosis

NZW Rabbits

Atherosclerotic area in abdominal aorta and cholesterol deposition in thoracic aorta of HC group were 37.2±4.4% and 28.7±1.9 mg/g, respectively. SM caused a 56% decrease ($p<0.05$) in atherosclerotic area and a 50% decrease ($p<0.05$) in cholesterol deposition (Table II). Probucol treatment (PB group) resulted in 74% decrease of atherosclerotic area (abdominal aorta) and 56% decrease of cholesterol deposition (thoracic aorta). There was a linear correlation between atherosclerotic area and cholesterol deposition.

ApoE(-) mice

Atherosclerotic lesion area was determined in the arch, thoracic, and abdominal aortas. Significant lesions were observed in CT group ($3342 \pm 648 \times 10^{-4}$ cm^2). DPPD treatment reduced the lesion area by 46% ($p<0.01$) and SM-EW-1 treatment (SaB group) reduced the lesion area by 52% as compared with that of CT group ($p<0.01$) (Table II). The difference in reduction of lesion area between SaB and DPPD groups was not statistically significant.

Conclusion

Both pathological animal models demonstrated that the water-soluble fraction of *S. miltiorrhiza* (SM-EW-1), which contained Sal B as a potent antioxidant, reduced atherosclerosis (*43*). LDL from SM-EW-1 treated hypercholesterolemic NZW rabbits contained more α-tocopherol and was more resistant to oxidation *ex vivo*. A similar trend was also observed in apoE(-) mice. SM-EW-1 treatment reduced endothelial damage in cholesterol-fed NZW rabbits.

One of the potential mechanisms of SM-WE-1 to protect LDL from oxidation *ex vivo* is the protection of α-tocopherol in LDL (*12,47*). It has been shown that ascorbic acid preserves lipophilic antioxidants in LDL (*48*). The α-

Table II. Reduction of Atherosclerotic Lesion in Experimental Animals

	Atherosclerotic areas (%)	*Atherosclerotic area (10^{-4} cm^2)*	*Reduction (%)*
NZW rabbits			
ND (n=5)	0.3±0.1	-	-
HC (n=7)	37.2±4.4[a,b]	-	0
PB (n=7)	9.6±2.9[a]	-	-74%
SM (n=7)	16.5±2.7[b]	-	-56%
ApoE(-) mice			
CT (n=10)	-	3342±648[c,d]	-
DPPD (n=10)	-	1823±643[c]	-46%
SaB (n=10)	-	1620±430[d]	-52%

NOTES: In NZW rabbits, atherosclerotic area (expressed in area %) of abdominal aorta was determined. In apoE(-) mice, atherosclerotic lesion areas in the arch, thoracic, and abdominal aortas were determined. Values with the same superscript in the same column are significantly different by using Student's t test.

[a] $p<0.01$

[b] $p<0.01$

[c] $p<0.01$

[d] $p<0.01$

tocopherol contents of LDL from SM-EW-1 treated animals were significantly higher than those of control groups (*43*). Since SM-EW-1 contained no α-tocopherol or ascorbic acid, we suggest that SM-EW-1 may indirectly protect LDL form oxidation by preventing the loss of α-tocopherol. It is also likely that SM-EW-1 scavenges ROS in the blood stream and spares α-tocopherol in LDL from being oxidized. Lipophilic antioxidants, such as probucol and DPPD, may compete with α-tocopherol within LDL and are not able to preserve α-tocopherol. In PQRST study, probucol treatment has been found to decrease serum concentrations of diet-derived lipophilic antioxidants (*48*). ApoE(-) mice model is a good model of lipoprotein oxidation causing by elevated plasma cholesterol and delayed VLDL and LDL catabolism (*8,9,46*). Increase of LDL α-tocopherol was also observed in apoE(-) mice treated with a Sal B-enriched fraction (SM-EW-1).

Endothelial dysfunction is an early event in atherogenesis. Antioxidants, including α-tocopherol and probucol are reported to restore endothelial function in cholesterol-fed rabbits. This study demonstrated that SM-EW-1, a Sal B-enriched fraction of *S. miltiorrhiza*, reduces endothelium damage effectively in NZW rabbit model. We suggest that water-soluble antioxidants may on one hand scavenge ROS in blood stream and reduce the direct injury to aortic endothelium, while on the other hand preserve α-tocopherol in LDL to resist further oxidation in the subendothelial space.

From previous and current studies, it is concluded that *S. miltiorrhiza*, which contains salvianolic acid B as a potent antioxidant, increases α-tocopherol content in LDL, prevents LDL from oxidative modification, and reduces atherosclerosis in hypercholesterolemic NZW rabbits and apoE(-) mice.

Acknowledgements

We thank Profs. Chuang-Ye Hong, Shing-Jong Lin, and Yuh-Lien Chen, National Yang Ming University and Prof. Lee-Young Chao, Academia Sinica for advice and technical support. A large-scale preparation of Sal B was provided by Dr. Yun-Lian Lin, National Research Institute of Chinese Medicine. This study was supported by the Department of Health and Veterans General Hospital-Taipei, Taiwan.

References

1. Steinberg, D.; Parthasarathy, S.; Carew, T. E.; Khoo, J. C.; Witztum, J. L. *N. Engl. J. Med.* **1989**, *320*, 915-924.
2. Steinberg, D. *Circulation* **1997**, *95*, 1062-1071.
3. Ross, R. *Nature* **1993**, *362*, 801-809.
4. Lusis, A. J.; Navab, M. *Biochem. Pharmacol.* **1993**, *46*, 2119-2126.
5. Parthasarathy, S.; Santanam, N.; Ramachandran, S.; Meilhac, O. *J. Lipid Res.* **1999**, *40*, 2143-2157.
6. Mao, S. J. T.; Yates, M. T.; Rechtin, N. E.; Jackson, R. L.; Van Sickle, W. A. *J. Med. Chem.* **1991**, *34*, 298-302.
7. Sparrow, C. P.; Doebber, T. W.; Olszewski, J.; Wu, M. S.; Ventre, J.; Stevens, K. A.; Chao, Y. S. *J. Clin. Invest.* **1992**, *89*, 1885-1891.
8. Tangirala, R. K.; Casanada, F.; Miller, E.; Witztum, J. L.; Steinberg, D.; Palinski, W. *Arterioscler. Thrmob. Vasc. Biol.* **1995**, *15*, 1625-1630.
9. Uhrman, B.; Buch, S.; Vaya, J.; Belinky, P. A.; Coleman, R.; Hayek, T.; Aviram, M. *Am. J. Clin. Nutr.* **1997**, *66*, 267-275.
10. Zhang, S. H.; Reddick, R. L.; Avdievich, E.; Surles, L. K.; Jones, R. G.; Reynolds, J. B.; Quarfordt, S. H.; Maeda, N. *J. Clin. Invest.* **1997**, *99*, 2858-2866.
11. Halliwell, B. *Am. J. Clin. Nutr.* **1995**, *61*(suppl), 670S-677S.
12. Kagan, V. E.; Serbinova, E. A.; Forte, T.; Scita, G.; Packer, L. *J. Lipid Res.* **1992**, *33*, 385-397.
13. Lin, C.-C. The Pharmacognostical Therapy in Treatment of Blood Stasis. Kaohsiung Medical College, Kaohsiung, Taiwan, pp 1-97.
14. Wang, N.; Luo, H. W.; Niwa, M.; Ji, J. *Planta Medica.* **1989**, *55*, 390-391.

15. Wu, H.; Li, J.; Peng, L.; Teng, B.; Zhai, Z. *Chin. Med. Sci. J.* **1996**, *11*, 49-52.
16. Yu, W. G.; Xu, L N. *Acta Pharmaceutica Sinica* **1994**, *29*, 412-416.
17. Li, C. P.; Yung, K. H.; Chiu, K. W. *Am. J. Chin. Med.* **1990**, *18*, 157-166.
18. Wang, J. Z.; Chen, M. E.; Xu, Y. Q. *Chin. J. Modern Develop. Traditional Med.* **1991**, *11*, 420-421.
19. Lei, X. L.; Chiou, G. C. *Am. J. Chin. Med.* **1986**, *14*, 26-32.
20. Lei, X. L.; Chiou, G. C. *Am. J. Chin. Med.* **1986**, *14*, 145-152.
21. Cheng, Y. Y.; Fong, S. M.; Chang, H. M. *Chin. J. Modern Develop. Traditional Med.* **1990**, *10*, 609-611.
22. Zhu, P. *Chin J Surgery.* **1990**, *28*, 9-12.
23. Yagi, A.; Fujimoto, K.; Tanonaka, K.; Hirai, K.; Takeo, S. *Planta Medica* **1989**, *55*, 51-54.
24. Wu, T. W.; Zeng, L. H.; Fung, K. P.; Wu, J.; Pang, H.; Grey, A. A.; Weisel R. D.; Wang, J. Y. *Biochem. Pharmacol.* **1993**, *46*, 2327-2332.
25. Zhao, B. L.; Jiang, W.; Zhao, Y.; Hou, J. W.; Xin, W. J. *Biochem. Mol. Biol. Int.* **1996**, *38*, 1171-1182.
26. Liu, G.-T.; Zhang, T.-M.; Wang, B.-E.; Wang, Y.-W. *Biochem. Pharmacol.* **1992**, *43*, 147-152.
27. Lin, T.-J.; Zhang, K.-J.; Liu, G.-T. *Biochem Pharmacol.* **1996**, *51*, 1237-1241.
28. Bao, G. *Acta Acad. Med. Sin.* **1993**, *15*, 320-324.
29. Fung, K. P.; Zeng, L. H.; Wu, J.; Wong, H. N.; Lee, C. M.; Hon, P. M.; Chang, H. M.; Wu, T. W. *Life Sci.* **1993**, *52*, 239-244.
30. Cheng, Y. Y.; Fong, S. M.; Hon. P. M. *Chung-Kuo Chung His i Chieh Ho Tsa Chih* **1992**, *12*, 424-426.
31. Yang, Y.; Wang, L.; Li, L.; Chen, H. J *West China Univ. Med Sci.* **1993**, *24*, 143-146.
32. Yuan, Q. L.; Li, Y.; Chang, L. F.; He, G. B. *China J. Chin. Materia Medica* **1993**, *18*, 433-435.
33. Cheng, Y. Y.; Fong, S. M.; Hon, P. M.; Li, C. M.; Chang, H. M. *Chin. J. Modern Develop. Traditional Med.* **1991**, *11*, 543-546.
34. Xue, Q. F. *Chin. Med. J.* **1986**, *66*, 334-337.
35. Yu, G. R. *Chin. J. Modern Develop. Traditional Med.* 1984, 4, 546-547.
36. Dong, J. C.; Xu, L. N. *Acta Pharm. Sin.* **1996**, *31*, 6-9.
37. Xu, H. T.; Chen, S. L.; Li, L. S. *Chin. J. Modern Develop. Traditional Med.* **1990**, *10*, 737-739.
38. Zhou, X. M.; Lu, Z. Y.; Wang, D. W. *Chung-Kuo Chung Hsi i Chieh Ho Tsa Chih* **1996**, *16*, 480-482.
39. Lee, A. R.; Wu, W. L.; Chang, W. L.; Lin H. C.; King, M. L. *J. Nat. Prod.* **1987**, *50*, 157-160.
40. Kasimu, R.; Basnet, P.; Tezuka, Y.; Kadota, S.; Namba, T. *Chem. Pharm. Bull.* **1997**, *45*, 564-566.

41. Ai, C.-B.; Li, L.-N. *J. Nat. Prod.* **1988**, *51*, 145-149.
42. Li, L.-N.; Tan, R.; Chen, W.-M. *Planta Medica* **1984**, *50*, 227-228.
43. Wu, Y.-J.; Hong, C.-Y.; Lin, S.-J.; Wu, P.; Shiao, M.-S. *Arterioscler. Thromb. Vasc. Biol.* **1998**, *18*, 481-486.
44. Regnstrom, J.; Nilsson, J.; Tornvall, P.; Landou, C.; Hamsten, A. *Lancet* **1992**, *339*, 1183-1186.
45. Wallin, B.; B. Rosengren, H. G. Shertzer and G. Camejo. 1993. *Anal. Biochem.* **1993**, *208*, 10-15.
46. Palinski, W.; Ord, V.; Plump, A. S.; Breslow, J. L.; Steinberg, D.; Witztum, J. L. 1994 *Arterioscler. Thromb.* **1994**, *14*, 605-616.
47. Jialal, E.; Grundy, S. M. *J. Clin. Invest.* **1991**, *87*, 597-601.
48. Elinder, L. S.; Hadell, K.; Johansson, J.; Molgaard, J.; Holme, I.; Olsson, A. G.; Walldius, G. *Arterioscler. Thromb. Vasc. Biol.* **1995**, *15*, 1057-1063.

Chapter 24

Suppressive Effects of 1′-Acetoxychavicol Acetate on Superoxide and Nitric Oxide in Cell Culture Systems, and on Hydrogen Peroxide in Mouse Skin

A. Murakami[1], Y. Nakamura[2,3], K. Koshimizu[1], and H. Ohigashi[2]

[1]Department of Biotechnological Science, Faculty of Biology-Oriented Science and Technology, Kinki University, Iwade-Uchita, Wakayama 649-6493, Japan
[2]Division of Applied Life Sciences, Graduate School of Agriculture, Kyoto University, Kyoto 606-8502, Japan
[3]Current address: Laboratory of Food and Biodynamics, Nagoya University Graduate School of Bioagricultrial Sciences, Nagoya 464-8601, Japan

Antioxidative compounds can be divided into two main categories: those scavenging reactive oxygen and nitrogen species and those suppressing the *de novo* synthesis or activity of free radical-generating enzymes, including NADPH oxidase and inducible nitric oxide (NO) synthase, in inflammatory leukocytes. In search for effective chemopreventive agents in vegetables and fruits, some compounds that selectively suppress free radical-generating pathways have been recognized to be preventive for cancer in several rodent models. For example, 1'-acetoxychavicol acetate (ACA) markedly blocked superoxide and NO generation in cell culture systems, and suppressed hydrogen peroxide and edema formation in double 12-*O*-tetradecanoylphorbol-13-acetate (TPA)-treated mouse skin. These results suggest that our approach, directed towards free radical generation inhibitors, may lead to the discovery of promising inhibitors of carcinogenesis.

The close relationships between food constituents and their efficacies for prevention of "life style-related diseases" including cancer have been proven in
The close relationships between food constituents and their efficacies for prevention of "life style-related diseases" including cancer have been proven in
recent decades. While attention has been frequently paid to anti-oxidative vitamins or polyphenols for the prevention of carcinogenesis in rodents, the above-mentioned ubiquitous phytochemicals do not exclusively represent the beneficial, cancer-protective effect of vegetable intake when the great diversity of plant secondary metabolites is taken into account.

Southeast Asia is one of the appropriate sites in the world for collecting diverse and uncommon edible plants containing various and abundant biologically active substances (1). In fact, there is considerable utilization of commonly eaten plants in the traditional folk medicines of Southeast Asia (2). The above background leads to the premise that edible plants from Southeast Asian countries have a higher potential for chemoprevention as compared with those found in temperate zones.

Accordingly, our continuing screening tests for in vitro anti-tumor promoting activity, from more than 500 species of vegetables and fruits from Southeast Asian countries, have suggested that those from Thailand (3), Indonesia (4), and Malaysia (5) have greater potentials for cancer prevention.

In the course of our studies, (1'S)-1'-acetoxychavicol acetate (ACA, Figure 1) was indicated to be a particularly promising compound, which potently suppressed tumor promoter-induced Epstein-Barr virus activation (6). ACA was originally isolated from the rhizomes of Languas galanga (Zingiberaceae) as an anti-ulcer agent, and was also shown to inhibit xanthine oxidase activity (7). In this study, we describe the suppressive effect of ACA on superoxide (O_2^-) and nitric oxide (NO) in cell culture systems as well as on hydrogen peroxide (H_2O_2) formation in mouse skin.

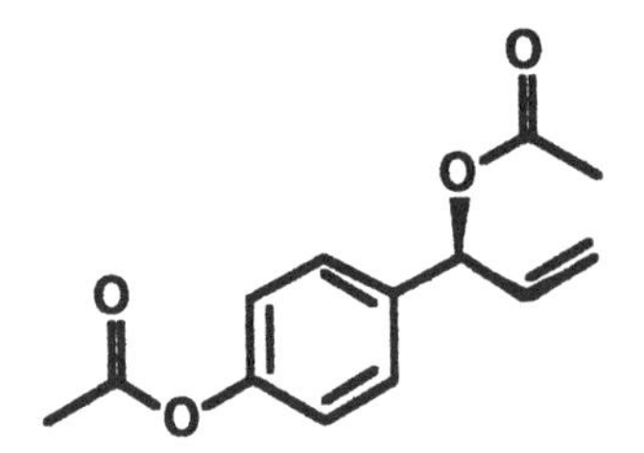

Figure 1. Chemical structure of ACA.

Materials and Methods

Superoxide Generation From Differentiated HL-60 Cells

Differentiated HL-60 cells, suspended in 1 mL of Hank's buffer, were treated with each test compound dissolved in 5μL DMSO or the vehicle (5μL DMSO).

After preincubation at 37 °C for 15 min, the suspension was centrifuged and the extracellular compounds were removed by washing twice with 1% bovine serum albumin. Then, the cells were suspended in 1 mL of Hank's buffer, and incubated with 100 nM 12-*O*-tetradecanoylphorbol-13-acetate (TPA) or vehicle and 1 mg/mL cytochrome *c* at 37 °C for 30 min. The reaction was terminated by adding a superoxide dismutase solution (10,000 U/mL) and by palcing the sample tubes in ice. After centrifugation, the level of extracellular O_2^- was measured by the cytochrome *c* reduction method, in which reduced cytochrome *c* was quantified by measuring the visible absorption of the supernatant at 550 nm. Cells treated with the compound, cytochrome *c*, and vehicle without TPA, and cells with the vehicle without the compound, cytochrome *c*, or TPA were used as negative and positive controls, respectively. Cells treated with the vehicle without the compound, cytochrome *c*, or vehicle without TPA were used as a blank. Inhibitory rates (IRs) were calculated by the following formula: {1 – [(sample, $Abs._{550}$) - (negative, $Abs._{550}$)/(positive, $Abs._{550}$) – (blank, $Abs._{550}$)]} x 100 (%). Cell viability was determined by a trypan-blue dye exclusion test (*8*). Each experiment was done independently in duplicate twice, and the data are shown as mean ± standard deviation (SD) values.

Nitric Oxide Generation From RAW 264.7 Cells

Murine macrophage cell line RAW 264.7 cells, grown confluent in 1 mL of Dulbecco's Modified Eagle Medium(DMEM) on a 24-well plate, were treated with lipopolysaccharide (LPS) (100 ng/mL), tetrahydrobiopterin (10 mg/mL), interferon (IFN)-γ (100 U/mL), L-arginine (2 mM), and the test compound (5 μL in DMSO solution) or vehicle. After 24 h, the NO_2^- concentrations were measured by the Griess assay (*9*). Cells treated with the compound and vehicle without stimulation, and cells with the vehicle without the compound or stimulation were used as negative and positive controls, respectively. Cells treated with the vehicle without the compound, and the vehicle without stimulation were used as blanks. IRs were calculated by the following formula: {1 – [(sample, $Abs._{530}$) - (negative, $Abs._{530}$)/(positive, $Abs._{530}$) – (blank, $Abs._{530}$)]} x 100 (%). Cytotoxicity was measured by a 3-(4,5-dimethylthiazol-2-yl)-2,5-diphenyltetrazolium bromide (MTT) assay (*10*). Each experiment was done independently in duplicate twice, and the data are shown as mean ± SD values.

Hydrogen Peroxide Generation and Edema Formation In Mouse Skin

Each experimental group consisted of 5 mice. ACA (810 nmol/100 μL acetone), or acetone (100 μL) was topically applied to the shaved area of the dorsal skin 30 min before application of a TPA solution (8.1 nmol/100 μL in acetone). After 24 h, the same dose of the test compounds, or acetone and TPA, were applied to the same region. Mice were sacrificed by cervical dislocation 1 hr after the second TPA treatment, and then biochemical parameters were measured as described below. We divided the mice into six groups as follows: group 1 (acetone x 2/acetone x 2); group 2 (acetone-TPA/acetone-TPA); group 3 (ACA-TPA/ACA-TPA); group 4 [ACA (priming)-TPA/acetone-TPA]; and group 5 [acetone-TPA/ ACA (activation)-TPA]. Skin punches (epidermis and dermis) were obtained from excised dorsal skins with an 8-mm-diam. cork borer and weighed with an analytical balance. The skin punches were minced in 3 mL of 50 mM phosphate buffer (pH 7.4) containing 5 mM sodium azide, and then homogenized at 4 °C for 30s twice. The homogenate was centrifuged at 10,000 g for 20 min at 4 °C. The H_2O_2 content was determined by the phenol red-horseradish peroxidase method (*11*). The data are shown as mean ± SD values.

Results and Discussion

Suppressive Effects of ACA on Free Radical Generation in Cell Culture Systems

Inflammation is a universal and physiological response in the processes of carcinogenesis. Treatment of the tissues with ultraviolet light, endotoxins, proinflammatory cytokines, or chemical tumor promoters are known to lead to chemotaxis, differentiation, and infiltration of inflammatory leukocytes. This includes neutrophils and macrophages that produce reactive oxygen and nitrogen intermediates, prostaglandins, and cytokines. In particular, neutrophil-induced O_2^- generation has been reported to be involved in the activation of certain procarcinogens, including polyaromatic hydrocarbons (12). In addition, leukocyte-derived free radicals are known to modify protein residues and DNA bases (13).

ACA and genistein, a soybean isoflavone, were tested for their inhibitory effects on tumor-promoter TPA-induced O_2^- generation in human promyelocytic leukemia HL-60 cells being differentiated into granulocyte-like cells by treatment with 1.3% DMSO, for 6 days (14). The extracellular O_2^- production was detected by measuring the levels of the reduced form of cytochrome c. In the current experiments, O_2^- scavenging or quenching activity was considered to be negligible, since we removed the extracellular compounds by washing the cells with a 1% BSA solution.

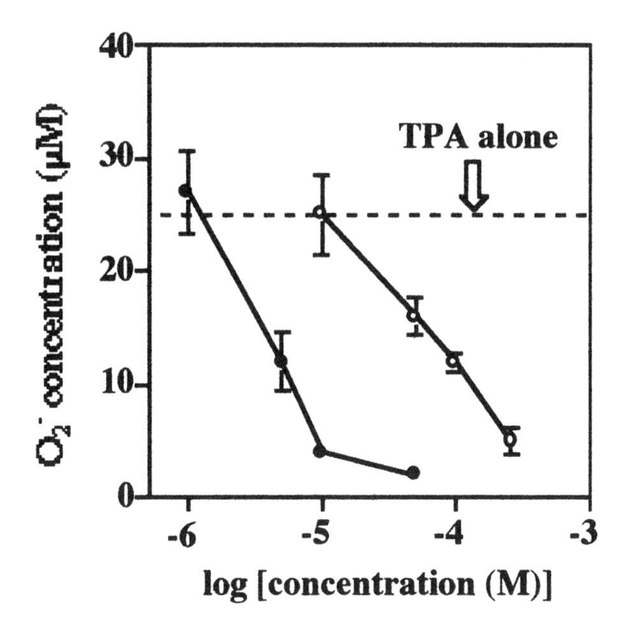

Figure 2. Suppressive effects of ACA and genistein on TPA-induced O_2^- generation in differentiated HL-60 cells. (●), ACA; (○), genistein.

As shown in Figure 2, treatment of HL-60 cells with TPA resulted in marked O_2^- generation (26 μM in the media). No detectable cytotoxicity was observed in each experiment. ACA was found to attenuate O_2^- generation in a concentration dependent manner (50% inhibitory concentration: IC_{50} = 4.3 μM), and its suppressive potency is much higher than that of genistein (IC_{50} = 100 μM). ACA may suppress the assembly and/or activity of the NADPH oxidase that is expressed in differentiated HL-60 cells and is responsible for O_2^- generation because ACA showed no O_2^- scavenging activity in the xanthine-xanthine oxidase system up to a concentration of 500 μM (data not shown), and, as mentioned above, we removed the extracellular compounds from the reaction buffer. A chemical characteristic of ACA, namely, bearing no free hydroxyl group in its structure, should support the idea that ACA has no radical scavenging potentials.

Suppressive Effect of ACA and L-NIO on NO Generation in RAW 264.7 Cells.

NO, a gaseous free radical, is synthesized in biological systems by a family of enzymes, constitutive NO synthase (cNOS) and inducible (iNOS) (15). In the latter, NO is readily released by iNOS function which is induced by stimulation of LPS and/or IFN-γ to form stoichiometric amounts of L-citrulline from L-arginine in some cell lines such as macrophages (16). While NO was primarily reported as an endothelium-derived relaxing factor and is thought to act as an intra- and intercellular messenger, excess generation of NO by iNOS has attracted attention on account of its relevance to carcinogenesis (17).

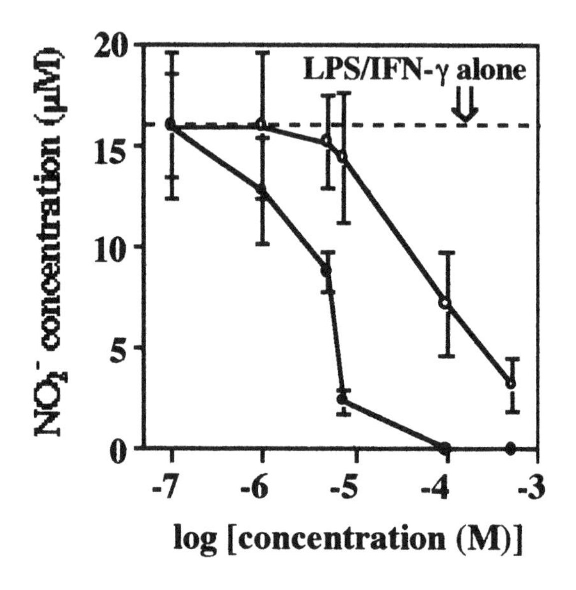

Figure 3. Suppressive effects of ACA and L-NIO on LPS/IFN-γ-induced NO_2^- generation in RAW 264.7 cells. (●), ACA; (○), L-NIO.

NO has been reported to cause mutagenesis (18), deamination of DNA bases (19), and more importantly, to form carcinogenic N-nitroso compounds through its simultaneous conversion to NO_2^-, which covalently binds to primary and a secondary amines under acidic conditions (20). Moreover, an important chemical property of NO is that it reacts rapidly and spontaneously with a O_2^- to form a peroxynitrite anion ($ONOO^-$) (21), which is more toxic than O_2^- or NO to biological systems, causing modification of proteins (e.g., 3-nitrotyrosine) (22) or nucleic acids (e.g., 8-nitroguanine) (23). Collectively, suppression of the iNOS-induced NO generation at excess amounts is now widely accepted as a new paradigm for the chemoprevention of cancer.

As shown in Figure 3, ACA (IC_{50} = 5.3 μM) showed a substantial suppression of NO_2^- generation and its potency is comparable to or higher than that of L-N-iminoethyl-L-ornithine (L-NIO, IC_{50} = 9.7 μM), synthetic iNOS inhibitor. No cytotoxicity was observed in each experiment.

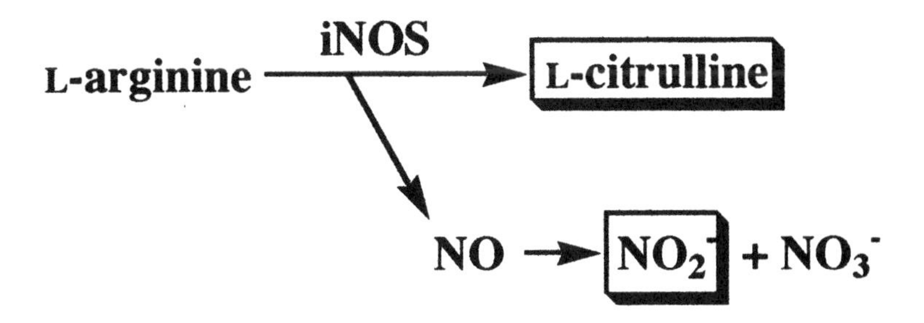

Figure 4. Production of L-citrulline and NO_2^- by iNOS.

We have recently reported that ACA is a potent suppressor of LPS- or IFN-γ-induced iNOS messenger RNA and protein expression in RAW 264.7 cells (24). Thus, suppression by ACA of NO_2^- production is, at least in part, attributable to iNOS induction attenuation. We can rule out the possibility for ACA to scavenge NO in RAW 264.7 cells, because it inhibited both NO_2^- and L-citrulline in a quite similar concentration-dependent manner (data not shown). If ACA had any NO scavenging potential, the inhibitory rate of L-citrulline would be lower than that of NO_2^- (Figure 4).

Suppression by ACA of TPA-induced H_2O_2 Production and Edema Formation in Mouse Skin

Wei et al. reported that double applications of TPA to mouse skin led to excessive free radical production (25). Ji and Marnett (26) designated these two application stages as "priming" (the first stage, as illustrated by leukocyte recruitment, maturation, and infiltration of inflammatory leukocytes into inflamed lesions) and "activation" (the second stage, as illustrated by free radical production from accumulated leukocytes), respectively.

In the present study, we attempted to determine whether ACA inhibits the priming and/or activation stages when conducting this double TPA application model. In groups 2-6, TPA (8.1 nmol) was topically applied to the dorsal skin of mice twice within a time interval of 24 h. Treatment with double dose TPA (group 2, 810 nmol each) caused a dramatic elevation of edema weight and H_2O_2 generation in the dermis and epidermis 1 h after the second TPA application, as compared with the vehicle control (group 1) (Figures 5 and 6). ACA was also topically applied 30 min prior to each TPA treatment. Double pretreatment with ACA substantially suppressed both edema and H_2O_2 formation (27). Subsequently, we addressed a question regarding which stage, either priming (leukocyte infiltration) or activation (oxidative burst), can be blocked by a single ACA pretreatment. Although a single pretreatment with ACA (810 nmol) during the activation stage remarkably suppressed edema and H_2O_2 formation, the same treatment during the priming stage was much less effective, indicating that ACA is defined as a specific suppressor of the activation stage. This result is consistent with that observed in cell culture studies (14).

As summarized in Table I, ACA has been observed to be an effective suppressor of free radical generating enzymes such as xanthine oxidase, NADPH oxidase and iNOS, in biological systems. This anti-oxidative profile is distinct from those of polyphenolic compounds, and vitamins C and E.

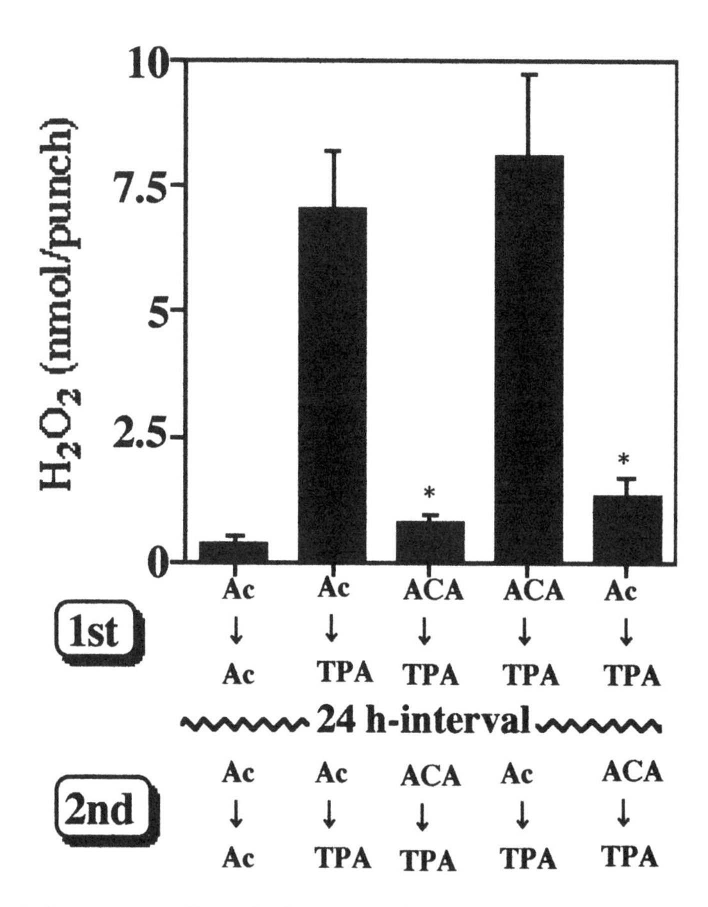

*Figure 5. Suppressive effect of ACA on double TPA-induced H_2O_2 generation in mouse skin. Ac, acetone. * $P < 0.001$ in Student t-test.*

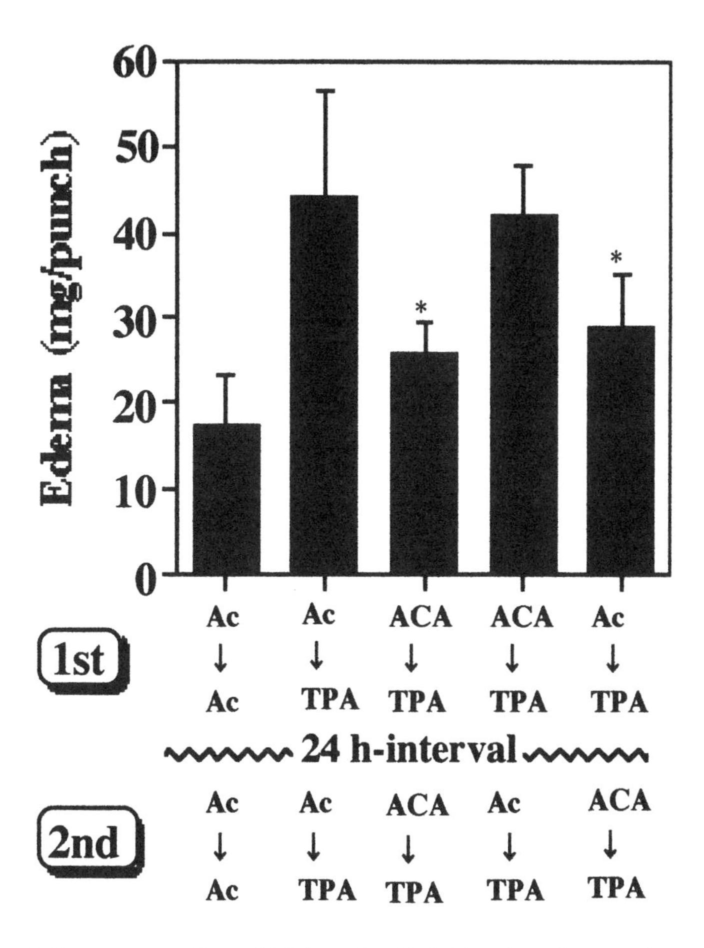

*Figure 6. Suppressive effect of ACA on double TPA-induced edema formation in mouse skin. Ac, acetone. * $P < 0.001$ in Student t-test.*

Table I. Research on the activity of ACA

Activity	Reference
Suppression of:	
Xanthine oxidase activity	*7*
TPA-induced EBV activation	*6*
TPA-induced O_2^- generation (NADPH oxidase)	*14*
Lipid peroxidation	*27*
TPA-induced hydroperoxide formation	*14*
LPS- or IFN-γ-induced NO generation	*24*
LPS-induced NF-κB activation	*24*
TPA-induced H_2O_2 and edema formation	*27*
Cellular proliferation markers	*28-30*
Enhancement of:	
Glutathione S-transferase activity	*30*
Quinone reductase activity	*30*

Table II. Chemopreventive activities of ACA in rodent models

Model	Reference
DMBA[a]/TPA-induced papilloma in mouse skin	*14*
4-NQO[b]-induced squamous cell carcinoma in rat tongue	*28*
AOM[c]-induced aberrant crypt foci in rat colon	*29*
AOM[c]-induced adenocarcinoma in rat colon	*30*
DEN-induced GST-P[d] in rat liver	*31*
Choline-deficient diet-induced GST-P[d] in rat liver	*31*
NMBA[e]-induced squamouse cell carcinoma in rat esophagus	*32*

[a]Dimethylbenz[a]anthracene. [b]Nitroquinoline-1-oxide, [c]Azoxymethane. [d]Glutathione-S-transferase-P. [e]N-Nitrosomethylbenzylamine.

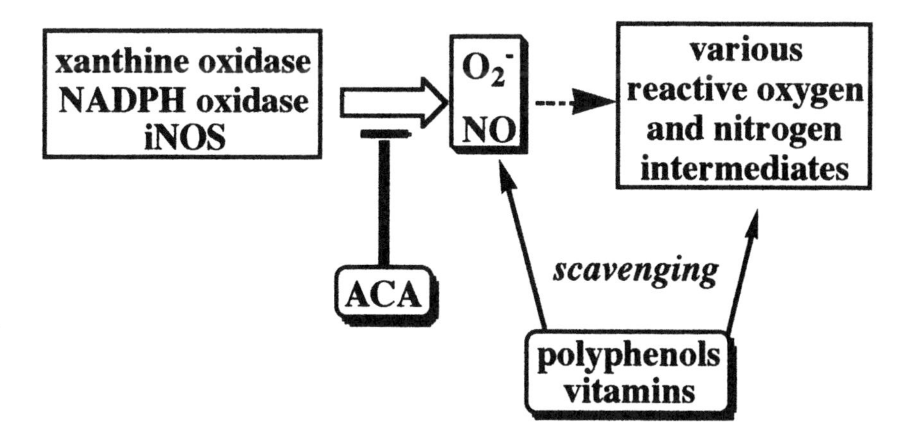

Figure 7. The mechanisms by which ACA and polyphenols inhibit oxidative stress.

Conclusion

The present study has shown that ACA, in contrast to general polyphenolic compounds, is an effective and unique anti-oxidant, which can suppress free radical-generating pathways in vitro and in vivo (Figure 7). So far, we have found biological activities of ACA relating to chemopreventive activity (Table I), and some of which may be involved in the mechanisms of action by which ACA inhibited chemical carcinogenesis in rodent models (Table II).

Acknowledgement

This study was supported by grants-in-aid from the Japan Society for the Promotion of Science and from the Ministry of Education, Science, Sports and Culture of Japan. The authors thank Yoshimi Ohto and Oe Kyung Kim for their excellent technical assistance.

References

1. Jacquat, C. and Bertossa, G. In Plants from the Markets of Thailand Duang Kamol, Bangkok, Thailand, 1990.
2. Siemonsma, J. S, Piluek K. In Plant Resources of South-East Asia 8 (Vegetables), Prosea Education, Bogor, Indonesia, 1994.

3. Murakami, A.; Jiwajinda, S.; Koshimizu, K.; and Ohigashi, H. *Cancer Lett.* **1995**, *95*, 139-146.
4. Murakami, A.; Morita, H.; Safitri, R.; Ramlan, A.; Koshimizu, K.; and Ohigashi, H. *Cancer Detec. Prev.* **1988**, *22*, 516-525.
5. Murakami, A.; Manaf, A. A.; Kamaruddin, M.-S., Koshimizu, K.; and Ohigashi H. *Biosci. Biotechnolnol. Biochem.* **2000**, *64*, 9-16.
6. Kondo, A.; Ohigashi, H.; Murakami, A.; Suratwadee, J.; Koshimizu, K. *Biosci. Bietechnol. Biochem.* **1993**, *57*, 1344-1345.
7. Noro, T.; Sekiya, T.; Katoh, M.; Oda, Y.; Miyase, T.; Kurayanagi, M.; Ueno, A.; and Fukushima, S. *Chem. Pharm. Bull..* **1988**, *36*, 244-248.
8. Brus, I.; and Glass, G. B. *Stain Technol.* **1973**, *48*, 127-132.
9. Green L. C.; Wagner D. A.; Glogowski P. L.; Skipper P. L.; Wishnok J. S.; and Tannenbaym S. R. *Anal. Biochem.*, **1982**, *126*, 131-138.
10. Sladowski, D.; Steer, S. J.; Clothier R. H.; and Balls M. *J. Immunol. Methods.*, **1992**, *157*, 203-207.
11. Wei H.; and Frenkel. *Cancer Res.*, **1992**, *52*, 2298-2303.
12. Trush, M. A.; Esterline, R.L.; Mallet, W.G.; Mosebrook, D.R.; and Twerdok, L.E. *Adv. Exp. Med. Biol.* **1991**, *283*, 399-401.
13. Lamb, N. J.; Gutteridge, J. M.; Baker, C.; Evans, T. W.; Quinlan, G. J. *Crit. Care Med.* **1999**, *27*, 1738-1744.
14. Murakami, A.; Ohura, S.; Nakamura, Y.; Koshimizu, K.; Ohigashi, H. *Oncology* **1996**, *53*, 386-391.
15. Vanvaskas, S; Schmidt, H. H. H.W. *J. Natl. Cancer Inst.* 1997, *89*, 406-407.
16. Nathan, C; Xie, Q. *Cell* **1994**, *78*, 915-918.
17. Ohshima, H; Bartsch H. *Mutat. Res.* **1994**, *305*, 253-264.
18. Arroyo, P. L.; Hatch-Pigott, V.; Mower, H. F.; Cooney, R. V. *Mutat. Res.* **1992**, *281*, 193-202.
19. Wink, D. A.; Kasprazak, K. S.; Maragos, C. M.; Elespuru, R. K.; Misra, M.; Dunams, T. M.; Cebula, T. A.; Koch, W. H.; Andrews, A. W.; Allen, J. S.; Keefer, L. K. *Science* **1991**, *254*, 1001-1003.
20. Miwa, M.; Stuehr, D. J.; Marletta, M. A.; Wishnok, J. S.; Tannenbaum, S. R. *Carcinogenesis* **1987**, *8*, 955-958.
21. Ischiropoulos, H.; Zhu, L.; Beckman, J. S. *Arch. Biochem. Biophys.* **1992**, *298*, 446-451.
22. van der Vliet, A.; Eiserich, J. P.; O'Neil, C. A.; Halliwell, B.; Cross, C. E. *Arch. Biochem. Biophys.* **1995**, *319*, 341-349.
23. Yermilov, V; Rubio, J.; Becchi, M.; Freisen, M.D.; Pignatelli, B.; Ohshima, H. *Carcinogenesis* **1995**, *16*, 2045-2050.
24. Ohata, T.; Fukuda, K.; Murakami, A.; Ohigashi, H.; Sugimura, T.; and Wakabayashi, K. *Carcinogenesis* **1998**, *19*, 1007-1012.

25. Wei, H.; Wei, L.; Frenkel, K.; Bowen, R.; and Barnes, S. *Nutr. Cancer* **1993**, *20*, 1-12.
26. Ji, C.; Marnett, L.J. *J. Biol. Chem.* **1992**, *267*, 17842-17878.
27. Nakamura, Y.; Murakami, A.; Ohto, Y.; Torikai, K.; Tanaka, T.; and Ohigashi, H. *Cancer Res.* **1998**, *58*, 4832-4839.
28. Ohnishi, M.; Tanaka, T.; Makita, H.; Kawamori, T.; Mori, H.; Satoh, K.; Hara, A.; Murakami, A.; Ohigashi, H.; and Koshimizu, K. *Jpn J Cancer Res* **1996**, *87*, 349-356.
29. Tanaka, T.; Makita, H.; Kawamori, T.; Kawabata, K.; Mori, H.; Murakami, A.; Satoh, K.; Hara, A.; Ohigashi, H.; and Koshimizu, K. *Carcinogenesis* **1997**, *18*, 1113-1118.
30. Tanaka, T.; Kawabata, K.; Kakumoto, M.; Makita, H.; Matsunaga, K.; Mori, H.; Satoh, K.; Hara, A.; Murakami, A.; Koshimizu, K.; and Ohigashi, H. *Jpn. J. Cancer Res.* **1997**, *88*, 821-830.
31. Kobayashi, Y.; Nakae, D.; Akai, H.; Kishida, H.; Okajima, E.; Kitayama, W.; Denda, A.; Tsujiuchi, T.; Murakami, A.; and Koshimizu, K.; Ohigashi, H.; and Konishi, Y. *Carcinogenesis* **1998**, *19*, 1809-1814.
32. Kawabata, K.; Tanaka, T.; Yamamoto, T.;Ushida, J.; Hara, A.; Murakami, A.; Koshimizu, K.; Ohigashi, H.; Stoner, S. D.; and Mori, H. *Jpn. J. Cancer Res.* **2000**, *91*, 148-155.

Author Index

Subject Index

D

E

F

G

H

I

J

L

M

N

O

P

S

Q

R

T

U-V

W

X